Tropical Plant Species and Technological Interventions for Improvement

Edited by Muhammad Sarwar Khan

Published in London, United Kingdom

Tropical Plant Species and Technological Interventions for Improvement
http://dx.doi.org/10.5772/intechopen.98109
Edited by Muhammad Sarwar Khan

Contributors
Laura Graciela Mereles, Mario Smidt, Edelira Velázquez, Karen Patricia Martínez, Laura Correa, Eva Eugenia Soledad Coronel, Mahfut, Shinichi Enoki, Yoshinori Takahara, Kayode Paul Baiyeri, Kolawole Olajide, Raufu Olusola Sanusi, Mohammad Babakatcha Abubakar, Oluwole Olalekan Oke, Azeez Olalekan Ibrahim, Deola-Tayo Lordbanjou, Margaret Ikhiwili Oniha, Sharon Oluwatobi Akinpelu, Eze Frank Ahuekwe, Parichart Burns, Pimpilai Saengmanee, Uthaiwan Doung-Ngern, John Samuel Kennedy, Jeeva Kasiviswanathan Lekshmi, Rafiq Ahmad Shah, Parshant Bakshi, Hamidullah Itoo, Gaganpreet Kour, Hortensia Brito-Vega, José Manuel Salaya-Domínguez, Edmundo Gómez-Méndez, Armando Gomez-Vazquez, Maritza Alejo Jeronimo, Edward Manuel Arevalo de la Cruz, Muhammad Sarwar Khan, Irfan Ali, Faiz Ahmad Joyia, Ghulam Mustafa, Safdar Ali Mirza, Wakshum Shiferaw, Sebsebe Demissew

First published in London, United Kingdom, 2023 by IntechOpen
IntechOpen is the global imprint of INTECHOPEN LIMITED, registered in England and Wales, registration number: 11086078, 5 Princes Gate Court, London, SW7 2QJ, United Kingdom

British Library Cataloguing-in-Publication Data
A catalogue record for this book is available from the British Library

Additional hard and PDF copies can be obtained from orders@intechopen.com

Tropical Plant Species and Technological Interventions for Improvement
Edited by Muhammad Sarwar Khan
p. cm.
Print ISBN 978-1-80356-074-8
Online ISBN 978-1-80356-075-5
eBook (PDF) ISBN 978-1-80356-076-2

For EU product safety concerns:
IN TECH d.o.o., Prolaz Marije Krucifikse Kozulić 3, 51000 Rijeka, Croatia,
info@intechopen.com or visit our website at intechopen.com.

Meet the editor

Muhammad Sarwar Khan has been a full professor at the University of Agriculture, Faisalabad, since 2008. He has over 25 years of experience in teaching, research, and administration. He obtained his BSc and MSc from the University of Agriculture, Faisalabad, followed by a Ph.D. from the University of Cambridge, UK, and postdoctoral training at the Waksman Institute for Microbiology, Rutgers University, and the University of Central Florida, USA. He has occupied various positions at the National Institute of Biotechnology and Genetic Engineering (NIBGE), Faisalabad. His publications in high-impact factor journals have earned an impressive impact factor of 435. He has authored eight international books and 30 book chapters. He is a fellow of several international societies, including the Cambridge Commonwealth Society, the Cambridge Philosophical Society, and the Royal Society of Biology. The government of Pakistan has awarded him the President's Medal for Technology in recognition of his achievements.

Contents

Preface

Tropical plants include fruit, flowering (indoor and outdoor), climbers, and herbaceous (including perennials, annuals, and bulbous plants), as well as palms, ferns, and other species. Several factors need to be taken into account to grow these plant species successfully in tropical and subtropical regions of the world. Soil, nutrition, and water management; insect and disease management; and genetic improvement through novel approaches are among these factors. This makes these plants more important to farmers, breeders, and technologists looking to increase their sustainable output.

Tropical Plant Species and Technological Interventions for Improvement consists of thirteen chapters organized in an easy-to-follow manner. In the introductory chapter titled "Integrative Technologies for Sustainable Plant Improvement," Khan comprehensively highlights the use of technological advances in improving plant genetics for agronomy and value addition. Chapter 2, "Emerging Trends to Improve Tropical Plants: Biotechnological Interventions", by Ali and colleagues, discusses the importance of conventional and modern scientific approaches for the conservation and improvement of tropical plant species. They further explain how the information retrieved from pan-genome, super-pan-genome, and pan-transcriptome has enriched marker-assisted selection, molecular breeding, and transgenic approaches in the sustainable development of tropical plants. Enoki and Takahara in Chapter 3, "Applications of Biotechnological Approaches in the Product and Breeding of *Phalaenopsis* Orchids", discuss the role of biotechnology in improving the orchid industry and highlight how it will benefit researchers, producers, and fanciers of *Phalaenopsis* orchids. In Chapter 4, "Lesser Known African Indigenous Tree and Fruit Plants: Recent Evidence from Literatures and Regular Cultivation Culture", Baiyeri and Olajide suggest exploiting the indigenous plant species as a key resource for ensuring healthy food systems in Africa. Chapter 5, "Challenges and Advances in the Production of Export-Quality Macadamia and Its Integral Use with Green Technologies", by Mereles and colleagues, highlights the technologies used to process macadamia nuts for the development of products of high nutritional quality. Burns and colleagues in Chapter 6, "Papaya: The Versatile Tropical Fruit", explain the pros and cons of papaya production technologies and the benefits of multidisciplinary approaches to enhancing the effective farm management and production of quality fruits for consumption and processing. Shah and team in Chapter 7, "Tamarillo (*Cyphomandra betacea* (Cav.)) Origin, Cultivation, Breeding and Management", discuss breeding, production, and consumption technologies of the Tamarillo fruit plant. In Chapter 8, "The Production and Marketing Issues of Pineapple (*Ananas comosus*) under Humid Tropical Conditions in the State of Tabasco and Way-out", Jeronimo and colleagues describe pineapple production technologies for domestic and foreign markets. In Chapter 9, "Cassava Production Enterprise in the Tropics", Raufu Olusola Sanusi et al, explain the production technology and integrated weed and pest management of cassava. Chapter 10 by Mahfut, "Identification of Native *Dendrobium* Based on Morphological and Anatomical Characters in Liwa Botanical Garden", describes the development of a key for the identification of orchid plants. Kennedy and Lekshmi highlight key technological advances in tropical pest management using host resistance, natural enemies, selective pesticides, ecological engineering, and habitat management strategies in Chapter 11, "Holistic Pest Management Strategies in Tropical Plant Species". In Chapter 12, "Phytochemical Contents of Essential Oils from *Cymbopogon* Species: A Tropical Medicinal Plant", Oniha and colleagues explain improvements in extraction and quantification techniques to harvest a pure yield of essential oils from lemongrass. In Chapter 13, "Effects of the Invasive Alien *Prosopis juliflora* (Sw.) DC and Its Management Options in Ethiopia: A Review", Shiferaw and Demissew discuss the invasion of *Prosopis juliflora* and its management options in Ethiopia.

Muhammad Sarwar Khan, Ph.D.
Center of Agricultural Biochemistry and Biotechnology (CABB),
Faculty of Agriculture,
University of Agriculture,
Faisalabad, Pakistan

Chapter 1

Introductory Chapter: Integrative Technologies for Sustainable Plant Improvement

Muhammad Sarwar Khan

1. Introduction

Agriculture systems around the world must produce more food while producing less waste. Sustainable agricultural practices and food systems, which cover both production and consumption, must be evaluated holistically and fully to be effective. Tropical plants include fruit, climbers, flowering perennials, annuals, and bulbous plants as well as indoor and outdoor floral plants. Also included in the list of tropical plant species are ferns, palms, and other plants. To grow these plant species successfully in tropical and subtropical regions of the world, several factors are needed to be taken care of. Because they are crucial to food production and are getting harder to get in many parts of the world, healthy soils, clean water, and plant genetic resources must all be used and maintained sustainably. With the expectation that there would be 9 billion people on the planet by the year 2050, an increase in crop output and quality would be necessary to satisfy the needs. An expanded discussion on how to grow fruit and ornamental plants for esthetic, dietary, and nutraceutical benefits may be found in the introductory chapter. The impact of technological advancements on the sustainable cultivation of plants is also examined. Further, the integrative strategies including OMICS and reliable methods for gene discovery and genome sequencing, as well as the application of CRISPR/Cas and gRNA/Cas, organelles transformation innovations, and value-adding strategies such as biopharming to create superior plants for agronomic, industrial, and value-adding features are highlighted.

2. Tropical plants: economic and nutritional importance

Tropical plants are essential for providing food, fiber, and shelter, but they also have esthetic value because of their lovely hues, fragrances, and greens, which also have a cooling effect from evapotranspiration. They are crucial for people living not only in tropical areas but also in other parts of the world's economies, cultures, and spiritual life. For example, there are over 100 different varieties of the 400 different species of bananas that are grown in Africa alone, nearly 1 million hectares of coffee plantations in Brazil, 2 million hectares of sugarcane plantations in Hawaii, and over 3 million hectares of rice paddies in tropical regions of Asia. Compared with plants found in temperate climates, tropical plants are significantly different. They thrive in a variety of environments, but particularly in rainforests, and have evolved to

IntechOpen

withstand extreme heat and heavy humidity. Additionally, as long as adequate water is available, many tropical plants may thrive on nutrient-deficient soils [1].

The economic value of tropical plants comes from their use in silviculture and timber production. Moreover, various species are used as sources of food, medicine, and raw materials (**Figure 1**). The esthetic value is derived from their beauty and usefulness in preserving the environment. On the other hand, they provide a vital service in maintaining soil fertility through their nutrient-rich soils and preventing erosion by providing habitats for microflora that act as decomposers of organic matter, root-dwellers, and rhizospheric antagonists [2]. Tropical plants also contribute to a variety of other services such as carbon sequestration (sequestering atmospheric carbon dioxide into biomass) by high photosynthetic rates, and water purification (through high evapotranspiration rates), and carbon capture and sequestration (through deposition of carbon on forest floors). In addition, they also provide a variety of amenities such as shade, food, and shelter for humans and animals; recreation opportunities for people; cultural heritage for indigenous civilizations; environmental education for children; protection from natural hazards such as floods and fires, and above all, biodiversity conservation.

Tropical plants contribute to the food supply, provide a source of nonfood raw materials, provide medicines, and are an essential part of human life in many cultures. They also have high nutritional content, making them an essential nutrition source for the tropics. Among nonfood products, tropical trees are rich in essential oils and resins used for cosmetics and medicines [3]. Medicinal plants such as the soapberry (*Sapindus mukorossi*) are used for treating skin ailments such as acne because it is rich in antimicrobial agents, which help to prevent bacterial or fungal infections. Other important medicinal plants include *Croton zaeus*, which has been used to treat diarrhea and other digestive disorders such as stomach aches; Passion vine (*Passiflora quadrangularis*), which contains compounds that suppress pain and Trumpet tree (*Cecropia peltata*), is known as a sort of cure-all in Caribbean folk medicine because of the long list of its uses.

Figure 1.
Tropical plants, their sustainable cultivation, and value-added products are presented in the graphical abstract of the introductory chapter.

The edible plants are used to flavor foods as well as some bitter medicines and to make dyes, perfumes, medicines, and cosmetics. The medicinal plants provide relief from pain and suffering. The flowers of these plants are used to produce dyes for dying cloths or other materials. Some flowers contain potent alkaloids that can be extracted for use in therapeutics. The leaves of some tropical plants serve as fodder for animals, while others serve as fertilizer for agriculture. Being valuable food sources, some tropical plants have been grown worldwide for thousands of years including peanuts, potatoes, cocoa beans, and rice, and now have become essential crops owing to a source of nutrition because of richness in a wide range of vitamins and minerals in addition to having high amounts of protein, dietary fiber, and low-fat levels. Further, these plants help to prevent chronic diseases such as heart diseases and cancer by reducing free radicals that cause oxidative stress due to the presence of antioxidants in them. High fiber contents help in digestion and prevent constipation that reducing the risk of colon cancer by preventing inflammations of the digestive tract. High levels of vitamin C in many tropical fruits such as papaya, mangoes, mangosteen, etc., help to boost immunity against common infections. Vitamin C also protects against oxidative stress caused by free radicals, which may lead to cell damage or death if left unchecked [4].

3. Tropical ecosystem

Tropical plants grow in a wide range of habitats, from the equator to the tropics. They thrive in warm weather and grow in rainforests, deserts, and even on mountainsides. The problems of deforestation, soil degradation, and changing climate pose serious threats to tropical ecosystems including forestry and agriculture sectors. This resulted in an average decline of 15.8 million hectares of tropical trees since 2017. If the issue remains unattended, it will erode social-ecological resilience in the tropics and will ultimately result in self-propagating feedback and regime shifts. Various factors including climate change and soil erosion pose a negative impact on the stability of natural ecosystems. Moreover, the seasonal, interannual, and decennial climatic fluctuations badly affect the vegetation dynamics in these areas. Land degradation is also more noticeable in the tropics, and it affects biodiversity and soil characteristics. The major causes of land degradation are anthropogenic [5]. Hence, sustainable management of natural ecosystems has been recommended as the only way to control factors affecting the stability and preservation of tropical ecosystems. Sustainable use encompasses the management and use of natural resources including tropical natural resources. They can sustain their natural biodiversity, yield, renewal capacity, vitality, and capacity to satisfy the relevant ecological, economic, and social functions at local, national, and global levels, and that they do not cause damage to other ecosystems. This reflects contemporary discourse in sustainable development and governance, which emphasizes the importance of public-private and civil society partnerships, with the potential to bridge multilateral norms and local action by drawing on a diverse number of actors in civil society, government, and business [6].

4. Evolution and diversification

The tropics comprise the geographic regions of the earth centered on the equator. They exhibit substantially variable landscapes ranging from deserts to rainforests and from hot lowlands to snow-capped mountains. Hence, we can find a variety of ecosystems in tropical regions with extreme climatic conditions along with a rich diversity of living organisms including plants, animals, and microbes.

Tropical plants are proposed to be originated 93–115 million years ago yet the rate of diversification increased dramatically during the last 15 million years. Hence, the tropics are considered a diverse rich region with a huge number of crop plants including cereals such as maize, rice, sorghum, and millet; tuberous crops such as potato, sweet potato, and cassava; vegetables such as tomato, peppers, many cucurbits; cash crops such as cocoa, rubber, tobacco, cotton; fruits such as banana, pineapple, mango, papaya; and other crops such as peanut, common beans, oil palm, coconut, sugarcane, and coffee. It is estimated that two-thirds of all angiosperms including crop as well as non-crop plant species are found within the tropics. Moreover, rich biodiversity of plants other than angiosperms such as Ferns [7], bryophytes, and liverworts are found highly concentrated in the tropics. The dispersal of plant species in tropical areas is not uniform. The Neotropical areas and the region of Asia Pacific are the most biodiverse while Africa and oceanic islands contain the least biodiversity. The tropical hyper-diversity is not well understood, and it has variously been attributed to be a *museum of diversity*, showing constant speciation and low extinction [8], or a *cradle of diversity*, referring to more recent and rapid speciation [9].

5. Tropical plant species—as an ornament

Tropical plants, with unique colors, shapes, and fragrances, have a great esthetic value and hence are very popular among culturists, collectors, hobbyists, and even enthusiasts. They not only lift the spirit of the room and indoor environment, uniquely and refreshingly but also make us feel alive, lively, and fresh. The craze for growing these housing plants has accelerated to multifold during the COVID pandemic particularly. Tropical flowering plants have enchanting colors and are key elements in the beautification of any landscape. The most popular tropical flowers are Amaryllis, Paper flower, Garden cosmos, Flamingo flower, Bush lily, and Ohe naupaka, etc. Though, most of the indoor plants available in nurseries are either grown by cuttings or seeds. However, these conventional techniques are slow and the resultant plants are susceptible to diseases. A valuable alternative is the use of tissue culture techniques for the mass production of disease-free plants [10]. The plants developed through tissue culture have uniform growth, hence providing rapid mass production of disease-free plants, off-season propagation, mass production in limited space, and can overcome the challenges of cultivating plants by traditional propagation techniques.

Researchers have worked out different protocols for the mass multiplication of these flowering plants. Different combinations of growth hormones and culture conditions were optimized for the establishment of *in vitro* regeneration and micropropagation. Different explants (leaf disc, mature embryos, and bulb scale) were tested in Amaryllis and other tropical flowers. Different varieties of hardy ornamental tropical plants have been explored, i.e., Bamboo (Goldstripe, Ghost bamboo, Slender weavers, and Chinese dwarf), Colocasia (Pink china), Hibiscus (Berry awesome, Perfect storm, and Cranberry crush), Bird of Paradise (Mexican bird of paradise), Palms (Dwarf Palmetto, Kumaon, Chusan, Miniature Chusan palm and Windmill), Canna (Peach, Stuttgart, Gigantum and Skyhawk), Hostas (Abiqua drinking gourd, Lakeside shore master and Dancing queen), and Ferns (Lady fern, Western sword fern, and

Christmas fern). Over the years, numerous researchers have tried to manipulate bamboo flowering and seed sets under *in vivo* and *in vitro* conditions [11]. Two different polyploidy groups were found to be present in woody bamboo. Tropical bamboo is hexaploid, whereas temperate bamboo is tetraploid. Likewise, different molecular markers have been employed to explore genetic diversity. Using these markers, it has been elucidated that lowland *bamboo* is associated with distinct geographic regions. Genetic diversity among the ornamental palm accessions helped to explore their origin, genetic identification, and conservation [12]. Some other prominent indoor plants, i.e., spider plants, African violet, baby rubber plant, and weeping fig had also been worked out to overcome existing bottlenecks in propagation.

6. Tropical plant species—as a source of fruits and lumber

In addition to flowering and herbaceous plants, tropical plant species are at the forefront to fulfill the fruit and lumber demands of the rapidly increasing population. The most valuable tropical fruits are jackfruit, dragon fruit, lychee, banana, passion fruit, papaya, acai, rambutan, coconut, and mango. They are not only an important source of nutrients, bioactive compounds, and primary as well as secondary antioxidants but also bring high economic returns to the growers. So, tropical fruit plants are in high demand worldwide owing to their nutritional and nutraceutical value. Recent updates in cultivation techniques, efforts to develop climate-resilient genotypes, control of fruit diseases, exploitation of postharvest physiology, and the discovery of bioactive compounds have further promoted the adoption and usage of tropical fruits. As a result, numerous fruit species have been explored that have the potential to be transformed from minor tropical fruits to major tropical fruits and can be introduced elsewhere [13].

Tropical lumber trees are also very important for domestic as well as industrial applications resulting in the substantial economic growth of lumber-producing areas. Furthermore, such areas not only contribute to the conservation of animal and plant biodiversity but are of central importance in the global trade of sawn wood, roundwood, and plywood too. They are so important that only rosewood has a market of 26 billion USD in China per annum. More than 600 species of lumber trees have been explored so far, in different tropical regions, worldwide. Though numerous tree species are on the verge of extinction, researchers are striving hard to protect and promote them using various interventions. Advancements in clonal propagation, micrografting, nodal culture, genetic transformation, development of DNA databases for the forensic identification of lumber [14], and establishment of seed banks have helped not only to secure endangered plant species but have also helped to improve their production and quality.

7. Tropical plants—a source of medicine

Tropical plants are second major source of oxygen on the earth after oceanic phytoplankton. They are valuable indoor plants that can help to restore oxygen balance in the closed space and are an exclusive part of the top-five indoor plant species having the ability to produce maximum oxygen. These include Boston Fern (*Nephrolepis exaltata*), Peace Lily (*Spathiphyllum spp.*), Snake Plant (*Sansevieria trifasciata),* and Areca Palm (*Dypsis lutescens*), and Gerber Daisy (*Gerbera jamesonii).* These tropical plants play a critical role to moderate the level of carbon dioxide in the atmosphere. The modern world is highly reliant on fossil fuels to fulfill ever-increasing energy needs. This has led to an alarming increase in carbon accumulation in the atmosphere. Here comes the importance of carbon sinks for balancing carbon concentration. Grasslands, peat bogs, coastal ecosystems, coral reefs, wetlands, boreal forests, and tropical rainforests are important carbon sink ecosystems playing their role in balancing oxygen–carbon levels.

Tropical plants have also been explored as a source of valuable industrial products and drugs. Owing to the enriched biodiversity, they provide 60% of the chemical entities all over the world. They have been called the largest pharmacy in the world because more than 70% of the drugs are derived from these plants, directly or indirectly. Most synthetic drugs are also derivatives of tropical plant products. Half of the best 25 pharmaceutical agents also come from tropical forests. They have been identified as a valuable source of anticancer agents. Further, the first known antimalarial drug "quinine" was also derived from a neotropical tree [15].

8. Technological interventions

Indoor tropical plants not only act as oxygen balancing agents, but also have positive psychological effects, help in reducing indoor pollution, purifying indoor air, and absorbing volatile compounds such as formaldehyde, benzene, and trichloroethylene. Indoor air often contains volatile organic compounds such as formaldehyde, benzene, and chloroform. These toxins come from different sources including cooking, showering, furniture, and smoking. House plants can remove some toxins from the air, but they aren't very efficient: A homeowner would need more than 20 plants to remove formaldehyde from a typical room. Developing improved plants through recent innovative approaches can be of great help in this context (**Figure 2**). Researchers have worked out that the detoxification ability of the plants can be improved by the expression of the mammalian gene(s). Stuart strand and colleagues introduced a rabbit gene (*CYP2E1)* into a common houseplant, pothos ivy (*Epipremnum aureum*), and resultant plants were able to remove injected chloroform and benzene from the vial containing transgenic ivy plants [16].

Developing fancy indoor plants has gotten attention and different research groups have attempted to produce glowing plants. Mitiouchkina et al. [17] engineered tobacco plants with a fungal bioluminescence system that converts caffeic acid (present in the plants) into luciferin for the production of self-sustained luminescence, visible to the naked eye. Researchers have also engineered metabolic pathways to divert the natural supply of caffeic acid resulting in their ability to glow. Hence, these plants can glow throughout their life cycle. Likewise, transgenic papaya has been developed to stand against viral infection and is successfully grown in Hawaii. SunUp and Rainbow are the commonly grown varieties of virus-resistant transgenic papaya.

With the advent of next-generation techniques and advancements in DNA sequencing, it has become much more feasible to explore the genome of any plant for the desired traits. Evolutionary biology has got a great pace with these advancements. Phylogenetic and macroevolutionary analysis has been employed to define their genetic relatedness, thus helping out to track better plant species. Non-coding and repetitive DNA sequences play critical roles in determining the phenotype and genome evolution. The pan-genome analysis offers a valuable platform to evaluate the genetic diversity of species via investigation of their entire genome repertoire.

Figure 2.
Integrative strategies and their applications to improve plants for better nutrition, medicine, and vaccines.

It is now feasible to make multiple high-quality genomes that can be used to construct high-resolution pan-genomes making it possible to track all of the variations. However, high-throughput new tools would be required for the assembling, displaying, and interaction studies of such high-resolution pan-genomes.

Genome editing is one of the emerging innovations in the current millennium that has proved its potential to develop plants of desire. This innovative technology has been employed for stacking gene mutations, manipulating gene expression, and improvement of yield. Further, CRISPR-edited plants are not taken as conventional GMOs so, need not get approval from the regulatory bodies provided they are free from exogenous DNA. This has emerged as a user-friendly tool to address challenges in the production of tropical plants and improve their nutritional value. Numerous tropical plants have been targeted to improve through CRISPR/Cas9 and have achieved phenomenal successes during the last decade. Researchers at Cold Spring Harbor Laboratory precisely edited tomato genes involved in fruit size and shape, flowering time, self-pruning, and growth habitat, hence generating new alleles for valuable traits and improved plant architecture [18]. RAS-PDS1 and RAS-PDS2 (phytoene desaturase genes) were mutated in bananas to improve carotenoid biosynthesis, and it was reported to be improved by 59% [19]. Mutants of cassava plants were developed by targeting the MePDS gene and more than 95% of mutants exhibited partial albino or albino phenotype in cotyledonary-stage somatic embryos. Mutant embryos developed into plantlets indicating that 22–47% of the mutants were stable [20]. CRISPR-mediated editing of nCBP-1 and nCBP-2 (elF4E isoforms) in cassava resulted in improved resistance against cassava brown streak disease. Differential expression and genome-wide studies revealed numerous genes involved in salinity, drought, cold, and oxidative stresses. These genes can be targeted for improved tolerance against abiotic stresses in cassava. EgWRKY genes in African oil palm appeared to be upregulated in response to abiotic stresses. These findings revealed the crucial role of EgWRKY in abiotic stress tolerance, hence providing a great opportunity to edit the palm genome for enhanced abiotic stress tolerance. Likewise, S-genes mutations in papaya boosted its defense response against insect pests and pathogens by increasing the accumulation of papain (a cysteine protease). Targeted mutation of TcNPR3NPR3 resulted in upregulation of PR gene expression and increased resistance against pathogens in Theobroma cacao. Though different tropical plants have been mutated through CRISPR/Cas9 system, certain limitations are there, which need to be addressed for the widespread applications of technology.

Acknowledgements

The author acknowledges Drs. Faiz Ahmad Joyia, Ghulam Mustafa, and Rimsha Riaz for their significant contributions to the chapter.

Author details

Muhammad Sarwar Khan
Centre of Agricultural Biochemistry and Biotechnology (CABB), University of Agriculture, Faisalabad, Pakistan

*Address all correspondence to: sarwarkhan_40@hotmail.com

References

[1] Engler A. Plants of the northern temperate zone in their transition to the high mountains of tropical Africa. Annals of Botany. 1904;**8**(72):523-540

[2] Flores BM, Staal A, Jakovac CC, Hirota M, Holmgren M, Oliveira RS. Soil erosion as a resilience drain in disturbed tropical forests. Plant and Soil. 2020;**450**(1):11-25

[3] Sharmeen JB, Mahomoodally FM, Zengin G, Maggi F. Essential oils as natural sources of fragrance compounds for cosmetics and cosmeceuticals. Molecules. 2021;**26**(3):666

[4] Saranraj P, Behera SS, Ray RC. Traditional foods from tropical root and tuber crops: Innovations and challenges. In: Innovations in Traditional Foods. United Kingdom, United States: Woodhead Publishing, Elsevier; 2019. pp. 159-191

[5] Allen K, Dupuy JM, Gei MG, Hulshof C, Medvigy D, Pizano C, et al. Will seasonally dry forests be sensitive of resistant to future changes in rainfall regimes? Environmental Research Letters. 2017;**12**:023001

[6] Delabre I, Alexander A, Rodrigues C. Strategies for tropical forest protection and sustainable supply chains: Challenges and opportunities for alignment with the UN sustainable development goals. Sustainability Science. 2020;**15**:1637-1651

[7] Kreft H, Jetz W, Mutke J, Barthlott W. Contrasting environmental and regional effects on global pteridophyte and seed plant diversity. Ecography. 2010;**33**:408e419

[8] Wallace AR. Tropical Nature, and Other Essays. London, UK: Macmillian; 1878

[9] Pennington RT, Hughes M, Moonlight PW. The origins of tropical rainforest hyperdiversity. Trends in Plant Sciences. 2015;**20**:693-695

[10] Chin CF, Tan HS. The use of proteomic tools to address challenges faced in clonal propagation of tropical crops through somatic embryogenesis. Proteomes. 2018;**6**:21. DOI: 10.3390/proteomes6020021

[11] Yuan JL, Yue JJ, Gu XP, Lin CS. Flowering of woody bamboo in tissue culture systems. Frontiers in Plant Sciences. 2017;**8**:1589. DOI: 10.3389/fpls.2017.01589

[12] El-Banna AN, El-Denary ME, Dewir YH, Naidoo Y. Genetic relationship and diversity in some ornamental palms based on proteins and randomly amplified polymorphic DNA markers. HortScience. 2017;**52**(3):338-342

[13] Galan SV. Potential of minor tropical fruits to become important fruit crops. Acta Horticulturae. 2013;**975**:581-591. DOI: 10.17660/ActaHortic.2013.975.74

[14] Ng CH, Ng KKS, Lee SL, et al. DNA databases of an important tropical timber tree species *Shorea leprosula* (Dipterocarpaceae) for forensic timber identification. Scientific Reports. 2022;**12**:9546. DOI: 10.1038/s41598-022-13697-x

[15] Kinghorn AD, Balandrin MF. Human medicinal agents from plants. In: Kinghorn AD, Balandrin MF, editors. ACS Symposium Series 534. Washington, DC: ACS Publishers; 1993

[16] Zhang L, Routsong R, Strand SE. Greatly enhanced removal of volatile organic carcinogens by a genetically modified houseplant, Pothos Ivy

(*Epipremnum aureum*) expressing the mammalian cytochrome P450 2e1 gene. Environmental Sciences and Technology. 2019;**53**(1):325-331. DOI: 10.1021/acs.est.8b04811

[17] Mitiouchkina T, Mishin AS, Somermeyer LG. Plants with genetically encoded autoluminescence. Nature Biotechnology. 2020;**38**:944-946

[18] Rodríguez-Leal D, Lemmon ZH, Man J, Bartlett ME, Lippman ZB. Engineering quantitative trait variation for crop improvement by genome editing. Cell. 2017;**171**:470-480.e8. DOI: 10.1016/j.cell.2017.08.030

[19] Kaur N, Alok A, Shivani N, Pandey P, Awasthi P. CRISPR/Cas9-mediated efficient editing in phytoene desaturase (PDS) demonstrates precise manipulation in banana cv. rasthali genome. Functional & Integrative Genomics. 2018;**18**:89-99. DOI: 10.1007/s10142-017-0 577-5

[20] Odipio J, Alieai T, Ingelbrecht I, Nusinow DA, Bart R, Taylor NJ. Efficient CRISPR/Cas9 genome editing of phyteone desaturase in cassava. Frontiers in Plant Sciences. 2017;**8**:1780. DOI: 10.3389/fpls.2017.01780

Chapter 2

Emerging Trends to Improve Tropical Plants: Biotechnological Interventions

Irfan Ali, Faiz Ahmad Joyia, Ghulam Mustafa, Safdar Ali Mirza and Muhammad Sarwar Khan

Abstract

Tropical plants are an integral part of the ecosystem and are of significance for the well-being of humanity. Since their domestication in 10,000 BC, conventional breeding has played a crucial role in their conservation and widespread adaptation worldwide. Advancements in multi-omics approaches, that is, genomics, metabolomics, transcriptomics, proteomics, whole genome sequencing, and annotation, have led to the identification of novel genes involved in crucial metabolic pathways, thus helping to develop tropical plant varieties with desirable traits. Information retrieved from the pan-genome, super-pan-genome, and pan-transcriptome has further uplifted marker-assisted selection and molecular breeding. Tissue culture techniques have not only helped to conserve endangered plant species but have also opened up new avenues in terms of mass-scale propagation of ornamental plants. Transgenic technology is increasingly contributing to the betterment of tropical plants, and different plant species have been engineered for valuable traits. Likewise, genome editing is appearing to be a promising tool to develop tropical plants having the potential to fulfill future needs. Hence, this chapter highlights the importance of conventional and modern scientific approaches for the conservation and improvement of tropical plant species.

Keywords: tropical plants, ornamental, medicinal, value addition, biotechnological interventions, genetic engineering, domestication, tissue culture, endangered species

1. Introduction

Tropical plants are those that grow in warm climates. They typically grow best in areas where the temperature ranges consistently between 75 and 85 degrees Fahrenheit (24 and 29 degrees Celsius). The term "tropical" is used for a variety of plants, from palms to orchids. Tropical plants often have thick leaves that are waxy or shiny to help reduce evaporation and protect them from too much light. These plants also have flowers with colorful petals or leaves that attract pollinators like butterflies [1].

Tropical plants are among the most colorful and beautiful of all plant species. They are very diverse, ranging from small herbaceous plants to tall trees. Many tropical plants have been introduced in areas outside their natural range because they can tolerate colder temperatures than other species of the same genus or family. Tropical plants grow in a wide range of habitats, from the equator to the tropics. Tropical plants thrive in warm weather and grow in rainforests and deserts, and even on mountainsides [2]. These plants have adapted to their environment over time by developing unique features that help them survive in hot climates without suffering from heat stress or lack of moisture. Some tropical plants have thick stems that provide support during strong winds, while others still use their roots as storage organs called tubers or rhizomes that store nutrients until they are needed by the plant itself or by animals or humans [3].

The tropics have been a source of many new food crops that have been introduced to other regions of the world: sorghum (which originated in Africa, Winchell, [4]), soybean (from China, Singh, [5]), mung bean (from China, Shahrajabian, [6]), peanut (from South America, Ozias-Akins, [7]), cassava (from Brazil, Bester, [8]), lima bean, and fava bean (from Peru, [9] Santos, 2008). Several plant families have their center of diversity in the tropics. These include Arecaceae (palms), Lamiaceae (mint family), Lauraceae (laurel family), and Rubiaceae (coffee family). The largest family by the number of species is the Orchidaceae, with about 25,000 species.

Because most living forms thrive here, the tropical zone is home to the majority of plant and animal species in the world. The domestication of food crops is one of the most important developments in human history that made it possible for people to settle down and live in communities rather than wandering around with their herds and flocks.

2. Tropical plants and ecosystem

The tropics are usually defined as the area between the Tropic of Cancer and Tropic of Capricorn bounded by lines of latitude running through the Arctic and Antarctic circles, which is 23.5 degrees north and south of the equator, respectively. This area spans among Central America and South America, Africa, India, and Southeast Asia. Some tropical plants can also be found in subtropical regions such as California. The region lies beneath the intertropical convergence zone (ITCZ), a belt where tropical cyclones often form during their lifetime before moving into more temperate latitudes where they die out. However, they can be found year-round in some areas as well as at higher altitudes inside the tropics—like in Indonesia or Ecuador [10].

Tropical climates generally do not experience marked changes in temperature throughout the year, with average temperatures ranging from 18 to 24°C (64 to 75°F). The average annual rainfall varies greatly from region to region within this climate zone, with some areas receiving more precipitation annually, while other areas may receive less than that. Humidity is high throughout these regions, and climatic conditions are relatively constant throughout the year. Within tropical regions, distinct wet and dry seasons are caused by changes in the direction of wind flow that result in periods of more intense (wet) or weaker (dry) rainfall. The length and onset dates for wet and dry seasons vary by the geographic location within tropical zones [11].

The climate in the tropics is characterized by high temperatures and a nearly constant rate of evaporation. The evaporation results in a higher concentration of water vapor in the air, which produces high relative humidity. The high temperature reduces the capacity of the air to hold water vapor so that even though there is high humidity, the air remains very dry. High temperatures also cause a lot of water to evaporate from the ocean's surface, which make it rain more on land and make the air even more humid [12]. The tropics are characterized by wide diurnal temperature variations with large day–night differences (diurnal range) and small annual ranges. The diurnal range is due to a strong daily heating-and-cooling cycle as well as large daily fluctuations in solar radiation on land surfaces (high solar intensity). Also, there are no big changes in the way the atmosphere moves, like seasonal winds or changes in cloud cover, that would have a long-term effect on the surface temperature, like months [13].

Tropical plants have a significant impact on the ecosystem. The tropics contain more than half of the world's species, and the majority of plants and animals live in this region. Tropical rain forests are vital to the planet because they act as carbon sinks and produce oxygen, which helps to regulate the global climate [14]. Tropical rain forests are also home to endangered species such as the Sumatran tiger, Bornean orangutan, Sumatran rhinoceros, sun bears, and Asian elephants, which need preservation. These animals depend on tropical rain forests for food and shelter. Tropical rain forests are also a valuable economic resource because they provide many useful things that people use every day [15]. Tropical plants are essential to the survival and preservation of other living organisms. One of the most important traits of tropical plants is that they purify water through their roots. This filtering of water through soil acts as a safeguard against many different diseases, such as cholera, dysentery, and typhoid. The roots grow deep into the ground and hold onto small amounts of water that can sustain plant life. Tropical plants also prevent soil erosion by acting as a natural buffer against strong winds, storms, floods, and wildfires. Due to their deep root systems, this allows for proper drainage of water from floods and droughts alike. Tropical plants also provide food for animals and people by acting as a natural canopy for fruits and nuts, which are used for both food and trade [16].

3. Improvement of tropical plants using conventional approaches

The Agricultural Revolution began when people started cultivating crops on a large scale to feed larger populations. This allowed them to grow more food per unit of land than hunting and gathering. Enhanced agricultural productivity led to population growth and an increase in urbanization. Traditional methods for the improvement of tropical plants have been developed over thousands of years. Plants were selected and bred by farmers and gardeners, who then shared their knowledge with other growers. Because of these changes, there are now many beautiful plants that can be used as decorations or as food.

3.1 Domestication

Crop domestication is the process of selecting and breeding plants or animals to produce food and other desired products. It is a human-mediated artificial selection process that gives rise to domesticated organisms with certain desirable characteristics. Plants were first domesticated around 10,000 BC when early farmers selected and bred the best edible plants they found in their fields. They later moved on to domesticating animals, cultivating grain, and storing food [17].

3.2 Selective breeding

One of the most common ways to improve plants is through selective breeding. This is a process where desirable traits are selected and passed on to offspring generation after generation. The selection of naturally occurring varieties in the wild and, subsequently, in cultivated areas was the first type of plant breeding. Planting-harvesting cycles exert selection pressure on genetic diversity. Some plant phenotypes were profoundly altered as a consequence of this process, as shown by the derivation of maize from teosinte. The result is a plant that has better survival characteristics or greater productivity than its ancestors [18]. The earliest evidence for selective breeding dates back to 7000–7500 BC in Jiahu, China; here, rice was bred from wild rice, *Oryza rufipogon,* and domesticated through artificial selection with a combination of harvesting and replanting the plant's seeds. Other early examples of selective breeding are emmer wheat, barley, flax, and cotton [19].

Selective breeding requires careful selection of individuals with desirable traits, and they must be further propagated by vegetative means (such as cuttings) so that all their descendants have these same traits. It takes time and patience but can be very rewarding if done right! Another common method is simply growing out large numbers of seedlings until one appears that has the desired trait(s). This type of selection does not require vegetative propagation or any genetic engineering—just patience! [20].

3.3 Intuitive farmer selection

One way that farmers have traditionally developed new varieties is through intuitive farmer selection (IFS). This is a form of plant breeding where farmers select plants that exhibit desirable traits and save seeds from those plants to plant the following year. This process can be repeated for several generations until the desired characteristics are fixed in the population. The first step of IFS is to observe what happens to plants over time, including which ones express certain traits and which ones do not. Farmers then select the best examples of these traits and save their seeds for planting the next season. This process can be repeated for many years before any new varieties are created [21].

3.4 Pure line breeding

Pure line selection (PLS) is a method in which a new variety is created by selecting an individual with desirable characteristics from an existing population (often consisting of many different varieties). It involves repeated cycles of crossing or self-pollination between related plants or clones that are genetically identical to each other and then selecting for one or more traits. This process can be used to produce a wide range of new varieties, including dwarfing rootstocks, disease-resistant plants, fruits with improved flavor and color, etc. PLS has been used in agriculture since ancient times, but it became more important after the nineteenth century when it helped breeders create new varieties with desirable traits such as yield, drought tolerance, frost resistance, disease resistance, and so on. PLS is also known as single gene selection or monogenic selection. It involves selecting a plant that has one desired characteristic and eliminating all other plants that do not possess that trait. So, it is possible to make a variety with only one type of flower or fruit. This makes it easier for farmers to buy seed stock of the right variety [22].

3.5 Mass selection breeding

Mass selection is the process by which farmers select the best seeds from their crops and save them for the next year. This was the main way that new varieties were developed before modern times. To select seeds, farmers would have to grow them out in their fields and observe which ones were the most productive and hardy and had good taste. Once they had chosen a few plants to grow again, they would save their seeds for the following season's crops. Over many generations, this process gradually produced new varieties of plants that were well suited to local conditions and tastes. Mass selection is a form of artificial selection that allows only those plants that exhibit the desired trait to reproduce. This method does not require any knowledge about genetics or the mechanism by which genes are inherited. The breeder just picks a desirable trait and grows out plants with that trait over and over again to choose the best ones for breeding [23].

4. Improvement of tropical plants using innovative approaches

4.1 Cell and tissue culture

Tropical plant communities contribute an enormous proportion of the global plant species and represent more than 42% of the total carbon reserves throughout the world [24]. They play a critical role in the well-being of humanity. Our daily dependence on tropical plants and/or their products is outstanding. They are not only major contributors to food and feed but also a valuable source of spices, essential oils, fruits, sugar, and beautiful hardwoods. In addition, tropical regions also produce different types of fibers, resins, gums, plant essences, and dyes, which are extensively used in therapeutics and various industrial by-products. They are the dominant source of trade among the continents; for example, Africa and Latin America are the major suppliers of cacao and coffee, South America is the largest producer of sugar and bioethanol, whereas Asia produces most of the natural rubber.

With the increased demand for tropical plants, that is, rubber, oil palm, cocoa, banana, pepper, and pineapple, their large-scale production is direly needed. High-quality planting material can only be produced through tissue culture, thus the only potential strategy to fulfill the increased demand for ornamental tropical plants and plants of economic value. So, different types of explants, cultural media, and conditions have been worked out by different research groups. Micropropagation by somatic embryogenesis has been established through direct or indirect routes. The number of clonal plants is usually lower in the case of direct embryogenesis as compared with indirect somatic embryogenesis, so indirect embryogenesis is preferred for the commercial-scale production of ornamental plants, particularly in the case of endangered plant species, where only limited explant material is available. However, callus induction and somatic embryogenesis are dependent on exogenous auxin and culture conditions [25]. Exogenous auxin is involved in the downregulation of essential genes involved in embryogenesis perhaps by DNA methylation or other cellular processes, which is still to be explored. Removal of auxin from the culture media supports embryogenesis in most plant species [26]. Exploring the developmental trajectory of callus induction, indirect somatic embryogenesis, and direct somatic embryogenesis will be of great help for the commercial-scale production of indoor tropical plants and edible tropical plant species. Different types of explants including immature zygotic embryos, mature zygotic embryos, immature female inflorescence, immature male inflorescence, immature leaves, mature leaves, young plantlets, and shoots were tested for the micropropagation of oil palm *via* indirect somatic embryogenesis [27].

Advancements in molecular biology have not only explored the critical pathways and genes involved in different phases of *in vitro* growth but also led to manipulation of certain genes to improve somatic embryogenesis and regeneration. Though most of the research has been reported on *Arabidopsis*, yet these findings may be extended for the betterment of other plant species including tropical plants. LEAFY COTYLEDON genes (LEC1 and LEC2) were found to be involved in the key pathways of somatic embryogenesis, and their overexpression triggered the upregulation of YUC genes, which are responsible for the increase of endogenous auxin levels. Other valuable genes involved in somatic embryogenesis are BBM, WOX, and SERK. Genetic manipulation of these genes can be of great help to further explore and understand fundamental cellular processes involved in somatic embryogenesis and clonal propagation [28]. In addition, proteomics studies have helped to elucidate the fundamental processes involved in these crucial cell growth phases, thus helping out to promote plant growth under *in vitro* conditions. Three valuable proteins linked with somatic embryogenesis were identified as osmotin-like proteins, chitinase, and β-1,3-glucanase. In *Picea glauca*, 48 differentially expressed proteins were observed to be of crucial importance during different stages of the development of somatic embryos. Efforts have been made to explore the proteome profile of oil palm embryogenic lines, embryogenic cell suspensions of coffee [29], auxin-induced embryogenic and non-embryogenic tissues of tamarillo, secondary somatic embryogenesis in cassava, and somatic embryos of avocado, which could further be exploited to improve clonal propagation of these valuable tropical plant species [30].

Though efforts have been made to unravel the fundamental cellular processes involved in key regulatory pathways of cell differentiation, dedifferentiation and commercial scale production of tissue culture plants have been possible. Still, certain impediments are there that need to be addressed for further improvement of existing clonal propagation systems. These include the genotype-dependent nature of the cultures, slow response of the cultured tissues, lower conversion rate of embryonically competent tissues, and heterogeneity of the cultured samples.

4.2 Genomics approaches

Breeding of the tropical tree plants is complicated owing to polyembryony, parthenocarpy, long juvenile phase, polyploidy, generation cycle, heterozygosity, and insufficient genomic resources. Advancements in multi-omics approaches, that is, genomics, metabolomics, transcriptomics, proteomics, whole genome sequencing, and annotation, have led to the identification of novel genes involved in crucial metabolic pathways responsible for sugar metabolism, fruit development, fruit ripening, stress tolerance, shelf life, etc. Interventions in genome-wide association (GWAS), genomic selection (GS), genetic transformation, and genome editing through CRISPR/Cas9 have helped to develop tropical plant varieties with desirable traits.

Developments in sequencing techniques have not only helped researchers devise molecular markers but also paved the way to explore genetic diversity in different plant species. The world's largest germplasm collection of bananas has been maintained at Biodiversity International Transit Centre (ITC), Belgium [31]. Diversity arrays technology (DArT) was employed for the selection of carotenoid-rich bananas, thus helping out to promote nutrient-enriched bananas [32]. In papaya, 21,231 SSR markers were developed from genic regions, of which 73 SSR markers were validated for fruit ripening. The SCAR marker (CPFC1) was developed for the fruit flesh color in papaya, facilitating the identification of progenies based on pulp color. Likewise, AFLP markers were developed for the characterization of pink-fleshed and white-fleshed guava genotypes. Comparative mapping and germplasm characterization were performed by SSR markers in Musa species. These studies helped out to shortlist the Musa accessions with relatively higher content of minerals including calcium (111.1-fold higher), potassium, and magnesium (4.7-fold higher) [33]. Jackfruit draft genome assembly has helped to explore numerous gene families involved in starch synthesis and fruit development. SSR and ISSR markers have also successfully been employed for the identification of dragon fruits with white pulp and pink pulp [34].

Molecular breeding, that is, marker-assisted selection (MAS), marker-assisted backcrossing (MAB), and marker-assisted introgression (MAI), is quite helpful in the efficient development of new genotypes, mapping population, phenotyping, and genotyping. QTLs (quantitative trait loci) are of pivotal importance in this context, to track particular desired traits. In papaya, 21 QTLs were identified for the key quality traits of fruits, that is, fruit weight, fruit width, fruit length, flesh thickness, flesh sweetness, fruit firmness, and skin freckle. Similarly, 460 SNPs were predicted as potential molecular markers for the selection of particular fruit traits and diversity studies [35]. In mandarin, four QTLs were identified to be associated with fruit weight, three with peel puffing, and one with sugar content. The whole genome assembly of mango has helped to explore polyembryony, identification of non-coding RNAs, and QTLs for flavonoid biosynthesis and fruit weight [36].

The discovery of genome-wide SNPs has opened up new avenues in high-throughput genotyping and marker-assisted breeding, thus helping out to develop the novel genotypes. These SNPs have been used for the GWAS and identification of QTLs relevant to fruit traits in tropical plants, that is, mango, papaya, avocado, cassava, banana. These QTLs can effectively be used for the betterment of traditional breeding in terms of reduced time and cost. In addition, information retrieved from the pan-genome, super-pan-genome, and pan-transcriptome has been employed in the mining of genetic determinants of various phenotypes, thus helping out the betterment of tropical plant species.

4.3 Genetic engineering

Transgenic technology is increasingly contributing to the betterment of tropical and sub-tropical plants. Numerous fruit trees, crop plants, and ornamental plant species have been engineered for valuable traits, esthetic value, and the cleanup of the environment. Compared with the annual crops, tree plants are tough targets as far as their genetic manipulation is concerned. The complexity of the genome, low transformation efficiency, complex cultivation environment, long breeding cycle, and recalcitrant nature of the plant tissues are the major impediments in the genetic transformation of tree plants. Researchers have worked out to resolve these bottlenecks, and protocols have been established for the genetic transformation of numerous tropical and subtropical plants including citrus, mango, banana, pineapple, litchi, passion fruit, plantain, longan, and avocado. These plant species have been engineered not only for valuable agronomic traits but also for the improved quality and quantity of the fruits.

Populus alba is taken as a model plant for the establishment of genetic transformation in tree plants. It has been engineered for insect resistance, herbicide tolerance, and decreased lignin content. Insect-resistant transgenic poplar plant has been approved for commercial-scale cultivation in China, wherein stable expression of the transgene was observed in 8- to 10-year-old transgenic plants, hence providing broad-spectrum resistance against insect pests [37]. Stable transformation of Cavendish banana cv. Grand Nain was also reported using *uidA* and the potential virus-resistance gene (BBTV) along with the *nptII* gene as a selectable marker expressed.

Diverse tropical plants including papaya, oil palm, cassava, Picea, Ulmus, and Pinus have been engineered for disease resistance, herbicide tolerance, and resistance against insect pests [38]. The first draft of the papaya genome sequence from the commercial virus-resistant transgenic fruit tree opened up new avenues for the genetic transformation of tree plants [39]. Papaya has a relatively small genome of 372 Mb, with nine pairs of chromosomes and diploid inheritance. Other desirable features of the papaya are short generation time (9–15 months), primitive sex chromosome system, and continuous flowering throughout the year. These features make papaya a promising system for the exploration of fruit tree genomics and tropical tree genomes.

The first commercialized transgenic papaya carrying coat protein for papaya ring spot virus, PRSV CP gene, was introduced in Hawaii in 1998. CP-transgenic papaya plants appeared to have variable levels of resistance against ring spot viral isolates from different geographical regions. Isolates from Florida, Bahamas, and Mexico have delayed and made symptoms mild, whereas isolates from Thailand and Brazil have delayed symptoms; as a result, the virus can overcome resistance and thus may cause pathogenicity. Rainbow, a hemizygous line, also appeared to be susceptible to viral isolates from Taiwan [40]. Hence, the level of resistance appeared to be different against isolates from different regions. Broad-spectrum resistance has also been attempted to develop through RNAi by targeting the conserved domain of the PRSV CP gene [41]. The disease-resistant transgenic papaya was reported to be environmentally safe, with no harmful effects on human health. Oil palm is another valuable tropical plant that has extensively been used for the production of edible oil. Parveez and Christou [42] reported its genetic transformation through a biolistic transformation using multiple gene(s) constructs, that is, gusA, bar, hpt under ubiquitin, and CaMV35S promoters. The resultant transformants were selected on 50 mg/L Basta. Bahariah et al. [43] published biolistic transformation, and resultant transformants were selected through a mannose selection system containing mannose @ 30 g/L. Oil palm genome has also been targeted for the production of polyhydroxybutyrate (biodegradable plastics). Three genes (phaB, bktB, and phaC) responsible for the bacterial PHB biosynthesis were expressed under the Ubi promoter. The expression of biodegradable plastic was detected to be in the range of 0.33 to 0.58 mg/g of dry weight. The transgenic oil palm plants showed normal growth; thus, no deleterious effects of transgene expression were observed on plant growth.

Cassava (*Manihot esculenta*) is the third major contributor of staple food in sub-Saharan Africa, where it is grown as a starch-storing root crop. It has been engineered for valuable agronomic traits as well as to boost its nutritional value. Zeoline protein has been expressed in roots wherein the total soluble protein was detected to be increased up to 12.5% of the dry weight, thus showing a fourfold increase in protein content as compared with non-transgenic plants. The Nigerian cultivars, TMS 91/02324 and TMS 95/0505, were engineered for resistance against CBSD and CMD.

The transformed cultivars showed an increased level of resistance against the noxious viral pathogens [44]. AtFER1 and AtIRT1were overexpressed in cassava for the increased accumulation of zinc and iron in roots (40 μg/g and 145 μg/g dry weight, respectively).

So, tropical plants have not only been engineered for improved agronomic performance and additive nutrients but also been engineered to clean up the indoor air and increase esthetic values. Indoor air often contains benzene, formaldehyde, chloroform, and other volatile organic compounds. Modern lifestyle has promoted the production of undesired molecules coming from furniture, smoking, and showering. A normal room needs at least 20 plants for the removal of these toxic molecules. The detoxifying ability of the plants can be increased through transgenic technology. Incorporation of mammalian cytochrome P450 2e1 (rabbit CYP2E1) in pothos ivy boosted its ability to detoxify the abovementioned hazardous molecules. The engineered plants were able to detoxify chloroform and benzene in the closed vials within 8 days of culture, thus showing great potential to detoxify the undesired molecules [45]. Likewise, sulfur metabolism can be engineered to improve resistance to SO_2. Transgenic tobacco plants overexpressing serine acetyltransferase and cysteine synthase gene(s) were highly tolerant to sulfite and SO_2 [46]. Engineering tropical plants with the said gene can uplift their ability to tolerate sulfite and sulfur dioxide.

4.4 Role of biotechnology to secure endangered plant species

Tropical plants occupy approximately 1/20th of the earth's surface [47]. They comprise 2/3rd of the terrestrial plant species globally. Most tropical regions are categorized into biodiversity hotspots, but they possess different challenges due to the increasing rate of the human population. The tropical biodiversity hotspots have increased habitat loss, species richness, and an increasing number of endemic species [48]. An important cause of these changes is deforestation due to agricultural and industrial expansion during the past 30 years. As a result, the diversity in the tropical ecosystem is endangered, which has high environmental concerns regarding biodiversity degradation and extinction of species. One example of a biodiversity hotspot is Sumatra, where ~0.84 million hectares of forest land has declined [49]. The same is happening in other developing countries. Intensified agriculture and changes in the use of tropical land have a durable impact on the global biodiversity, and their future consequences are just estimated.

It has been studied by Rodriguez-Echeverry et al. [50] that fragmentation, changes in land use, and habitat loss are causing the decline of biodiversity and the composition of species is changing. The ecosystem processes and invasive species are altering. In the past decade, various studies conducted to find its impact on the genetic composition of tropical plant species and genetic alterations were observed [51]. Habitat loss, population differentiation, and genetic diversity losses have consequences that are caused by inbreeding, genetic drift, and increased distances by isolation [52]. The outcome of these changes is also caused by different life traits such as dispersal strategy, the density of plant species, mating, and gene flow in tropical lands. The information on genetic resources collected from one of the few species cannot reflect the plant community. However, genetics and molecular biology provide efficient approaches to precisely calculate the genetic variability in tropical plants.

Conservation management is an important issue in the tropical ecosystem. Various human and social factors have been identified that are responsible for biodiversity loss [51]. Agriculture expansion, corruption, human growth, and incompetency in the development of genetic conservation strategies have increased the risk to the tropical ecosystem and sustainable management. The genetic information of the tropical species has increased the probability to maintain the genetic conservation of targeted plant species. The recommendations on the monitoring program and sustainable management in a particular fragment of tropical land are based on the genetic information of species, species richness, and processes of the ecosystem. Lack of sufficient genetic study along with population fragmentation data of different plant species could lead to the development of poor management practices for the conservation of plant species. The land-use change process is fast in the tropics, and there is a need for a robust method for identification of biodiversity hotspots and development of strategy in the identified area. Genetics alone does not provide a sufficient solution for the determination of the hotspots in the tropical plant community. Therefore, the emerging molecular and biotechnological techniques provide the required solutions to identify the genetic diversity in the high number of plants. These biotechnology-based tools provide the pre-requisite baseline information of the dominant species composition. It also helps to identify the high conservation value of different habitats that could be used in the conservative management practices of endangered species.

One important technique to investigate multiple plant species is amplified fragment length polymorphism (AFLP). After DNA extraction of the selected plant species from nearly 1 cm^2 leaf tissues of each plant, the AFLP protocol of Vos et al. [53] provides better results. The extra genetic material can be stored at −20°C for a longer period at optimized conditions. The samples were excised with the single enzyme to incubate overnight or to amplify with a single primer. In this regard, different PCR protocols are optimized for different plant species [54]. For the efficient reproducibility of the AFLP procedure, two to ten samples of each plant species provide reliable information. The fragments occurring from the restriction steps in the repetition of samples are considered important. The results obtained from the AFLP linked with PCR techniques can be visualized by transforming into the fragment presence-absence matrix. The analysis can be done by collecting samples of more than one hundred genotyped species. Different values of variation per species use the common genetic diversity indices [55].

In vitro conservation of germplasm included a large number of techniques involving the incubation of plant germplasm under controlled conditions. Although the nutrient requirement and conditions of growth vary for each tropical plant species, it provides a robust and reliable solution for the conservation of endangered species in the tropical region [56]. More commonly, the younger developing tissues of the explant can be used as the source. Genetic transformation has become an important tool in the additive one-point improvement in comparison with the mutation breeding that develops the subtractive one-point improvement. The genetically modified ornamental plants can be more acceptable to the consumers due to their vibrant colors as compared to food crops, where different ethical concerns are present regarding the use of recombinant technology [57]. Different GM technologies include zinc finger proteins, RNA interference, miRNAs, and CRISPR-Cas-9, which can be applied in the development of transgenic plants (**Figure 1**). The quality of the explant and type of tissue help to develop the biotechnology-based strategy for the regeneration of endangered species. The genome-editing technology has been vastly improved in the previous years. After the advent of genome-wide scan analysis and next-generation sequencing, varied information becomes available on tropical indoor and outdoor plants. The available information on the sequences of tropical plants could provide great help in breeding and basic research.

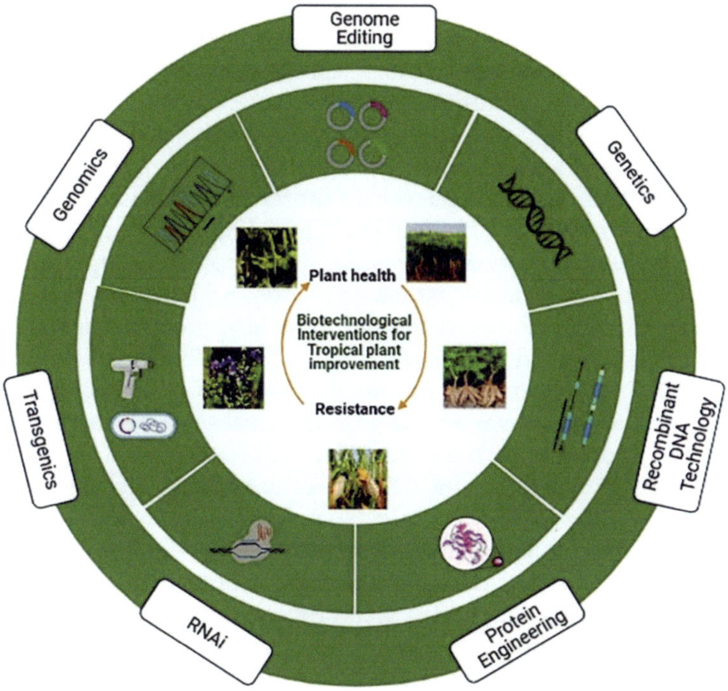

Figure 1.
Contribution of biotechnology to the improvement of germplasm and agriculture in tropical areas. The DNA of tropical plants can be genetically engineered for the improvement of different traits, conservation of endangered species, and resistance against pathogens.

The heterogeneity of the different biodiversity plots and land-use systems can be studied by using the fragment distance matrix that is based on the principal component analysis. The bioinformatics and statistical tools can be used by applying the function betadisper and R-package ggplot2. Precise estimation of the genetic diversity in an area of dominant species depends on the spatial scales of alpha, beta, and gamma. The alpha scales respond to the diversity within the plot, the beta scale is for the land-use system, while the gamma scale is the highest level of spatial scales. In the fragment pool approach, the differentiation at the alpha level is determined by using at least 10 fragment pairs within each plot. However, the β diversity level can be calculated by taking at least 40 pairwise fragments in every land system. On the other side, the γ diversity is based on using the 160 genetic fragments from the plots where the concerned species are dominant. To apply this technique, the Shannon Index can be applied within each plot.

5. Biotechnological challenges to improving tropical plants

The flora of tropical areas suffers from biotic and abiotic stresses. In the past few years, the effects of climatic changes are significant on tropical plants. It has been reported by the Intergovernmental Panel on Climate Change (IPCC) that the climatic conditions are a precursor to various stresses on plants and are considered the most important influencing factor in the decline of agricultural production in developing countries. Global warming negatively influences tropical and sub-tropical plants due to rapid alterations in the ecosystem, drought, rainfall patterns, floods, and biological outbreaks [58]. The balanced environment and lack of nutrients are the major constraints that enhance the losses of biological stresses. The losses due to abiotic stresses could be around 50 percent in tropical areas possibly due to a decline in plant metabolism [59].

The tropical plants of agricultural importance are suffering from viruses, bacteria, and insect attacks. Although there has been a substantial increase in food production in developing countries due to advances in breeding practices, the challenges of food security have never been met. The use of biotechnological approaches enables the genome alterations in plants to withstand abiotic and biotic stresses (**Figure 2**), which are usually difficult to achieve by conventional approaches. Advanced genome editing tools such as RNA-induced gene silencing, CRISPR-Cas, genome mapping, and next-generation sequencing have paved the way toward desirable genetic alterations in plants. However, the available biotechnological applications to agriculture in conventional crop species are the use of selective breeding approaches to bring improvement in genetic materials [66]. Some of the conventional crops of the tropical areas include maize, tomato, sweet potato, beans, pulses, nuts, cassava, sorghum, etc. These crops are present in the wet climate in tropical areas. On the other hand, the important crops of dry tropical climates are coffee, rubber, cotton, tobacco, tea, and sugarcane. The ornamental plants have their range such as hibiscus, plumeria, palms, ebony, teak, gardenia, fire brush.

Biotechnology has contributed to the solution of food security challenges by making robust genetic mutations. According to ISAAA infographics, more than 17 million farmers are benefiting from the cultivation of biotechnology crops. More

Figure 2.
(A) Coffee berry borer (modified from Benavides et al. [60]), (B & E) Coccinia grandis (modified from; Goebel et al. [61]), (C) coffee leaf rust (modified from Nelson, [62]; Gichuru et al. [63]), (D) stem borer larval caterpillar on sugarcane stalk (modified from Cilas et al. [64]), and (F) cacao swollen shoot virus (modified from; Kouakou et al. [65]).

than 28 countries are growing 14 biotechnology crops precisely, and major contributors are tropical countries such as Brazil, India, Argentina, China, Indonesia, and Paraguay. The area under biotech crops is 2.7 billion hectares, and 17 million farmers' earnings are associated with it [67]. The genetic alteration strategies are designed to increase crop production, improve quality, and develop resistance against biological agents such as pests, herbs, and bacterial and viral diseases. With the advancement of technology, biotechnological approaches have become robust, accurate, and more reliable, but there are some ethical and environmental concerns about their agricultural applications.

The first contribution of biotechnology to crop improvement came with the development of crops resistance to a broad range of selective herbicides. The biotechnological techniques proved valuable on a socioeconomic as well as on an ecological scale because they reduced the consumption of questionable herbicides. However, due to unorganized applications of herbicides over growing seasons, the resistance to herbicides is emerging that could overcome the advantage of selective tolerance. The scientific challenge to address this issue is diversification of herbicide consumption along with crop rotation. There is the need to make continuous improvements in resistance technologies and the identification of new biochemical pathways in plants to find new targets. A similar challenge resides in another trait of resistance to insect pests, which is also commercially exploited. Many commercially important crops in tropical climate regions are threatened by different insect pests such as coffee rust, Lepidopteran stem borers, Helicoverpa, coffee berry borers [64].

The most important biological challenges are plant viruses because of their complexity in their life cycle, replication, movement, diversity, and unique capability of developing mutations against resistance. Identification of a causative virus and its pathogen vector is crucial for estimating the epidemiology and economic losses that are essential for the development of management strategy. Viruses can only be detected by using molecular diagnostic techniques. Some potential virus threats of tropical crops are cacao swollen shoot virus, banana bunchy top virus, cassava mosaic virus, plum pox virus, potato virus X, cucumber mosaic virus, African oil palm ring spot virus, rice yellow stunt virus, etc. [68].

Since after their inception, biotechnological approaches have established a reputation as a potential way to overcome food shortages. The choice of biotechnological approaches depends on the targeted pathogen or abiotic stress [69]. For example, the short RNA-based RNAi approaches are efficient against RNA viruses because the viral genome can be targeted before replication and formation of proteins, but to target the DNA genome, CRISPR-Cas is a superior technology. Violation of biosafety measures could cause unwanted gene flow to the other plant or pathogen species.

Although biotech crops are becoming an important hope in the tropical region, some challenges also emerged with the increase in production of biotech crops. For example, with the continuous application of broad-spectrum herbicide on the GM crop plants, resistance also emerges in the weeds, which lose the selective advantage in genetically modified plants. Another area that has been addressed more by biotechnological approaches is the use of bacterial toxins to develop transgenic plants for durable resistance to insect pests. So far, the toxins from *Bacillus thuringiensis* are mostly applied for pest tolerance.

Plant biology is facing an additional level of complexity as compared to mammals because plants are sessile and grow in varied environmental conditions that are far from optimal climate conditions. There is evidence that changes in the ecosystem alter the pathogenicity of viruses and plant pests. In the past few years, various unpredictable changes have been reported in the tropical ecosystem, which need continuous investigation of the genome characterization of viruses, pests, and the response of host plants. Changes in humidity, temperature, and atmospheric pressure affect the growth of insect pathogens. The development of the post-transcriptional gene silencing strategy needs a very comprehensive study of the host, environment, and pathogen; otherwise, there are chances of horizontal gene transfer to other species [70]. Climate changes lead to ecosystem disturbances at different levels, which affect the efficiency of biotechnology techniques. The climatic changes affect the interaction of the ecological and biological communities including soil, natural habitats, plants, and biodiversity. For example, the plant susceptibility to disease increases due to high temperature and humidity [71]. However, these challenges can be overcome by combining conventional and biotechnological approaches to address biotic and abiotic stresses.

6. Future prospects

Tropical plants comprise two-thirds of the terrestrial plant species and have great importance in plant biodiversity. Due to increasing population, there is an urgency to double the food production globally. Thus, a series of biotechnological interventions can be made to conserve plant biodiversity (Van Montagu, 2020) and improve crop plants in tropical areas. Medicinal and fruit plants also need attention with regard to biotechnology-assisted breeding to combat biotic and abiotic stresses. Tropical forests can make an important contribution to the global demand for fruits, timber, and biomass. Biotechnological tools should be used to identify the potential of plants as new crops to mitigate malnutrition in tropical regions. Some important crops in tropical regions such as cassava, cowpea, sweet potatoes can be improved as cash crops by biotechnological efforts. Moreover, the nutritional contents of plants can be improved. For example, lathyrus is a leguminous protein-rich crop that is used after overnight soaking to remove the toxins on its split seeds. Engineering lathyrus genome could help in this regard. There are also gaps in the study of marker-assisted breeding on tropical plants, which can become an important tool in crop improvement. The tropical region has a range of medically important plants, and their properties can be broadened to increase diversity. Latest molecular biotechnology tools such as CRISP/Cas 9 have promising applications in gene activation, repression, gene mutation, and epigenetics. It has been effectively applied to citrus, apple, petunia, and various other plants. Similar research models can be applied to economically important tropical plants.

7. Conclusion

The sustainability of the food supply chain depends on tropical plants, which are also vital sources of medicine, fiber, wood, and energy. Tropical plants are present in more than 60% of all terrestrial plant species worldwide. In addition to domestication, attempts have been made to enhance these plants using both conventional and cutting-edge scientific interventions. Parthenocarpy, polyembryony, polyploidy, protracted juvenile phase, heterozygosity, and generation cycle problems have been helped by genomic techniques. Innovations in cell and tissue culture have enabled micro-propagation and the commercial plantation of various plants, in addition to aiding in the conservation of endangered tropical species. Additionally, transgenic technology has made it feasible to combine genes from multiple species, enabling plants to withstand biotic and abiotic challenges as well as other types of stress. Another useful intervention that has helped to create plants with the desired features and the capacity to mitigate potential climate change difficulties is genome editing. Therefore, scientific advancements are essential for both the preservation of tropical plant species and the cultivation of tropical plants that will meet future demands.

Author details

Irfan Ali[1], Faiz Ahmad Joyia[1], Ghulam Mustafa[1], Safdar Ali Mirza[2] and Muhammad Sarwar Khan[1*]

1 Center of Agricultural Biochemistry and Biotechnology (CABB), University Road, University of Agriculture, Faisalabad, Pakistan

2 Department of Botany, GC University, Lahore, Pakistan

*Address all correspondence to: sarwarkhan_40@hotmail.com

References

[1] Yadav A, Verma P, Raghubanshi AS. An overview of the role of plant functional traits in tropical dry forests. Research anthology on ecosystem conservation and preserving. Biodiversity. 2022;**2022**:92-117

[2] Winters G, Beer S, Willette DA, Viana IG, Chiquillo KL, Beca-Carretero P, et al. The tropical seagrass Halophila stipulacea: Reviewing what we know from its native and invasive habitats, alongside identifying knowledge gaps. Frontiers in Marine Science. 2020;**7**:300

[3] Liu C, Huang Y, Wu F, Liu W, Ning Y, Huang Z, et al. Plant adaptability in karst regions. Journal of Plant Research. 2021;**134**(5):889-906

[4] Winchell F, Brass M, Manzo A, Beldados A, Perna V, Murphy C, et al. On the origins and dissemination of domesticated sorghum and pearl millet across Africa and into India: A view from the Butana Group of the Far Eastern Sahel. African Archaeological Review. 2018;**35**(4):483-505

[5] Singh G, Dukariya G, Kumar A. Distribution, importance and diseases of soybean and common bean: A review. Biotechnology Journal International. 2020;**24**(6):86-98

[6] Shahrajabian MH, Sun W, Cheng Q. A short review of health benefits and nutritional values of mung bean in sustainable agriculture. Polish Journal of Agronomy. 2019;**37**:31-36

[7] Ozias-Akins P, Breiteneder H. The functional biology of peanut allergens and possible links to their allergenicity. Allergy. 2019;**74**(5):888-898

[8] Bester AU, Carvalho IR, da Silva JAG, Hutra DJ, Moura NB, Lautenchleger F, et al. Three decades of cassava cultivation in Brazil: Potentialities and perspectives. Revista Colombiana de Ciencias Hortícolas. 2021;**15**(2):e12087

[9] Santos JO, Araújo AS, Gomes RL, Lopes ACA, Figueiredo MV. Rhizobia Phaseolus lunatus Symbiosis: Importance and diversity and tropical soils-a review. Dynamic Soil, Dynamic Plant. 2008;**2**(2):56-60

[10] Duhn V. Tropical nephrology: A review. International Journal of Child Health and Human Development. 2021;**14**(2):211-222

[11] Sheldon KS. Climate change in the tropics: Ecological and evolutionary responses at low latitudes. Annual Review of Ecology, Evolution, and Systematics. 2019;**50**:303-333

[12] Bird MI, Haig J, Hadeen X, Rivera-Araya M, Wurster CM, Zwart C. Stable isotope proxy records in tropical terrestrial environments. Palaeogeography, Palaeoclimatology, Palaeoecology. 2020;**538**:109445

[13] Rada F, Azócar A, García-Núñez C. Plant functional diversity in tropical Andean páramos. Plant Ecology & Diversity. 2019;**12**(6):539-553

[14] Sakai S, Kitajima K. Tropical phenology: Recent advances and perspectives. Ecological Research. 2019;**34**(1):50-54

[15] Hasnat GT, Hossain MK. Global overview of tropical dry forests. Handbook of Research on the Conservation and Restoration of Tropical Dry Forests. 2020;**2020**:1-23

[16] Abdullah SRS, Al-Baldawi IA, Almansoory AF, Purwanti IF,

Al-Sbani NH, Sharuddin SSN. Plant-assisted remediation of hydrocarbons in water and soil: Application, mechanisms, challenges, and opportunities. Chemosphere. 2020;**2020**(247):125932

[17] Purugganan MD. Evolutionary insights into the nature of plant domestication. Current Biology. 2019;**29**(14):R705-R714

[18] Zhang F, Shi Y, Ali J, Xu J, Li Z. Breeding by selective introgression: Theory, practices, and lessons learned from rice. The Crop Journal. 2021;**9**(3):646-657

[19] Meyer RS, DuVal AE, Jensen HR. Patterns and processes in crop domestication: An historical review and quantitative analysis of 203 global food crops. New Phytologist. 2012;**196**(1):29-48

[20] Turner-Hissong SD, Mabry ME, Beissinger TM, Ross-Ibarra J, Pires JC. Evolutionary insights into plant breeding. Current Opinion in Plant Biology. 2020;**54**:93-100

[21] Louwaars NP. Plant breeding and diversity: A troubled relationship? Euphytica. 2018;**214**(7):1-9

[22] Saxena KB, Saxena RK, Hickey LT, Varshney RK. Can a speed breeding approach accelerate genetic gain in pigeonpea? Euphytica. 2019;**215**(12):1-7

[23] Nyouma A, Bell JM, Jacob F, Cros D. From mass selection to genomic selection: One century of breeding for quantitative yield components of oil palm (*Elaeis guineensis* Jacq.). Tree Genetics & Genomes. 2019;**15**(5):1-16

[24] Enri SR, Petrella F, Ungaro F, Zavattaro L, et al. Relative importance of plant species composition and environmental factors in affecting soil carbon stocks of Alpine Pasures. Agriculture. 2021;**11**:1047

[25] Namasivayam P. Acquisition of embryogenic competence during somatic embryogenesis. Plant Cell Tissue and Organ Culture. 2007;**90**:1-8

[26] Loschiavo F, Pitto L, Giuliano G, Torti G, Nuti-Ronchi V, Marazziti D, et al. DNA methylation of embryogenic carrot cell cultures and its variations as caused by maturation, differentiation, hormones and hypomethylation drugs. Theoretical and Applied Genetics. 1989;**77**:325-331

[27] Jayanthi M, Mohan NM, Mandal PK. Direct somatic embryogenesis and plantlet regeneration in oil palm. Plant Biochemistry and Biotechnology. 2011;**20**:249-251

[28] Tobias M, Marc G, Luis S. Correlation of mRNA and protein in complex biological samples. FEBS Letters. 2009;**583**:3966-3973

[29] Campos NA, Paiva LV, Panis B, Carpentier SC. The proteome profile of embryogenic cell suspensions of *Coffea arabica* L. Proteomics. 2016;**16**:1001-1005

[30] Garcia G, Romero SE, Panis C, Carpentier SC. The use of 2D-DIGE to understand the regeneration of somatic embryos in avocado. Proteomics. 2013;**13**:3498-3507

[31] Biswas MK, Bagchi M, Biswas D, Harikrishna JA. Genome-wide novel genic microsatellite marker resource development and validation for genetic diversity and population structure analysis of banana. Genes. 2020;**11**:1479

[32] Amorim EP, Vilarinhos AD, Cohen KO, Amorim VBO, Santos-Serejo JA, Oliveira S, et al. Genetic diversity of carotenoid-rich bananas

evaluated by diversity arrays technology (DArT). Genetics and Molecular Biology. 2009;**103**:96-103

[33] Devarajan R, Jayaraman JK, Somasundaram SM, Ragupathy S, Raman P, Sathiamoorthy K, et al. Genetic diversity in fresh fruit pulp mineral profile of 100 indian musa accessions. Food Chemistry. 2021;**361**:130080

[34] Tao J, Qiao G, Wen XP, Gao GL, Liu T, Peng ZJ, et al. Characterization of genetic relationship of dragon fruit accessions (Hylocereus spp.) by morphological traits and ISSR Markers. Scientia Horticulturae. 2014;**170**:82-88

[35] Bohry D, Christine H, Ramos C, Henrique P, Santana M, Boechat B, et al. Discovery of SNPs and InDels in papaya genotypes and its potential for marker assisted selection of fruit quality traits. Scintific Reports. 2021;**11**:292

[36] Bally ISE, Bombarely A, Chambers AH, Cohen Y, Dillon NL, Innes DJ, et al. The 'Tommy Atkins' mango genome reveals candidate genes for fruit quality. BMC Plant Biology. 2021;**21**:108

[37] Chang S, Mahon EL, MacKay HA, Rottmann WH, Strauss SH, Pijut PM, et al. Genetic engineering of trees: Progress and new horizons. In Vitro Cellular & Developmental Biology – Plant. 2018;**54**:341-376. DOI: 10.1007/s11627-018- 9914-1

[38] Ren Y, Zhou X, Dong Y, Zhang J, Wang J, Yang M. Exogenous gene expression and insect resistance in dual Bt toxin *Populus* × euramericana 'neva' transgenic plants. Frontiers in Plant Sciences. 2021;**12**:660226. DOI: 10.3389/fpls.2021.660226

[39] Ming R, Hou S, Feng Y. The draft genome of the transgenic tropical fruit tree papaya (*Carica papaya* Linnaeus). Nature Letters. 2008;**452**:24. DOI: 10.1038/nature06856

[40] Kung YJ et al. Generation of transgenic papaya with double resistance to papaya ringspot virus and papaya leaf-distortion mosaic virus. Phytopathology. 2009;**99**:1312-1320

[41] Jia R, Zhao H, Huang J, Kong H, Zhang Y, Guo J, et al. Use of RNAi technology to develop a PRSV-resistant transgenic papaya. Scientific Reports. 2017;7(12636)

[42] Parveez GKA, Christou P. Biolistic-mediated DNA delivery and isolation of transgenic oil palm (Elaeis guineensis Jacq.) embryogenic callus cultures. Journal of Oil Palm Research. 1998;**10**:29-38

[43] Bahariah B, Parveez GKA, Masani MYA, Masura SS, Khalid N, Othman RY. Biolistic transformation of oil palm using the phosphomannose isomerase (pmi) gene as a positive selectable marker. Biocatalysis and Agricultural Biotechnology. 2013;**2**:295-304

[44] Abhary M, Siritunga D, Stevens G, Taylor NJ, Fauquet CM. Transgenic biofortification of the starchy staple cassava (Manihot esculenta) generates a novel sink for protein. PLoS One. 2011;**6**(1):e16256. DOI: 10.1371/journal.pone.0016256

[45] Zhang L, Routsong R, Strand SE. Greatly enhanced removal of volatile organic carcinogens by a genetically modified houseplant, Pothos Ivy (*Epipremnum aureum*) expressing the mammalian cytochrome P450 2e1 gene. Environmental Science and Technology. 2018;**53**:1

[46] Noji M, Saito M, Nakamura M, Aono M, Saji H, Saito K. Cysteine synthase overexpression in tobacco

confers tolerance to sulfur-containing environmental pollutants. Plant Physiology. 2001;**3**:973-980. DOI: 10.1104/pp.126.3.973

[47] Gardner TA, Barlow J, Chazdon R, Ewers RM, Harvey CA, Peres CA, et al. Prospects for tropical forest biodiversity in a human-modified world. Ecology Letters. 2009;**12**:561-582

[48] Myers N. Threatened biotas: "Hot spots" in tropical forests. Environmentalist. 1988;**8**:187-208

[49] Margono BA, Potapov PV, Turubanova S, Stolle F, Hansen MC. Primary Forest cover loss in Indonesia over 2000-2012. Nature Climate Change. 2014;**4**:730-735

[50] Rodriguez J, Echeverria C, Oyarzun C, Morales L. Impact of land use change on biodiversity and ecosystem services in the Chilean temperate forests. Landscape Ecology. 2018;**22**:5982

[51] Sloan S, Jenkins CN, Joppa LN, Gaveau D, Laurance WF. Remaining natural vegetation in the global biodiversity hotspots. Biological Conservation. 2014;**177**:12-24

[52] Sampson JF, Byrne M, Yates CJ, Gibson N, Thavornkanlapachai R, Stankowski S, et al. Contemporary pollen-mediated gene immigration reflects the historical isolation of a rare, animal-pollinated shrub in a fragmented landscape. Heredity. 2014;**112**:172-181

[53] Vos P, Hoger R, Bleeker M, Reijans M, van de Lee T, Homes M, et al. AFLP: A new technique for DNA fingerprinting. Nucleic Acids Research. 1995;**23**:4407-4414

[54] Kuchma O. Genetic Processes in Scots Pine (*Pinus sylvestris* L.). Gottingen, Germany: Georg-August-University; 2010

[55] Breidenbach N, Rahayu S, Siregar IZ, Siregar UJ, Hamzah and Finkeldey R. Genetic diversity of dominant plant species in tropical land use systems in Sumatra, Indonesia. Tropical Conservation Service. 2018;**11**:1-14

[56] Ahn CH, Ramya M, An HR, Park PM, Kim Y, Lee SY, et al. Progress and challenges in the improvement of ornamental plants by genome editing. Plants. 2020;**9**:687

[57] Nishihara M, Nakatsuka T. Genetic engineering of flavonoid pigments to modify flower color in floricultural plants. Biotechnology Letters. 2011;**33**:433-441

[58] Pautasso M, Doring TF, Garbelotto M, Pellis L, Jeger MJ. Impact of climate change on plant disease – Opinions and trends. European Journal of Plant Pathology. 2012;**133**:295-313

[59] Onyekachi OG, Boniface OO, Gemlack NF, Nicholas N. The Effect of Climate Change on Abiotic Plant Stress: A Review. London, UK: Intechopen; 2019

[60] Benavides P, Gongora CE, Bustillo AE. IPM program to control Coffee berry borer hypothenemus hampei, with emphasis on highly pathogenic mixed strains of Beauveria bassiana to overcome insecticide resistance in Colombia. In: Insecticides – Advances in Integrated Pest Management. London, UK: Intechopen; 2012

[61] Goebel FR, Sallam N. New pest threats for sugarcane in the new bioeconomy and how to manage them. Current Opinion in Environmental Sustainability. 2011;**3**:81-89

[62] Nelson S. Ivy Gourd (*Coccinia grandis*): Powdery Mildew. Flickr; 2014. Available from: https://www.flickr.com/photos/scotnelson/14162084871

[63] Gichuru E, Alwora G, Gimase J, Kathurima C. Coffee leaf rust (Hemileia Vastatrix) in Kenya – A review. Agronomy. 2021;**11**:2590

[64] Cilas C, Goebel F, Babin R, Avelino J. Tropical crop pests and diseases in a climate change setting – A few examples. In: In Book: Climate Change and Agriculture Worldwide. USA: Springer Publishers; 2016. pp. 73-82

[65] Kouakou K, Kebe BI, Kouassi N, Ake S, Cilas C, Muller E. Geographical distribution ofcacao swollen shoot virus molecular variability in Côte d'Ivoire. Plant Disease. 2012;**96**:1445-1450

[66] Taylor NJ, Fauquet CM. Biotechnology's greatest challenge. Available from: http://www.agbioworld.org/biotech-info/topics/dev-world/challenge.html [Accessed: 27th October 2022]

[67] Dinoglay C. Infographic: Here's where GM Crops are grown around the world today. Ithaca, NY: ISAAA; 2021

[68] Geering ADW, Randles J. Virus Diseases of Tropical Crops: BeLS. Chichester, USA: John Wiley & Sons Ltd; 2012

[69] Bigini V, Camerlengo F, Botticella E, Sestilli F, Savatin DV. Biotechnological resources to increase disease resistance by improving plant immunity: A sustainable approach to save cereal crop production. Plants. 2021;**10**:1146

[70] Casacuberta JM, Devos Y, du Jardin P, Ramon M, Vaucheret H, Nogue F. Biotechnological uses of RNAi in plants: Risk assessment considerations. Trends in Biotechnology. 2015;**33**:145-147

[71] Velasquez AC, Castroverde CDM, He SY. Plant and pathogen warfare under changing climate conditions. Current Biology. 2018;**28**:R619-R634

Chapter 3

Applications of Biotechnological Approaches in the Product and Breeding of *Phalaenopsis* Orchids

Shinichi Enoki and Yoshinori Takahara

Abstract

Phalaenopsis orchids native to the tropics are called "Moth Orchids". It is one of the most commercially popular orchids because of its beautiful, colorful, and long-lasting variety of flowers. Biotechnology used in the production and breeding of *Phalaenopsis* was reviewed in this chapter. In the commercial production of *Phalaenopsis*, biotechnologies, such as methods of aseptic sowing and tissue culture, have been used for a long time. Recently, molecular phylogenetic analysis of original species and molecular breeding by the transformation of *Phalaenopsis* has been actively studied. The role of biotechnology in the *Phalaenopsis* orchid industry is significant, and the development of the technology in this field will bring further benefits to researchers, producers, and fancier of *Phalaenopsis* orchids.

Keywords: orchids, *Phalaenopsis*, classification, micropropagation, molecular breeding

1. Introduction

The genus *Phalaenopsis* consists of approximately 60 species and the various traits of hybrids are due to easy interspecific and intergeneric crossing compared to other higher plants. Germination and propagation of *Phalaenopsis* in nature (or under natural conditions) are very difficult since their seeds contain no endosperm storing nutrients for germination. Therefore, "micropropagation" (mass proliferation) by aseptic culture technologies has been used for the research and industrial production of *Phalaenopsis* for a long time before. In recent years, molecular breeding by biotechnology has been extensively studied. In this chapter, we will review the latest knowledge of classification, proliferation methods, and molecular breeding of *Phalaenopsis* by biotechnology.

2. *Phalaenopsis* and related genera

2.1 Classification

The genus *Phalaenopsis* (Orchidaceae) is classified as subfamily *Epidendroideae*, tribe *Vendeae*, and subtribe *Aeridinae* [1]. The native species of *Phalaenopsis* are distributed throughout northern Australia to southern India, China, and Taiwan in tropical Asia. Cultivars called moth orchids (*Phalaenopsis* and *Doritaenopsis*) are mainly generated by crossing native species of genus *Phalaenopsis* and genus *Doritis* (*Doritis pulcherrima*). Because *Phalaenopsis* is amenable to artificial crossing with other species and even other genera, such as *Doritis*, *Ascocentrum,* and *Vanda*, many of the cultivars have been produced as interspecific and intergeneric hybrids [2].

2.1.1 Morphological classification

Phalaenopsis Orchids have been classified morphologically by unique features, such as pollen. *Phalaenopsis* are epiphytic orchids, which live sticking to trees. They are monopodial plants with a short stalk and three to six widely and fleshy leaves. Their flowers consist of sepal, petal, lip, and column, which are flower structures particular in Orchids. The genus *Phalaenopsis* are defined by Blume in 1825 and has been classified by many taxonomists mainly based on morphological features of flower structure [3] and the number of pollens [2, 4] and based on cytogenetic features, [5, 6] such as a number of chromosomes, chromosome shape, and permissibility of crossing.

Christenson [7] defined the genus *Phalaenopsis* which consists of 62 original species. He divided the genus into five subgenera by morphological classification. Of these, two subgenera also were subdivided into four sections. He integrated the genus *Doritis*, which has been treated as an independent genus by other taxonomists, into the genus *Phalaenopsis* (section *Esmeralda*) in the broad sense (**Table 1**). He systematically described species characteristics, habitat, history of discovery, etc. in this work. Currently, his work is one of the most referenced in the classification of *Phalaenopsis* orchids.

2.1.2 Molecular phylogenetic classification

Differences in opinion on the importance of morphological features, such as pollen numbers caused disagreement among taxonomists. Therefore, molecular phylogenetic analyses based on DNA information independent from morphology have been actively studied. Molecular phylogenetic analysis [8–11] supports Christenson's proposal that the closely related genus *Doritis* and *Kingidium* should be included in the genus *Phalaenopsis* (section *Esmeralda* and subgenus *Aphyllae*, respectively). However, because there were many consequences that classifications under subgenera do not match his classifications, reexaminations were proposed. Although many results that genus *Doritis* is classified into genus *Phalaenopsis* are shown, some taxonomists proposed that genus *Doritis* should be used, considering the established *Phalaenopsis* intergeneric hybrids of *Doritaenopsis* in the past.

Recently, researchers reported that distantly related genera *Lesliea*, *Nothodoritis*, *Ornithochilus*, *Hygrochilus*, etc. should be included in the genus *Phalaenopsis* [11, 12]. In the genus *Phalaenopsis* subgenus *Hygrochilus*, a new species, *Phal. pingxiangensis* were discovered in China [13]. Due to proposals for revision of classification criteria based

DOI: http://dx.doi.org/10.5772/intechopen.104597

Genus	Subgenus	Section	Species
Phalaenopsis	*Proboscidioides*	*Proboscidioides*	*lowii*
	Aphyllae	*Aphyllae*	*taenialis*
			braceana
			minus
			wilsonii
			stobartiana
			haiananensis
			honghenensis
	Parishianae	*Parishianae*	*appendiculata*
			gibbosa
			parishii
			lobbii
	Polychilos	*Polychilos*	*mannii*
			cornu-cervi
			borneensis
			pantheriana
		Fuscatae	*cochlearis*
			viridis
			fuscata
			kunstleri
		Amboinenses	*pulchra*
			violacea
			bellina
			micholizii
			fimbriata
			floresensis
			robinsonii
			gigantea
			fasciata
			doweryensis
			luteoka
			modesta
			maculata
			javanica
			mariae
			amboinensis
			luddemanniana
			venosa

Genus	Subgenus	Section	Species
			reichenbachiana
			pallens
			bastianii
			hieloglyphica
		Zebrinae	*inscriptioshinensis*
			speciosa
			tetraspis
			corningiana
			sumatrana
	Phalaenopsis	*Phalaenopsis*	*philippinensis*
			stuartiana
			amabilis
			aphrodite
			sanderiana
			schilleriana
		Deliciosae	*chibae*
			deliciosa
			mysorensis
			buyssoniana
		Esmeralda	*pulcherrima*
			regnieriana
		Stauroglottis	*equestris*
			celebensis
			lindenii

Table 1.
Classification of Phalaenopsis. *Created with reference to Christenson [7].*

on the molecular phylogeny of *Phalaenopsis* and related genera, the classification of *Phalaenopsis* orchids will become more diverse than ever.

2.2 Cultivars

The registration system for new cultivars of Orchids was established by Sander (Sander's Complete List of Orchid Hybrids [14]), and now the Royal Horticultural Society in the United Kingdom (RHS) is taking over the system. Thus, the history of hybridization of orchid cultivars (horticultural varieties) can be traced to their original species. Today, the database of Sander's list makes it easy for us to search for the ratio of each original species constituting orchid hybrids.

Major cultivars on the current market of moth orchids are divided into two groups (standard or novelty) (**Figure 1**). Standard types include traditional cultivars with white, pink, semi-alba (white flower with a red lip), and striped big flowers. Novelty

Figure 1.
Examples of cultivar types (standard, novelty).

types include cultivars with new colorful flowers, such as red, orange, yellow, multiple flowers, flowers of dots (spotted) or mottle (harlequin), and flowers with fragrance. Phylogenetic analysis of recent most popular cultivars revealed their original species composition as ancestors of these hybrids [15]. In standard cultivars, original species of subgenus *Phalaenopsis* were the most important ancestors. Most white flower hybrids were the progeny of *Phal. amabilis*, *Phal. aphrodite*, *Phal. schilleriana*. The pink flower hybrids were progeny of *Phal. amabilis*, *Phal. schilleriana* and *Phal. sanderiana*. *Phal. equestris* and *Phal. stuartiana* was important for the creation of semi-alba and striped hybrid cultivars, respectively. In novelty cultivars, original species in such subgenus *Polycilos* other than subgenus *Phalaenopsis* were important ancestors. Of spotted/harlequin cultivars, the genetic contribution in the generation of red spots of the famous hybrid *Phal.* Golden Poker "Brother" was 25, 18.75, 12.5, and 6.25% from *Phal. gigantea*, *Phal. leuddemanniana*, *Phal. Amboinensis,* and *Phal. faciata*, respectively.

3. Micropropagation

Phalaenopsis orchids are very difficult to germinate in nature because their seeds have no endosperm with nutrients for germination and to vegetatively propagate such as the method of bulb division. Therefore, propagation thorough aseptic culture has been desired. This section reviews the mass propagation methods using the tissue culture technic (micropropagation) and its problems in *Phalaenopsis*.

3.1 Aseptic sowing method

The aseptic sowing method greatly affected the industrial production of *Phalaenopsis* Orchids. A method of aseptically germinating the *Phalaenopsis* seeds with no endosperm on a medium that can artificially supply nutrients was developed. Many new cultivars have been created and produced in this method using various mediums, such as Knudson medium [16, 17], Vacin and Went (VW) medium [18], Murashige and Skoog (MS) medium [19], and Hyponex (Kano) medium [20]. However, the characteristics and quality of mature plants derived from seedlings are tending to vary genetically. Therefore, the development of a method to propagate moth orchids vegetatively using tissue culture and the production of clonal plantlets with the same traits has been desired.

3.2 Micropropagation

Protocorm-like bodies (PLBs) are generically used in the micropropagation of *Phalaenopsis*. PLBs are cell masses similar to protocorm, which is the state of enlarged embryos during orchid seed germination. PLBs are somatic embryos induced from somatic cells of Orchids [21]. Since a PLB can form a number of new secondary PLBs on the surface by culturing on an appropriate medium, the proliferation efficiency is very high (**Figure 2**) and then they grow to plantlets. On the other hand, callus induction is difficult in *Phalaenopsis* Orchid and embryogenic callus (EC) induction was first reported by Sagawa [22]. Although studies on proliferation using, such as EC, have also been conducted [23–25], the methods using PLBs are still mainstream in *Phalaenopsis* micropropagation because PLB is easier to grow to plantlet than a callus. To date, PLB induction methods using a variety of plant tissue have been established, as shown in **Figure 3.**

3.2.1 Flower stalk culture

Flower stalk culture is firstly performed in a vegetative propagation system of *Phalaenopsis* for PLB induction. In other Orchidaceae plants, PLB induction from shoot apex (shoot apical meristem) has been established. However, in monopodial *Phalaenopsis* orchids, varieties of alternate culture methods have been studied since only one shoot apex can be obtained from one strain and the removal of the shoot apex means the disappearance of the mother plant. Thus, flower stalk buds were firstly used for vegetative propagation of *Phalaenopsis* orchids [26]. Flower stalk culture is a method for obtaining the plantlets from dormant buds on the flower stalk. Although vegetative propagation systems that do not damage mother plants have been established by many researchers [27–30], the propagation efficiency of this method is

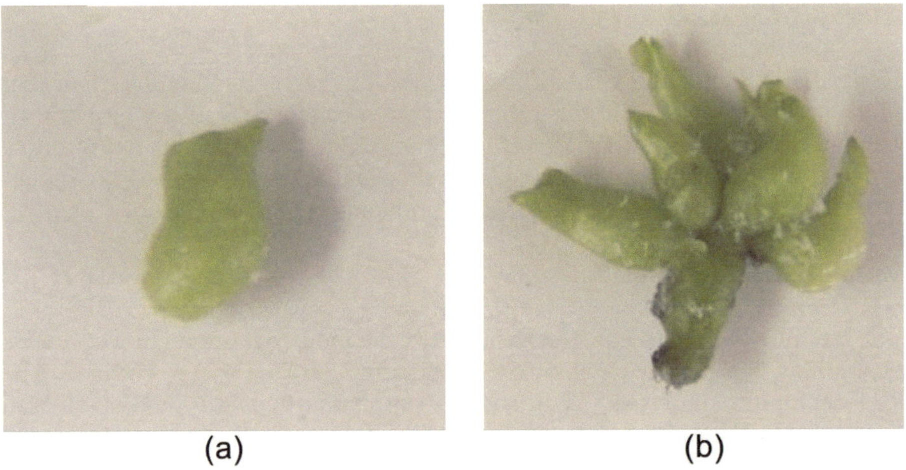

Figure 2.
PLB proliferation. a: PLB. b: Secondary PLBs formed on the original PLB.

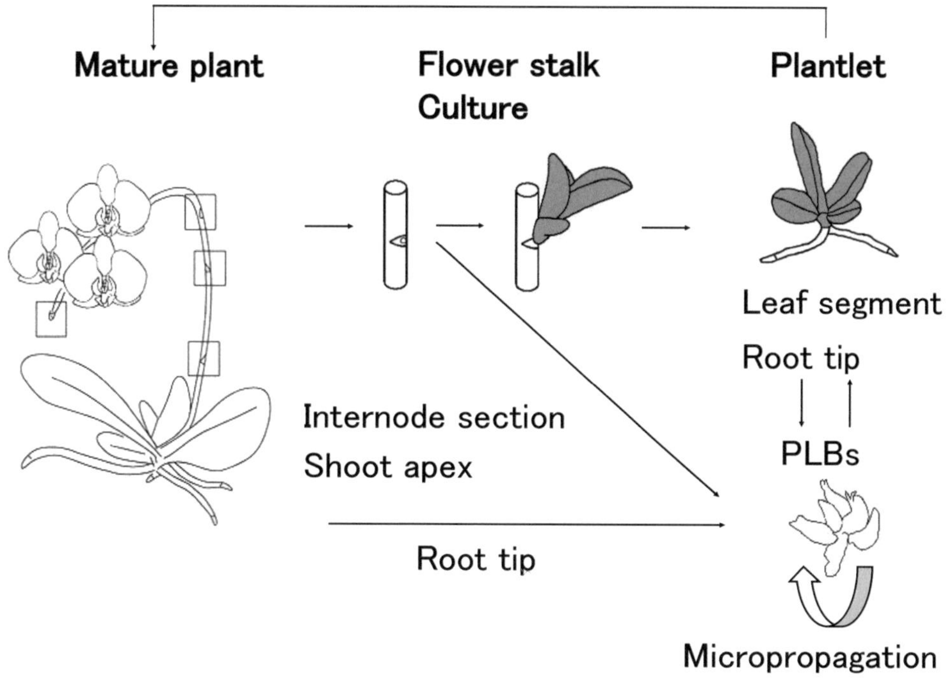

Figure 3.
The process of micropropagation in Phalaenopsis. *Micropropagation of* Phalaenopsis *orchids is performed using PLB derived from various tissues.*

still lower because only one plantlet can be obtained from one flower stalk bud. Therefore, reproduction of shoots from these plantlets [31, 32] or PLB induction from these shoots/plants (as described below) was conducted in practice.

3.2.2 PLB induction from plantlets

PLB induction using leaf segments of plantlets obtained by flower stalk culture has been studied in detailed conditions, such as medium, plant growth regulator, plantlets condition, temperature, lighting intensity, and subculture interval, and practically used since early times by Tanaka et al. [33, 34]. Also, many PLB induction methods are being studied because the leaves are easy to obtain and use as explants throughout the year [35, 36]. Hyponex, VW, and 1/2 strength MS medium are often used in this culture method. Since PLB induction from leaves is adventitious, the use of plant growth regulators, such as α -naphthalene acetic acid (NAA) and 6-benzylaminopurine (BAP) is essential. Highly active Thidiazuron (TDZ) instead of BAP is often used. Recently, efficient induction by leaf thin-section culture [37] and PLB induction using original species of *Phal. bellina* [38] and *Phal. cornu-cervi* [39], which are difficult to induce the PLB, have been studied.

Roots on plantlets are also easy to use without losing the mother strains and ideal tissue for PLB induction [40]. Park et al. [41] reported that highly efficient PLB induction from root tip on a modified MS medium supplemented with 2.3 mM TDZ. On the other hand, although it is necessary to sterilize, PLB induction is also possible from the aerial roots exposed to the air of potted mature plants [42].

3.2.3 Direct PLB induction

PLB also can be directly induced from flower stalk tissue on the mother strain. Internode segments from flower stalks were cultured for PLB induction. PLBs were formed at the bottom of the section with 50–80% after transferring the segments to a culture medium. Thomale GD medium supplemented with 10% coconut milk, 5 mg/l NAA, and 20 mg/l BAP was effective for PLB formation [43]. PLBs were also induced on the VW medium as a basic medium. Green PLBs with high proliferative efficiency were induced from the shoot apex of flower stalk bud with one or two leaf primordia on ND medium (NDM) supplemented with 0.1 mg/l NAA and 1 mg/l BAP [44].

3.2.4 PLB proliferation

The proliferation efficiency of PLBs induced from the tissues remarkably increases by adding cutting treatment. The upper part (tip) is apt to differentiate the shoot and the middle and bottom (base) parts tend to form new secondary PLBs on dividing PLBs [33, 45]. Protocorms with the trimmed base were formed secondary PLBs efficiently [46]. The survival rate tends to decrease with the division of PLBs. However, the PLB proliferation rate could be increased without decreasing the survival rate by partially incising the top of PLBs after removing the tip part of the PLB (partial incision treatment) [47]. Enoki and Takahara [48] developed a highly efficient PLB proliferation system by combining this treatment with elongated PLBs showing skotomorphogenesis in the dark.

3.3 Problems with micropropagation

3.3.1 Browning and death

Browning and death during tissue culture are critical problems for plant species, such as Orchids, including *Phalaenopsis*, fruit trees, etc. Although tissue culture technologies with cutting are essential for micropropagation of Orchids, these plant species are very sensitive to injury. Injured tissues elute a large amount of secondary metabolites, such as phenol-like substances into the medium [49] and it is thought that oxidative condensation of these substances destroys the physiological balance of the plant and then causes the death of tissues. There is a positive correlation between the exudation of phenolic compounds to medium and the survival rate of tissue explants in Mango [50]. In *Phalaenopsis*, phenolic compounds exudation causes poor regeneration from cultured plant tissues [34].

This phenomenon is reaction called wound responses, which are known in many plant species. Injury on plants causes plant defense system to production of antibacterial active substances, such as phenolic compounds or their own programmed cell death by hypersensitivity reactions due to production of reactive oxygen species, to prevent wounds from additional infection of fungi or insects [51]. Browning and death will occur in the tissue culture since these reactions may be excessive in *Phalaenopsis* orchids. Of these reactions, phenol is synthesized by phenylalanine ammonia-lyase (PAL), polyphenol oxidase (PPO), etc. in phenylpropanoid synthesis pathway. In fact, enzyme activities including PPO are higher in browning tissues of *Phalaenopsis* [52]. Therefore, activated charcoal adsorbing phenol [53, 54], antioxidants such as ascorbate acid (vitamin C) [55], L-2-aminooxy-3-phenylpropionic acid (AOPP, inhibitor of PAL) [56], and cycloheximide (inhibitor of

PPO) [57] were added to the medium in tissue culture of Orchids. A semisynthetic *Phalaenopsis* Shoot Reproduction (PSR) medium was developed that relieves the effects of phenolic compounds and enhances the survival rate of the explants of *Phalaenopsis* [31].

Recently, transcriptome analysis of *Phalaenopsis* during tissue browning provided comprehensive information on genes involved in browning and death other than the phenylpropanoid synthesis pathway [58]. However, the complex molecular mechanisms of browning and death are still unclear in *Phalaenopsis* orchids. Further elucidation of this molecular mechanism will make it possible to propose some more effective solutions to browning and death, and contribute to the commercial production of *Phalaenopsis*.

3.3.2 Interspecific and varietal differences

In the difficulty of micropropagation such as flower stalk [31], PLB [59], and callus [23] cultures of *Phalaenopsis*, there are large interspecific and varietal differences. This is probably because the moth orchid is a generic name for hybrids produced from various original species shown in **Table 1**. In fact, the ease of micropropagation in *Phalaenopsis* cultivars is due to characteristics of the original species involved in the creation of the cultivars [60], and thus there are few micropropagation methods that can be applied to all cultivars. Therefore, it is important to evaluate and estimate the proliferation difficulty of the original species in the development of the micropropagation method. Choice of proliferation methods based on the original species composition of the cultivars on Sander's list and information about their propagation difficulties from these investigations will be necessary for breeding of *Phalaenopsis* cultivars.

4. Molecular breeding

Various transformation methods have been studied as tools for molecular breeding. To date, a number of high-quality cultivars have been produced by traditional crossbreeding since interspecific and intergeneric hybrids are easy to obtain in Orchids, compared to other plant groups. However, it takes a lot of time and labor in improvement by traditional breeding, because the vegetative growth periods and reproductive cycle of the *Phalaenopsis* orchids are very long. Furthermore, genetic resources for new traits which are important in commerce have limitations found within only *Phalaenopsis* and closely related, crossbreeding possible genera. The transformation methods are one of the molecular breeding methods capable of solving these problems. In this section, we summarize the transformation methods and the application examples in practice.

4.1 Genetic transformation methods used in *Phalaenopsis*

4.1.1 Major methods

Genetic transformation methods are powerful tools for introducing useful genes of other plant species into target plant species. Transformation is advantageous in breeding because it can modify only specific traits of target plant species. Crossbreeding with the aim of improvement of only a particular trait is not suitable for

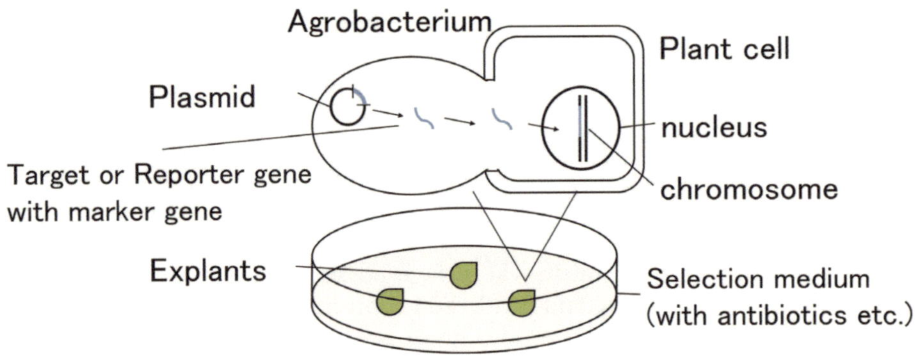

Figure 4.
Agrobacterium-*mediated transformation.*

Phalaenopsis orchids that have long reproductive cycles because multiple backcrossing at various times is required. To date, two transformation methods of *Agrobacterium*-mediated transformation (AT) and particle bombardment (PB) have been mainly used in *Phalaenopsis* orchids. The former is a method utilizing *Agrobacterium tumefaciens* (synonym: *Rhizobium radiobacter*) having the property of infected plant cells and sending their own genes into the infected plant genome (**Figure 4**). This gene part, the T-DNA region, is replaced with a useful target gene to be introduced by molecular biology techniques in practice. In this method, transgenic plants are obtained by the process of infection of *Agrobacterium* to explants, gene transfer by co-cultivation, sterilization of *Agrobacterium*, selection of transformed cells, and regeneration from the transformed cells to plants. The latter is a method of directly shooting gene-coated gold particles into cells using a gun device.

Many AT methods rather than PB have been studied in the examination of efficient transformation conditions in *Phalaenopsis* (**Table 2**). The first reported transformation in *Phalaenopsis* orchids was using the PB method by Anzai (1996) [70]. Belarmino and Mii (2000) [61] reported the first transformation of *Phalaenopsis* Orchids by AT. Thereafter, the success of transformation by AT was reported one after another [62–69]. Although PB has advantages, such as easy operation, and can be applied to a wide range of plant species and tissues, there is the largest bottleneck in the high cost of equipment and maintenance. The AT has lower maintenance costs and higher transformation efficiency than PB. In addition, gene silencing would occur less frequently and the later inheritance pattern of the transformed cultivar is also simple since a smaller number of copies of the gene are introduced in AT than in PB. The AT method has the disadvantage that it is difficult to use in monocotyledonous plants. However, the use of AT method in monocotyledonous plants, including *Phalaenopsis* orchids has also increased due to improved methods, such as the discovery of inducers for gene transfer into monocotyledonous plants in rice [81].

4.1.2 Target explants

The key to successful transformation depends on the ability of the tissue to regenerate since *Agrobacterium* particularly tends to infect cells that are active in cell division and since desired good cultivars cannot be created if the regeneration from the transformed cell to the mature plant is impossible. PLBs are often used as the target tissues for transformation rather than callus because a series of regeneration

Method	Explant	Marker genes	Reporter/Target genes	References
Reporter genes				
AT	Callus	*hpt, nptII*	*gus*	[61]
AT	PLBs	*hpt*	*gus, GFP*	[62–64]
AT	PLBs	*nptII*	*gus*	[65]
AT	protocorm	*BP/KNAT1, nptII*	*GFP*	[66, 67]
AT	protocorm	*hpt*	*GFP*	[68]
AT	protocorm	*hpt, nptII*	*gus*	[69]
PB	PLBs	*bar, nptII*	*gus*	[70]
Target genes				
AT	Callus	*hpt, nptII*	Wasabi defensin gene	[71]
AT	Callus	*nptII*	*LTP*	[72]
AT	PLBs	*hpt*	*PaFT*	[73, 74]
AT	PLBs	*nptII*	GAFP-NPI genes	[75]
AT/PB	PLBs	*hpt*	*CP, pflp*	[76, 77]
PB	flower	—	*F3'5'H*, CYP78A2 gene	[78, 79]
PB	PLBs	—	*PeUFGT3*	[80]

Abbreviations: AT, Agrobacterium-mediated transformation bar bialaphos resistance BP/KNAT1, Arabidopsis class 1 KNOX; CP, CymMV coat protein; F3'5'H, flavonoid-3, 5-hydroxylase; GAFP, Gastrodia Antifungal Protein; GFP, green fluorescent protein; gus, β-glucuronidase; hpt, hygromycin phosphotransferase; LTP, lipid transfer protein; NPI, Neutrophils Peptide-I; nptII, neomycin phosphotransferase II; PaFT, Phal. amabilis Flowering locus T; PB, particle bombardment; PeUFGT3, Phal. equestris UDP glucose: flavonoid 3-O-glucosyltransferase; pflp, sweet pepper ferredoxin-like protein; PLB, protocorm-like body.

Table 2.
Examples of the transformation of Phalaenopsis.

processes from PLBs to plantlets has already been established in *Phalaenopsis* as shown in **Figure 3**. The protocorms are also sometimes targeted to perform crossbreeding in parallel with transformation.

4.1.3 Marker and reporter genes

Selectable marker genes with target/reporter genes are introduced into target explants. In general, an antibiotic resistance gene such as *neomycin phosphotransferase II* (*nptII*) (kanamycin resistance) or *hygromycin phosphotransferase* (*hpt*) (hygromycin resistance) is used as marker genes [82]. Since transformation in practice would not occur in all cells of target tissues, it is possible by culturing the explants infected with AT on a medium containing antibiotics responsible for marker gene to select and propagate only the transformed and survived cells, and then to regenerate the transformed plantlets.

At the stage of examining optimal transformation conditions, reporter genes are used instead of the desired target gene to be introduced. *β-glucuronidase* (*gus*) and *green fluorescent protein* (*GFP*) genes are popular as reporter genes [83]. Both genes are useful in calculating transformation efficiency because the success or not of transformation can be visually recognized in the introduced cells at an early stage in

Phalaenopsis. The GUS-transformed cells exhibit blue color by giving a substrate solution from the outside. The GFP-transformed cells emit green fluorescence when exposed to ultraviolet rays. Although GFP is convenient because it does not need a substrate and transformed cells are not destroyed unlike the use of the GUS solution, there is a problem that it is difficult to distinguish green fluorescence from tissue color in the case of green color tissues.

4.2 Applications for breeding

In recent years, molecular breeding of moth orchids using useful target genes derived from other species and gene functional analysis of moth orchid itself using genetic transformation technique have been performed in practice (**Table 2**). Traits, such as new flower color, plant-pathogen resistance, and cold tolerance, which are important in commercial cultivation, are poor in genetic resources within the genera *Phalaenopsis* and *Doritaenopsis*. It is difficult to introduce such a trait to *Phalaenopsis* cultivars through conventional breeding methods. Therefore, molecular breeding using transformation methods have been studied.

4.2.1 Flower traits

In many flower plants, including *Phalaenopsis*, flower traits such as flowering time and new colors are important for breeding. To accelerate the floral transition and shorten the reproductive cycle of *Phalaenopsis*, transformants were obtained by overexpression of *FT* (*Flowering locus T*), a floral transition-related gene derived from *Phal. amabilis* by AT method [73, 74]. Overexpression of FT is known to be involved in early flowering by promoting floral transition in *Arabidopsis thaliana* and other species. Currently, functional analysis of this gene for flowering has been continued in the transformed *Phalaenopsis*.

Regarding the flower color traits, functional analysis of pigment synthesis-related genes of *Phalaenopsis* itself using the transformation method has been performed. Hsu's group introduced *flavonoid-3, 5-hydroxylase* (*F3'5'H*) derived from *Phalaenopsis* into the petal of *Phalaenopsis*, confirming that the flower color changed from pink to magenta [78]. In addition, they revealed by the same method that new CYP78A2 in the Cytochrome P450 (CYP 450) group of *Phalaenopsis*, which is specifically expressed in the pollen tube, is also involved in anthocyanin pigment synthesis [79]. Functional analysis of *UDP glucose:flavonoid 3-O-glucosyltransferase* (*PeUFGT*)-suppressed transformants in *Phal. equestris* also proved that this gene plays a crucial role in the anthocyanin synthesis pathway [80]. Cultivars with a blue flower, which are rare in nature, have been produced by transformation technology in many flower plants without blue pigment synthesizing ability. The creation of a blue rose by the introduction of exogenous *F3'5'H* which is the key gene for the synthesis of delphinidin as blue pigment gave a great influence all over the world [84]. In addition to the previous blue carnation, blue chrysanthemums have also been produced in recent years by the same method. In *Phalaenopsis*, the first genetically engineered blue moth orchid using the same method was created by the group of Mii of Chiba University and Ishihara Sangyo Kaisha, Ltd. in Japan [85], and was exhibited for the first time in Japan in 2013.

F3'5'H itself exists in *Phalaenopsis*, although there is no report on the presence of delphinidin in *Phalaenopsis*. Furthermore, the presence of varieties of the original species *Phal. violacea* and *Dor. pulcherrima* exhibiting blue color has been known since the olden days and *Dtps*. Kenneth Schubert as the world first's blue moth orchid has been already produced by crossbreeding these original species. The moth orchid produced by the above transformation method and this cultivar is still not perfectly blue. It is known in many flower plant species that complex mechanisms due to some factors, such as pH, metal complex, and intramolecular stacking of anthocyanin, other than the kind of anthocyanin pigments are involved in the determination of blue flower color [86]. Although Griesbach [87, 88] revealed that some of these factors are involved in the blue flower color of *Phalaenopsis* by crossing test and chemical analysis of the hybrid and original species described above, the detailed molecular mechanisms which determine flower color are not clarified so far. Why does not the *Phalaenopsis* orchid with bright blue flowers still exist? Further elucidation of the molecular mechanism of blue coloration of the original species of *Phalaenopsis* may lead to perfect bluing of *Phalaenopsis* by molecular breeding using methods other than the introduction of pigment synthesis gene.

4.2.2 Plant defense

Disease resistance breeding is one of the important tasks in the breeding of *Phalaenopsis*. Infection of plant pathogens (bacteria and viruses) to plants causes serious damage to producers in the actual farm field. Recently, conferring pathogen resistance into *Phalaenopsis* by introducing foreign genes derived from other species is attempted. In transformed *Phalaenopsis* with GAFP (*Gastrodia* Antifungal Protein)—NPI (Neutrophils Peptide-I) genes, the disease resistance to *Colletotrichum gloeosporioides* causing anthrax disease was confirmed *in vitro* and *in vivo* [75]. The introduction of the Wasabi defensin gene derived from *Wasabia japonica* into *Phalaenopsis* increased the resistance to *Rrwinia carotovora* causing soft rot disease [71]. The research group of Chan et al. [76, 77] reported that double transformation with Cymbidium mosaic virus (CymMV) coat protein (CP) and sweet pepper ferredoxin-like protein (pflp) genes confer dual resistance to CymMV and *Erwinia carotovora* into *Phalaenopsis*. In the future, *Phalaenopsis* with further multiple resistances to pathogens might be produced.

4.2.3 Cold tolerance

The breeding of low-temperature stress tolerance is a serious issue in the moth orchids which are tropical plants. In general, *Phalaenopsis* orchids have poor cold tolerance and the structure of the cell membrane degenerates at 15 degrees or less, and it suffers irreversible damage from low temperature. The lipid transfer protein (LTP) gene is involved in the transfer of monomers, such as wax and cutin, and the stabilization of plasma membrane. The expression of this gene is known to confer various biotic (such as fungi) and abiotic (such as cold) stress tolerance upon plants [89, 90]. In fact, the introduction of LTP derived from rice (*Oryza sativa* cv. IAPAR9) into the callus of *Phal. amabilis* gave the regenerated transformed plants strong cold tolerance with growing healthy leaves at 10°C/7°C (day/night) [72].

5. Conclusion

The utilization of biotechnology such as micropropagation by tissue culture and transformation methods has played a very important role in the commercial production and breeding of *Phalaenopsis* orchids. The further development of such technologies in this field and the acquisition of new knowledge by many studies utilizing these technologies will contribute to the *Phalaenopsis* orchid industry.

Author details

Shinichi Enoki[1,2]* and Yoshinori Takahara[1]

1 Laboratory of Applied Plant BioTechnology, Nagaoka University of Technology, Niigata, Japan

2 Laboratory of Fruit Genetic Engineering, The Institute of Enology and Viticulture, University of Yamanashi, Yamanashi, Japan

*Address all correspondence to: esfirefly@gmail.com

References

[1] Dressler RL. Phylogeny and Classification of the Orchid Family. Cambridge: Cambridge University Press; 1993. DOI: 10.5860/choice.31-2679

[2] Sweet HR. The Genus *Phalaenopsis* (Vol. 1). California: Orchid Digest; 1980

[3] Shim PS. A new generic classification in the *Phalaenopsis* complex (*Orchidaceae*). Malayan Orchid Review. 1984;**18**:1-28

[4] Seidenfaden G. *Doritis*. Opera Bots. 1988;**95**:31-34

[5] Arends JC. Cytological observations on genome homology in eight interspecific hybrids of *Phalaenopsis*. Genetica. 1970;**41**(1):88-100. DOI: 10.1007/BF00958896

[6] Shindo K, Kamemoto H. Karyotype analysis of some species of *Phalaenopsis*. Cytologia. 1963;**28**(4):390-398. DOI: 10.1508/cytologia.28.390

[7] Christenson EA. Phalaenopsis: A Monograph. Oregon: Timber Press (OR); 2001

[8] Goh MWK, Kumar PP, Lim SH, Tan HTW. Random amplified polymorphic DNA analysis of the moth orchids, *Phalaenopsis* (*Epidendroideae*: *Orchidaceae*). Euphytica. 2005;**141**(1): 11-22. DOI: 10.1007/s10681-005-4814-y

[9] Padolina J, Linder CR, Simpson BB. A phylogeny of *Phalaenopsis* using multiple chloroplast markers. Selbyana. 2005;**26**: 23-27

[10] Tsai CC, Huang SC, Chou CH. Molecular phylogeny of *Phalaenopsis* Blume (*Orchidaceae*) based on the internal transcribed spacer of the nuclear ribosomal DNA. Plant Systematics and Evolution. 2005;**256**(1):1-16. DOI: 10.1007/s00606-005-0356-y

[11] Yukawa T, Kita K, Handa T, Hidayat T, Ito M. Molecular phylogenetics of *Phalaenopsis* (*Orchidaceae*) and allied genera: Re-evaluation of generic concepts. Acta Phytotaxonomica et Geobotanica. 2005; **56**(2):141-161. DOI: 10.18942/apg. KJ00004623240

[12] Kocyan A, Schuiteman A. New combinations in *Aeridinae* (*Orchidaceae*). Phytotaxa. 2014;**161**(1):61-85. DOI: 10.11646/phytotaxa.161.1.3

[13] Deng H, Zhang GQ, Liu ZJ, Wang Y. A new species and a new combination of *Phalaenopsis* (*Orchidaceae*: *Epidendroideae*: *Aeridinae*): Evidence from morphological and DNA analysis. Phytotaxa. 2015;**238**(3):243-254. DOI: 10.11646/PHYTOTAXA.238.3.3

[14] Sanders DF, Wreford MW. Sander's Complete List of Orchid Hybrids. England: David Sander's Orchids Limited; 1947

[15] Tang CY, Chen WH. Breeding and development of new varieties in *Phalaenopsis*. Orchid Biotechnology. 2007;**57**:1-22. DOI: 10.1142/ 9789812775900_0001

[16] Knudson L. Nonsymbiotic germination of orchid seeds. Botanical Gazette. 1922;**73**(1):1-25. DOI: 10.1086/ 332956

[17] Knudson L. A new nutrient solution for the germination of orchid seed. American Orchid Society Bulletin. 1946; **15**:214-217

[18] Vacin EF, Went FW. Some pH changes in nutrient solutions. Botanical

Gazette. 1949;**110**(4):605-613. DOI: 10.1086/335561

[19] Murashige T, Skoog F. A revised medium for rapid growth and bio assays with tobacco tissue cultures. Physiologia Plantarum. 1962;**15**(3):473-497. DOI: 10.1111/j.1399-3054.1962.tb08052.x

[20] Kano K. Studies on the Media for Orchid Seed Germination. Memoirs of Faculty of Agriculture. Kagawa: Kagawa University; 1964

[21] Lee YI, Hsu ST, Yeung EC. Orchid protocorm-like bodies are somatic embryos. American Journal of Botany. 2013;**100**(11):2121-2131. DOI: 10.3732/ajb.1300193

[22] Sagawa Y. Orchids, other considerations. In: Handbook of Plant Cell Culture. Vol. 5. New York: MacGraw-Hill Publishing Company; 1990. pp. 638-653

[23] Islam OM, Ichihashi S. Effects of sucrose, maltose and sorbitol on callus growth and plantlet regeneration in *Phalaenopsis*, *Doritaenopsis* and *Neofinetia*. Journal of the Japanese Society for Horticultural Science. 1999; **68**(6):1124-1131. DOI: 10.2503/jjshs.68.1124

[24] Tokuhara K, Mii M. Induction of embryogenic callus and cell suspension culture from shoot tips excised from flower stalk buds of *Phalaenopsis* (*Orchidaceae*). In Vitro Cellular & Developmental Biology. Plant. 2001; **37**(4):457-461. DOI: 10.1007/s11627-001-0080-4

[25] Tokuhara K, Mii M. Highly-efficient somatic embryogenesis from cell suspension cultures of *Phalaenopsis* orchids by adjusting carbohydrate sources. In Vitro Cellular & Developmental Biology. Plant. 2003; **39**(6):635-639. DOI: 10.1079/IVP2003466

[26] Roter G. A method of vegetative propagation of *Phalaenopsis* species and bybrids. Proceedings of the American Society for Horticultural Science. 1950; **55**:434

[27] Arditti J. Culture of flower-stald buds: A method for vegetative propagation of *Phalaenopsis*. American Orchid Society Bulletin. 1977;**46**:236-240

[28] Sagawa Y. Vegetative propagation of *Phalaenopsis* stem cutting. American Orchid Society Bulletin. 1961;**30**:808-809

[29] Scully RM. Stem propagation of *Phalaenopsis*. American Orchid Society Bulletin. 1966;**35**:40-42

[30] Tanaka M, Sakanishi Y. Factors affecting the growth of in vitro cultured lateral buds from *Phalaenopsis* flower stalks. Scientia Horticulturae. 1978;**8**: 169-178. DOI: 10.1016/0304-4238(78)90022-5

[31] Enoki S, Takahara Y. Application of a modified MS medium for tissue culture with cutting in *Phalaenopsis* -comparison with other conventional media with regard to the survival rate and varietal differences in cultural characteristics. Journal of Science and High Technology in Agriculture. 2014;**26**(2):109-117. DOI: 10.2525/shita.26.109

[32] Tse AT. Adventitious shoot formation on *Phalaenopsis* nodes. American Orchid Society Bulletin. 1971;**40**:807-810

[33] Tanaka M. Micropropagation of *Phalaenopsis* spp. Biotechnology in Agriculture and Forestry. 1992;**20**: 246-268

[34] Tanaka M. Clonal propagation of *Phalaenopsis* by leaf tissue culture.

American Orchid Society Bulletin. 1977; **1977**(46):733-737

[35] Chen JT, Chang WC. Direct somatic embryogenesis and plant regeneration from leaf explants of *Phalaenopsis amabilis*. Biologia Plantarum. 2006; **50**(2):169-173. DOI: 10.1007/s10535-006-0002-8

[36] Park SY, Murthy HN, Paek KY. Rapid propagation of *Phalaenopsis* from floral stalk-derived leaves. In Vitro Cellular & Developmental Biology. Plant. 2002; **38**(2):168-172. DOI: 10.1079/IVP2001274

[37] Park SYEC, Yeung E, Chakrabarty D, Paek K. An efficient direct induction of protocorm-like bodies from leaf subepidermal cells of *Doritaenopsis* hybrid using thin-section culture. Plant Cell Reports. 2002;**21**(1):46-51. DOI: 10.1007/s00299-002-0480-x

[38] Khoddamzadeh AA, Sinniah UR, Kadir MA, Kadzimin SB, Mahmood M, Sreeramanan S. *In vitro* induction and proliferation of protocorm-like bodies (PLBs) from leaf segments of *Phalaenopsis bellina* (Rchb. f.) Christenson. Plant Growth Regulation. 2011;**65**(2):381. DOI: 10.1007/s10725-011-9611-0

[39] Rittirat S, Klaocheed S, Thammasiri K. Enhanced efficiency for propagation of *Phalaenopsis cornu-cervi* (Breda) Blume & Rchb. F. Using trimmed leaf technique. International Journal of Agricultural and Biological Engineering. 2014;**8**(4):328-331. DOI: 10.5281/zenodo.1091904

[40] Tanaka M. Plantlet formation by root tip culture in *Phalaenopsis*. American Orchid Society Bulletin. 1976; **56**:1024-1222

[41] Park SY, Murthy HN, Paek KY. Protocorm-like body induction and subsequent plant regeneration from root tip cultures of *Doritaenopsis*. Plant Science. 2003;**164**(6):919-923. DOI: 10.1016/S0168-9452(03)00019-0

[42] Kobayashi M. Studies on the vegetative propagation of *Phalaenopsis* through root-tip culture. Journal of the Japanese Society for Horticultural Science. 1991;**59**(2):664-665

[43] Homma Y, Asahira T. New means of *Phalaenopsis* propagation with internodal sections of flower stalk. Journal of the Japanese Society for Horticultural Science. 1985;**54**(3):379-387. DOI: 10.2503/jjshs.54.379

[44] Tokuhara K, Mii M. Micropropagation of *Phalaenopsis* and *Doritaenopsis* by culturing shoot tips of flower stalk buds. Plant Cell Reports. 1993;**13**(1):7-11. DOI: 10.1007/BF00232306

[45] Soe KW, Myint KT, Naing AH, Kim CK. Optimization of efficient protocorm-like body (PLB) formation of *Phalaenopsis* and *Dendrobium* hybrids. Current Research on Agriculture and Life Sciences. 2014;**32**(4):179-183. DOI: 10.14518/crals.2014.32.4.032

[46] Murdad R, Hwa KS, Seng CK, Latip MA, Aziz ZA, Ripin R. High frequency multiplication of *Phalaenopsis gigantea* using trimmed bases protocorms technique. Scientia Horticulturae. 2006; **111**(1):73-79. DOI: 10.1016/j.scienta.2006.08.008

[47] Nii H, Kawamura H, Yoshihara H. Effect of the emarginate treatment on the formation of PLB in *Phalaenopsis* (In Japanese). The Japanese Society for Horticultural Science. 2004;**73**(1):191

[48] Enoki S, Takahara Y. Development of a highly efficient and simple micropropagation system for

Phalaenopsis using elongated Protocorm-like bodies induced by Skotomorphogenesis under dark conditions. Journal of the Japanese Society for Horticultural Science. 2014; **83**(2):149-155. DOI: 10.2503/jjshs1.CH-083

[49] Ishii M, Uemoto S, Fujieda K, Nonaka M, Shoyama Y, Miyahara Y, et al. A new biologically active phenolic from *Cattleya trianaei*. Phytochemistry. 1979;**18**(7):1211-1213. DOI: 10.1016/0031-9422(79)80137-5

[50] Krishna H, Sairam RK, Singh SK, Patel VB, Sharma RR, Grover M, et al. Mango explant browning: Effect of ontogenic age, mycorrhization and pre-treatments. Scientia Horticulturae. 2008; **118**(2):132-138. DOI: 10.1016/j.scienta.2008.05.040

[51] Enoki S, Suzuki S. Pathogenesis-related proteins in grape. Grape wine. Biotechnology. 2016;**43**:43-58. DOI: 10.5772/64873

[52] Chuanjun X, Ling L. Changes of total phenol content and the activities of PPO, POD and PAL during the browning in *Phalaenopsis* explant *in vitro*. Acta Horticulturae Sinica. 2006;**33**(3):671

[53] Wang PJ, Huang LC. Beneficial effects of activated charcoal on plant tissue and organ cultures. In Vitro. 1976: **12**(3):260-262

[54] Young PS, Murthy HN, Yoeup PK. Mass multiplication of protocorm-like bodies using bioreactor system and subsequent plant regeneration in *Phalaenopsis*. Plant Cell, Tissue and Organ Culture. 2000;**63**(1):67-72. DOI: 10.1023/A:1006420116883

[55] Arditti J, Ernst R. Micropropagation of Orchids. New York: John Willey and Sons; 1993

[56] Mitsukuri M, Johkan M, Yamasaki S, Tanaka H, Tezuka T, Mishiba K, et al. L-2-aminooxy-3-phenylpropionic acid (AOPP) controls Browning and Promotes adventitious bud formation in *Neofinetia falcata in vitro*. Journal of the Japanese Society for Horticultural Science. 2010;**79**(4):367-371. DOI: 10.2503/jjshs1.79.367

[57] Chuanjun X, Zhiwei R, Ling L, Biyu Z, Junmei H, Wen H, et al. The effects of polyphenol oxidase and cycloheximide on the early stage of browning in *Phalaenopsis* explants. Horticultural Plant Journal. 2015;**1**(3): 172-180. DOI: 10.16420/j.issn.2095-9885.2015-0030

[58] Xu C, Zeng B, Huang J, Huang W, Liu Y. Genome-wide transcriptome and expression profile analysis of *Phalaenopsis* during explant browning. PLoS One. 2015;**10**(4):e0123356. DOI: 10.1371/journal.pone.0123356

[59] Tanaka M. Studies on the clonal propagation of *Phalaenopsis* through *in vitro* culture. Memoirs of Faculty of Agriculture-Kagawa University (Japan). 1988;**49**:1-85

[60] Zhou TS, Tankaka M. Varietal differences of the ability of PLB formation on cultured leaf segments in *Phalaenopsis* and *Doritaenopsis* hybrids (In Japanease). The Japanese Society for Horticultural Science. 1995;**64**(2): 602-603

[61] Belarmino MM, Mii M. *Agrobacterium*-mediated genetic transformation of a *Phalaenopsis* orchid. Plant Cell Reports. 2000;**19**(5):435-442. DOI: 10.1007/s002990050752

[62] Julkifle AL, Rathinam X, Uma RS, Sreeramanan S. Optimisation of transient green fluorescent protein (GFP) gene expression in *Phalaenopsis*

violacea orchid mediated by *Agrobacterium tumefaciens*-mediated transformation system. Australian Journal of Basic and Applied Sciences. 2010;**4**(8):3424-3432

[63] Subramaniam S, Balasubramaniam V, Poobathy R, Sreenivasan S, Rathinam X. Chemotaxis movement and attachment of *Agrobacterium tumefaciens* to *Phalaenopsis violacea* orchid tissues: An assessment of early factors influencing the efficiency of gene transfer. Tropical Life Sciences Research. 2009;**20**(1): 39-41

[64] Subramaniam S, Rathinam X. Emerging factors that influence efficiency of T-DNA gene transfer into *Phalaenopsis violacea* orchid via *Agrobacterium tumefaciens*–mediated transformation system. International Journal of Biology. 2010;**2**(2):64. DOI: 10.5539/ijb.v2n2p64

[65] Chai ML, Xu CJ, Senthil KK, Kim JY, Kim DH. Stable transformation of protocorm-like bodies in *Phalaenopsis* orchid mediated by *Agrobacterium tumefaciens*. Scientia Horticulturae. 2002; **96**(1):213-224. DOI: 10.1016/ S0304-4238(02)00084-5

[66] Semiarti E, Indrianto A, Purwantoro A, Isminingsih S, Suseno N, Ishikawa T, et al. *Agrobacterium*-mediated transformation of the wild orchid species *Phalaenopsis amabilis*. Plant Biotechnology. 2007;**24**(3): 265-272. DOI: 10.5511/ plantbiotechnology.24.265

[67] Semiarti E, Indrianto A, Purwantoro YH, Martiwi INA, Feroniasanti YML, Nadifah F, et al. High-frequency genetic transformation of *Phalaenopsis amabilis* orchid using tomato extract-enriched medium for the pre-culture of protocorms. The Journal of Horticultural Science and Biotechnology. 2010;**85**(3):205-210. DOI: 10.1080/14620316.2010.11512655

[68] Hsing HX, Lin YJ, Tong CG, Li MJ, Chen YJ, Ko SS. Efficient and heritable transformation of *Phalaenopsis* orchids. Botanical Studies. 2016;**57**(1):1-12. DOI: 10.1186/s40529-016-0146-6

[69] Mishiba KI, Chin DP, Mii M. *Agrobacterium*-mediated transformation of *Phalaenopsis* by targeting protocorms at an early stage after germination. Plant Cell Reports. 2005;**24**(5):297-303. DOI: 10.1007/s00299-005-0938-8

[70] Anzai H, Ishii Y, Shichinohe M, Katsumata K, Nojiri C, Morikawa H, et al. Transformation of *Phalaenopsis* by particle bombardment. Plant Tissue Culture Letters. 1996;**13**(3):265-272. DOI: 10.5511/ plantbiotechnology1984.13.265

[71] Sjahril R, Chin DP, Khan RS, Yamamura S, Nakamura I, Amemiya Y, et al. Transgenic *Phalaenopsis* plants with resistance to *Erwinia carotovora* produced by introducing wasabi defensin gene using *Agrobacterium* method. Plant Biotechnology. 2006;**23**(2):191-194. DOI: 10.5511/plantbiotechnology.23.191

[72] Qin X, Liu Y, Mao S, Li T, Wu H, Chu C, et al. Genetic transformation of lipid transfer protein encoding gene in *Phalaenopsis amabilis* to enhance cold resistance. Euphytica. 2011;**177**(1):33-43. DOI: 10.1007/s10681-010-0246-4

[73] Mercuriani IS, Purwantoro A, Moeljopawiro S, Semiarti E. Insertion of a flowering gene, *PaFT*, into *Phalaenopsis amabilis* orchid using *Agrobacterium tumefaciens*. In: 5th Indonesia Biotechnology Conference. 2012. pp. 152-161

[74] Semiarti E, Mercuriani IS, Rizal R, Slamet A, Utami BS, Bestari IA, et al.

Overexpression of *PaFT* gene in the wild orchid *Phalaenopsis amabilis* (L.) Blume. AIP Conference Proceedings. 2015; **1677**(1):090005. DOI: 10.1063/1.4930750

[75] Li J, Kuang P, Liu RD, Wang D, Wang Z, Huang M. Transfer of GAFP and NPI, two disease-resistant genes, into a *Phalaenopsis* by *Agrobacterium tumefaciens*. Pakistan Journal of Botany. 2013;**45**(5):1761-1766

[76] Liao LJ, Pan IC, Chan YL, Hsu YH, Chen WH, Chan MT. Transgene silencing in *Phalaenopsis* expressing the coat protein of Cymbidium mosaic virus is a manifestation of RNA-mediated resistance. Molecular Breeding. 2004; **13**(3):229-242. DOI: 10.1023/B:MOLB.0000022527.68551.30

[77] Chan YL, Lin KH, Liao LJ, Chen WH, Chan MT. Gene stacking in *Phalaenopsis* orchid enhances dual tolerance to pathogen attack. Transgenic Research. 2005;**14**(3):279-288. DOI: 10.1007/s11248-005-0106-5

[78] Su V, Hsu BD. Cloning and expression of a putative cytochrome P450 gene that influences the colour of *Phalaenopsis* flowers. Biotechnology Letters. 2003;**25**(22):1933-1939. DOI: 10.1023/B:BILE.0000003989.19657.53

[79] Su V, Hsu BD. Transient expression of the cytochrome p450 CYP78A2 enhances anthocyanin production in flowers. Plant Molecular Biology Reporter. 2010;**28**(2):302-308. DOI: 10.1016/S0958-1669(00)00192-0

[80] Chen WH, Hsu CY, Cheng HY, Chang H, Chen H, Ger MJ. Downregulation of putative UDP-glucose: Flavonoid 3-O-glucosyltransferase gene alters flower coloring in *Phalaenopsis*. Plant Cell Reports. 2011;**30**(6):1007-1017. DOI: 10.1007/s00299-011-1006-1

[81] Hiei Y, Ohta S, Komari T, Kumashiro T. Efficient transformation of rice (*Oryza sativa* L.) mediated by *Agrobacterium* and sequence analysis of the boundaries of the T-DNA. The Plant Journal. 1994;**6**(2):271-282. DOI: 10.1046/j.1365-313X.1994.6020271.x

[82] Becker D, Kemper E, Schell J, Masterson R. New plant binary vectors with selectable markers located proximal to the left T-DNA border. Plant Molecular Biology. 1992;**20**(6): 1195-1197. DOI: 10.1007/BF00028908

[83] De Ruijter NCA, Verhees J, Van Leeuwen W, Van Der Krol AR. Evaluation and comparison of the GUS, LUC and GFP reporter system for gene expression studies in plants. Plant Biology. 2003;**5**(02):103-115

[84] Katsumoto Y, Fukuchi-Mizutani M, Fukui Y, Brugliera F, Holton TA, Karan M, et al. Engineering of the rose flavonoid biosynthetic pathway successfully generated blue-hued flowers accumulating delphinidin. Plant and Cell Physiology. 2007;**48**(11):1589-1600. DOI: 10.1093/pcp/pcm131

[85] Yuki S, Araki S, Suzuki T. Flavonoid-3′,5′-hydroxylase gene of *Commelina communis*. US Patent 8.440, 879 B2. [Accessed: 14-05-2013]

[86] Yoshida K, Mori M, Kondo T. Blue flower color development by anthocyanins: From chemical structure to cell physiology. Natural Product Reports. 2009;**26**(7):884-915. DOI: 10.1039/B800165K

[87] Griesbach RJ. The biochemical basis for the blue flower color mutations in *Doritis pulcherrima* and *Phalaenopsis violacea*. Lindleyana. 1997;**12**(2):64-71

[88] Griesbach R. A scientific approach to breeding blue orchids. Orchids. 2000;**74**: 376-379

[89] Hughes MA, Dunn MA, Pearce RS, White AJ, Zhang L. An abscisic-acid-responsive, low temperature barley gene has homology with a maize phospholipid transfer protein. Plant, Cell & Environment. 1992;**15**(7):861-865. DOI: 10.1111/j.1365-3040.1992.tb02155.x

[90] Choi AM, Lee SB, Cho SH, Hwang I, Hur CG, Suh MC. Isolation and characterization of multiple abundant lipid transfer protein isoforms in developing sesame (*Sesamum indicum* L.) seeds. Plant Physiology and Biochemistry. 2008;**46**(2):127-139. DOI: 10.1016/j.plaphy.2007.10.003

Chapter 4

Lesser Known African Indigenous Tree and Fruit Plants: Recent Evidence from Literatures and Regular Cultivation Culture

Kayode Paul Baiyeri and Kolawole Olajide

Abstract

Indigenous plant species native to Africa have numerous uses. They have a long and rich ethno-medicinal history with well-known native applications in different African countries. The effects of these indigenous underutilized crops in local traditional medicine differ. But they play an important role in enhancing food and nutrition security of the population. Tropical plant species have economic potential as they make great socio-economic impact on the livelihoods of rural dwellers. Despite their economic, food and nutritional values, these plants are still underutilized and have not been brought under regular cultivation culture due to inadequate information about their food values and their agronomic requirements for cultivation. Their potential values to the African food system could be enhanced if they are domesticated and prevented from going into extinction. Thus, the potential implications for long-term sustainable food security of these plants should not be neglected. Therefore, there is the need to recognize and enable indigenous foods from the indigenous plant species to serve as a key resource in ensuring healthy food systems in Africa. The inherent potential of the following tropical indigenous plant species African Walnut (*Plukenetia conophora* Muell Arg.), Saba (*Saba senegalensis* (A. DC.) Pichon), Baobab (*Adansonia digitata* L.) and Kapok (*Ceiba pentandra* (L.) Gaertn.) are discussed in this review.

Keywords: forest resources, nursery management, fertilizer use, nutritional quality, growth and yield

1. Introduction

The extinction of plant species resulting from human activities throughout the world has become a major concern [1]. Forest resources diminish as a result of deforestation, which has negative impact on agriculture, medicine and economic enterprises of man [1, 2]. Forest resources provide numerous goods and services to man such as food, medicine, wood, fiber and energy, they were taken for granted in the past because they were available almost everywhere but the situation has changed due to adverse effects of human activities [3, 4]. Food insecurity remains a major challenge in developing countries and insufficient nutrient intake causes severe malnutrition affecting the populace [5].

Worldwide, the problem of food security leads to calorie deficit of more than 700,000,000 underfed people [6]. The valuation of edible fruits and vegetables that are underutilized is one of the ways out of this impasse, a few examples of such extinct fruit and vegetable species with potential to address global undernourishment problems particularly those confronting the developing countries are *Plukenetia conophora, Saba senegalensis, Adansonia digitata* and *Ceiba pentandra*. They are potential sources of vitamins, minerals, antioxidants and phenols [7]. Indigenous plant species are outstanding plants due to their numerous benefits. Many of them are richer in protein and other nutrient contents [8]. They are good sources of macro and micronutrients for human consumption; many indigenous fruits and vegetables are characterized by a high nutritional value in comparison with global vegetables like tomato and cabbage [9]. Notably, many are potential sources of vitamins and macro and microelements with the ability to provide them to children and adult at levels higher than those recommended by the World Health Organization (WHO) [10]. Consumption of fruits and vegetables can improve health and prevent the risk of developing chronic diseases including cancer [11]. Consuming adequate quantity of food can be assured by utilizing nutrient-rich fruits and leafy vegetables accompanied with staple food. Generally, indigenous plant species are important as food, medicine and socio-economic value.

In spite of the numerous potentials of these indigenous crops, they have not been cultivated like other tropical plant species in Nigeria as a result of lack of adequate knowledge on their nutritional value, climatic requirement, fertilizer requirement and agronomic practices. Crop yield is to a large extent, associated with its fertilizer requirements, and maintaining the yield and quality of a newly introduced crop involves suitable crop management practices to improve soil productivity [12, 13]. Soil amendment could be done using organic or inorganic fertilizer and may be combined [14]. Therefore, investigating into food, nutritional, medicinal, climatic and fertilizer requirements of these indigenous crops can provide evidence-based information encouraging the cultivation of these wild species in order to remedy food insecurity, improve the diet of the people and prevent the crops from going into extinction. The information will also be useful to the food industries and pharmaceutical companies.

2. African walnut (*P. conophora*)

2.1 Description

P. conophora Muell Arg. known as African walnut belong to Euphorbiaceae family [15]. It is a climber that twines around cocoa and kola nut trees for support [16] (**Figure 1**). *P. conophora* is a small tropical flowering shrub, a woody perennial plant 6 m – 18 m long when it attains maturity stage; the stem can be as wide as 16 cm and turns dark gray as it ages, but it is green and glabrous at tender age [17]. The leaf is simple, crenate and having serrated margin. They are spherical at the base with the leaves arranged alternately [18, 19]. The seeds can be boiled and eaten as snacks [20]. Fruits not yet developed have green color but change from dark brown to black at

Figure 1.
Staked African walnut plants.

Figure 2.
African walnut capsules.

maturity [21]. The seed is white when cracked upon shell removal with a thin layer between split halves. After eating the nut, the presence of chemical substances such as alkaloids gives a bitter taste upon drinking water [22]. Walnut seeds are housed in a pod having: one seed (single), two seeds (double), three seeds (triple), four seeds (quadruple) and five seeds (quintuple) [23]. The walnut shells are usually black or brown in color (see **Figure 2**).

2.2 Origin and distribution

Plukenetia conophora originated from tropical western and central Africa and it is available in Nigeria, Cameroon, Congo, Central African Republic, Gabon, Niger, and Sierra Leone [18, 24]. The African walnut is cultivated in Western and Eastern regions of Nigeria [18]. It is found in Uyo, Akamkpa, Akpabuyo, Lagos, Akure, Kogi, Ajaawa, Ogbomosho, Ibadan [25, 26], Ife, Ekiti and Osun State. In the South and cocoa producing States in Nigeria, walnut is available [24, 27]. African walnut thrives on loamy soils that are deep, fertile, moist and well drained. Walnut does well on silty clay and loam soils [28, 29] and for optimum growth, walnut requires high solar radiation.

2.3 Food and economic importance

P. conophora Muell Arg. is a multipurpose crop used for food, nutritional and economic purposes in Africa. This plant is grown purposely for its nuts, boiled in water and eaten as a snack [25]. In Nigeria, Sierra Leone and Ghana, the fruits improve the livelihood of the rural people by providing income [30]. Extracted oil from the nut is used in making wood varnishes, vulcanized oil for rubber, stand oil and leather substitutes [31, 32]. Essential oils are usually extracted and used in food, cosmetics, perfumes, soaps and drinks as flavor. It can also be used in treating skin diseases and as remedy for cancer. Walnut peels (shell) combined with other materials are used as filler in dynamite [33]. The shell can be included in catfish meal with no negative effect on the performance and health status of the fish [34].

2.4 Medicinal value

P. conophora have numerous ethno-medicinal uses among the African rural populace. The leaves, root, bark and fruit are known for their medicinal values. Walnut leaves are used to treat venous insufficiency, hypoglycemia, hemorrhoids, indigestion, constipation, dysentery, diarrhea, syphilis, asthma, thrush, prolonged and constant hiccups, pruritus, eczema, fungal and microbial infections, psoriasis and parasitic skin conditions majorly among children, the elderly and immunosuppressed [35, 36]. African walnut can be used to expel worms; it can also treat rheumatism, kidney pain, cold, gout, cleaning of blood and abnormal menstrual bleeding [37]. The succulent leaves are used as vegetable and for treating cancers growing in the neck. They control inflammation of the gums and throat and mouth when used as tea [33]. Brown dye is extracted from the husk and leaf which is used to manage hiccups [38]. The root is effective in the treatment of piles. It lowers the risk of developing cancer and it controls high blood pressure [37] and it can be used as an antidote to snakebite, tonification of kidneys and strengthening of the back and knees [39]. The bark can be used in tea as laxative, chewed to reduce toothache and to treat high blood pressure while the root is used for frost bite and varicose ulcers [40, 41].

2.5 Nutritional qualities

The seed and leaf of *Plukenetia* are good sources of nutrients that can ensure food security and remedy malnutrition of the populace. Olajide *et al.* [42] evaluated the variability in proximate quality traits of 10 accessions of *Plukenetia conophora* from Southwestern Nigeria. They found that the seeds contained proximate contents and established that location of seed collection significantly affected ash, crude fat, crude fiber, dry matter, moisture content and nitrogen free extract, which suggests the need for selection and also gives way to improvement program. Agbo and Baiyeri [43] also reported variability in proximate and mineral qualities of five accessions of African walnut, which probably suggests genetic diversity or more probably, could be environmentally induced. Olajide *et al.* [42] also found that fresh and boiled walnut is an excellent food material with the ability to combat food insecurity in rural populace. They reported that the proximate composition in the nut include ash (6.40 and 6.34%), crude protein (17.04 and 19.20%) crude fat (40.77 and 39.74%), crude fiber (11.76 and 11.28%), dry matter (90.93 and 90.94%), moisture content (9.07 and 9.06%) and nitrogen free extract (14.97 and 14.40%) for fresh and boiled nuts respectively. Boiling positively influenced proximate qualities of African walnut as it increased protein content and dry matter while it reduced ash, crude fat, crude fiber, moisture content and nitrogen free extract. The proximate composition of *P. conophora* shows that it contains carbohydrate (4.17%), ash (3.32%), protein (29.14%), fat (54.14%) and various vitamin contents [44]. Suara *et al.* [45] reported 6.86% for moisture content, 11.78% for protein, 8.57% for total ash, 20.12% for crude fiber, 1.56% for total fat and 51.8% for total carbohydrate. In addition, Olajide *et al.* [46] evaluated the nutritional differences in 10 accessions of African walnut obtained from Southwestern Nigeria as affected by collection center and processing and the results suggested sufficient genetic variability in seeds of walnut obtained from Southwestern Nigeria, emphasizing the possibility for selection. The nutritional assessment of the seed revealed that the fresh and boiled seed contains iron (16.82 and 24.59 mg/kg), potassium (10781.0 and 10420.0 mg/kg), magnesium (5076.0 and 4621.0 mg/kg), phosphorus (162.7 and 229.7 mg/kg), zinc (65.2 and 54.54 mg/kg) and sodium (729.2 and 718.2 mg/kg). The results showed that processing had positively effect on iron, magnesium, phosphorus and zinc contents. Fresh seeds possessed higher quantity of zinc, potassium, magnesium and sodium compared to the boiled seeds. Conversely, iron and phosphorus were more in boiled seeds. Enujiugha [47] reported that walnut seeds contain mineral (465.95 mg/100 g of phosphorus, 57.37 mg/100 g of magnesium, 1.55 mg/100 g of iron and 6.84 mg/100 g of zinc). Phytochemical contents revealed 0.243 and 0.31% of phenol, 0.0142 and 0.0179% of phytate, 0.0851 and 0.0784% of tannin, 0.5419 and 0.5547% for alkaloid and 0.1396 and 0.1577% of glycoside for fresh and boiled nuts, respectively [48]. Concentration of phytate, alkaloids, phenol and glycosides were more pronounced in boiled seeds in comparison with the fresh nuts. On the other hand, higher value for tannin was obtained in fresh seeds. Ekwe and Ihemeje [19] found tannins of 0.89 mg/100 g, oxalate of 1.28 mg/100 g, phytic acid of 3.105 mg/100 g, trypsin inhibitors of 1.84 mg/100 g, saponin of 985.0 mg/100 g and alkaloid of 40.91 mg/100 g. However, higher concentration of alkaloid of 2.670 mg/kg and lower tannin of 0.56 mg/kg were recorded. Ayoola *et al.* [49] found that nutritional and elemental components are more in the nuts than in the leaves. The phytochemical contents observed in the seeds were also available in walnut leaves. The nut contained oil of 48–50%, the oil color is golden yellow, the taste and odor resembles that of linseed oil [50].

2.6 Climatic requirement

P. conophora is native to West Africa or Central Africa. It is abundant in Nigeria, Congo, Ghana and Cameroon. It occurs in the rain-forest region and plantations found at elevations from 250 to 1400 m [51]. The plant thrives on fertile, well—drained loam soils and can also grow on silt clay loam soils [29]. Generally, subsistence farmers grow *P. conophora* around gardens and backyards in humid and hot zones of tropical Africa [16]. It is also found in bottomlands, coves, rich woodlands and abandoned agricultural fields [52]. As a climber, the plant twines round the host plant to the apex in order to trap sufficient sunlight.

2.7 Cultivation

The plant thrives on deep, fertile, moist and well-drained loam soils [29]. The African walnut is majorly grown for subsistence consumption in the humid and hot regions of tropical Africa [16]. It does well in rich woodlands, fallow fields, bottomlands and coves [52]. The plant twines round the host plant to settle at the apex to receive more light from the sun, it may join trees to each other and hold a dead tree in position until it decays. Flowering occurs from November to early January and fruiting starts in February till September with the highest yield in July [16]. Walnut seed takes about 4–6 months to reach maturity stage [53].

2.8 Fertilizer requirement

Prior to the introduction of inorganic compounds, soil fertility has been improved through the breakdown of raw natural materials in the environment. This provided the soil with the needed nutrients for crop growth from only organic matter and this was enough in sustaining life [54]. Rob [54] noted that synthetic fertilizers do not add to the humus content of soil nor substitute it. Growth and dry matter yield of *Amaranthus cruentus* as affected by organic manure investigated by Daramola *et al* [55] showed that soil amendment using organic nitrogen sources produced the tallest plants, greater number of leaves, more branches and dry matter yield in comparison with the control. Olajide [23] conducted a study on influence of four poultry manure rates (0, 10, 20 and 30 t ha^{-1}) on early growth of African walnut and reported that morphological traits were positively affected with the application of PM at 10 t ha^{-1}, increase in PM beyond this rate resulted to a decline in growth of African walnut. The decline in growth at 20 t ha^{-1} and 30 t ha^{-1} of PM indicated that adequate amount of nutrients were released by the 10 t ha^{-1} of PM to complement the inherent nutrient in the soil. Sufficient nutrient supply produced high quality and better nutritious plants [56, 57]. As reported by Adebayo *et al.* [58], when manure is supplied at the required quantity, plants tend to grow at their optimal potential.

3. Saba (*Saba senegalensis*)

3.1 Description

S. senegalensis (A. DC.) Pichon is a large woody liana with white latex [59], from the Apocynaceae family [60]. The fruit is known as maad (Senegal), zaban (Mali), malombo (Congo Basin), wèda (Burkina Faso) and côcôta (Côte d'Ivoire) [61]. A climbing plant species that clings on other plants for support and growth (**Figures 3** and **4**). Saba trees are upwardly mobile plant found in tropical West Africa and the Western Sudan [62]. The plant can be grown in different ecological zones with rainfall from 100 mm to 1300 mm per annum and an altitude of 0–800 m [63]. Saba can withstand bush fire and has the ability to suppress

Figure 3.
Saba senegalensis plants.

Figure 4.
Staked Saba senegalensis plants.

weed (**Figure 5**). The bark has a dark gray color [64, 65] and it can reach up to 40 meters with the trunk above 40 cm in diameter [64, 66]. *S. senegalensis* fruit is a globulous envelope which contains seeds coated with yellow juicy pulp [67] (**Figure 6**). Juice from the fruit has become popular in urban areas of Côte d'Ivoire, Mali, Guinée, Burkina Faso, Senegal and the Gambia. *S. senegalensis* fruit has yellow pulp that is acidulous, tasty, sweet–sour when ripe and can be consumed directly or processed into other products [60, 67]. The fruits are often traded in towns and cities in most of the West African countries. In Nigeria, Mali, Burkina Faso, Ghana and Côte d'Ivoire, clusters of the fruits are been sold like oranges along the roadways. In Nigeria, the fruits of Saba are available from April to August.

Figure 5.
Saba senegalensis flower.

Figure 6.
Saba senegalensis fruit.

3.2 Origin and distribution

S. senegalensis is native of Gambia, Ghana, Guinea and other African countries [67]. It can thrive in different ecological zones having rainfall from 100–1300 mm annually, but mainly distributed along the river banks, open woodland and in rocky hills [60, 62, 67]. It is found in the Sudan savannas as well as in the Guinean savannas of Africa. It is a twining plant that normally needs staking. These areas are characterized by maritime trade winds with an average annual temperature range of 26–31°C, a dry climate with considerable variations in humidity. Maximum of precipitation in these areas occurs in the month of August with rainfall lasting from 2 to 4 months. The average annual rainfall in these locations varied from 400 to 1200 mm [68].

3.3 Food and economic importance

S. senegalensis fruit can be eaten in various ways; fresh or seasoned with sugar, salt or chilled [65, 67]. Saba fruit pulp is tart and pleasant to consume. In local communities, saba can be used to improve the taste of porridge made from cereals [59, 69]. The fruit pulp can also be made into nectar, preserves, jams and jellies [70]. The inner part of the shell is enveloped with superficial skin that can be eaten as chewing gum. The leaves can be made into sauces and condiments [59]. The inner materials that envelop the fruit pulp are dried and used to substitute lemon and tamarind [71]. Mechanical extractor can be used to change the form, making it possible to produce a refined puree that can be converted into different finished products like nectars, concentrated bases, syrups and marmalades. Saba can be included in food products like yoghurt. Saba fruits are highly cherished and highly prized in Africa, are the fruits are openly hawked in cities which results to improvement in the economy of the rural farmers [72]. The plant has the potential to suppress weeds and contributes to soil and water conservation [73].

3.4 Medicinal value

Saba has been used in herbal medicine with pronounced native applications. Ethnobotany alludes that the leaves, roots and fruit have the potential of treating certain diseases [74]. The fruits contained active compounds that could play a vital role in preventing and treating metabolic diseases and certain vitamin deficiencies [75]. Green fruits have the ability to fight against galactagogic, sterility and colic [59]. Ripen fruits are antiscorbutic, anorexic, stimulating and tonic [76]. The green fruits are preferred by the Fulani, which are prepared with salt; it is active in diuretic drug [59]. In cases of food poisoning, the leaves can be used to reduce the effect, and mashed leaves can as well be used in treating injuries [60]. When boiled, the vapor released can be inhaled to reduce coughing and headaches [77]. It can also be used in treating tuberculosis and pulmonary diseases; the leaves can prevent chronic headache and vomiting [78]. The whit latex can be used to treat pulmonary diseases and helps in fighting tuberculosis [68]. The powder from the dry root bark is effective in wound healing [64]. The roots of saba are used in treating infertility in females and skin burns. Root maceration, as a drink, is considered to be anti-hemorrhagic [59]. The latex is used as an adhesive in preparing poison for arrows. Saba leaves are made into sauces and spices as an appetizer having salty taste.

3.5 Nutritional qualities

The nutritional contents of the pulp are subject to very large variations, which are obviously linked to the differences in climate, nature of the soil and various analytical methods employed [59]. *S. senegalensis* fruit have high nutritional composition such as proximate, mineral, phytochemical and vitamin contents as established by previous studies. It can improve nutrition and health of the household. Olajide [23] evaluated the impact of four accessions on proximate, mineral, vitamin and phytochemical contents of *S. senegalensis* fruit pulp from Kogi State, Nigeria. He reported that ash content varied from 1.0–1.4%, percent carbohydrate values ranged from 11.60–34.90%, fat content ranged from 0.2–0.3%, fiber was in trace amount, moisture ranges from 63.6–86.5% and protein 0.09–0.18%. Oxalate value varied from 14.0–18.4 mg 100 ml^{-1}, phenol ranged from 11.2–13.4 mg 100 ml^{-1}, saponin varied from 0.4–2.5 mg 100 ml^{-1} and tannin ranges from 0.6–1.5 mg 100 ml^{-1}. The results also indicated that calcium ranged from 12.7–19.2 mg 100 ml^{-1}, iron varied from 0.02–0.08 mg 100 ml^{-1}, potassium ranges from 0.1–0.3, phosphorus ranged from 10.1–15.9 mg 100 ml^{-1} and zinc varied from 1.90–2.1 mg 100 ml^{-1}. He also found that vitamin B_{12} ranged from 0.05–0.07 mg 100 ml^{-1}, vitamin B_2 ranged from 0.7–4.8 mg 100 ml^{-1}, vitamin B_6 varied from 2.7–14.4 mg 100 ml^{-1}, vitamin C ranged from 34.3–66.9 mg 100 ml^{-1}, while β-Carotene and vitamin E contents were 0.5 and 0.01 mg 100 ml^{-1}, respectively. Like other fruits, saba has high amount of carbohydrates but the values varied widely (11–74.23 g/100 g) [60, 63, 70–72, 79, 80]. The oil content is 0.2 g/100 g [81] and the crude protein values range from 0.8 to 0.3 g/100 g [70, 79]. Boamponsem *et al.* [72] reported that *S. senegalensis* contains 47.5 ppm of magnesium, 810 ppm of calcium and 357.5 ppm of phosphorus. It also possessed phenol (264.76 mg/100 g), phytate (31.18 mg/100 g), oxalic acid (381.33 mg/100 g) and tannin (198.94 mg/100 g) [67].

Minerals such as calcium (51 ppm), phosphorus (357.5 ppm), magnesium (47.5 ppm) and potassium (152 ppm) were present in saba fruit pulp [72] but very low sodium content of <5 ppm. Nafan *et al.* [62] documented that the fruit is a potential source of vitamin C ranging from 34.8 to 67.5 mg/100 g. With acidity varying from 30 to 78.5 meq/100 ml; the acid taste of the fruit is high, hence the malic acid of 47.2 mg/100 g [72]. According to Kini *et al.* [77] the fruit contains β-carotene (1.559 mg 100 mg/100 g). Saba fruit has high water content of 80% [79]. All these components suggested that the fruits could supply the required nutrients and improve the health of consumers.

3.6 Climatic requirement

S. senegalensis can be found in West African countries and South Sudan [65, 72]. It can be seen growing along riverbanks [65], in woody savanna region and rocky areas [61, 77]. These locations have maritime trade winds, annual temperature of about 25–30°C and a dry climatic condition which varies in relative humidity. Maximum rainfall occurs in August and lasts for 2 to 4 months [70]. The species survives in different ecological conditions having rainfall between 100 mm to 1300 mm per annum with an altitude of 0–800 m [64, 65]. The plant is hardy and it resists bush fire.

3.7 Cultivation

Sales revenue for *S. senegalensis* fruits in Senegal are significant, accounting for 1/3 to 2/3 of farmers' income [80]. Programs meant at increasing *S. senegalensis* production and domestication are in place. Grafting of saba vine to encourage the domestication and reduce the time of fruiting has been carried out [82]. There is limited information or research work on fertilizer requirement for favorable growth of the seedlings in the nursery and field for its domestication [83]. *S. senegalensis* is predominant in regions that have sandy-loam to sandy-clay-loam soils [84]. It is not known whether physical and chemical properties of the soils are linked to the geographical distribution of Saba.

3.8 Fertilizer requirement of *S. senegalensis*

Food security, malnutrition and environmental degradation are being influenced by low soil fertility, inappropriate use and poor nutrient management strategies. Currently, there is dearth of information on the nutritional requirement and domestication of Saba in Nigeria. In maintaining the yield and quality of new crops, soil fertility management is of paramount importance [12]. The low fertility status of most tropical soils results in low crop production as most crops are nutrient demanding. Inorganic fertilizer such as NPK has strong effect on plant growth, development and yield [85]. Excessive use of NPK will result to loss of soil fertility, which has adverse effect on agricultural productivity, soil degradation and even cause water pollution. Conversely, regular use of organic fertilizers can improve organic matter content, water-holding capacity, enhance structure, nutrient cycling, helps in soil conservation, increase cation exchange capacity and encourage the activities of soil living organisms. Olajide *et al.* [86] evaluated early growth pattern of four accessions of Saba (*S. senegalensis*) in response to seven fertilizer rates (0 t ha^{-1}, 20 t ha^{-1} of PM + 200 kg ha^{-1} of NPK (20:10:10), 30 t ha^{-1} of PM, 30 t ha^{-1} of PM + 100 kg ha^{-1} of NPK, 30 t ha^{-1} of PM + 150 kg ha^{-1} of NPK, 40 t ha^{-1} of PM and 50 t ha^{-1} of PM) in the nursery and found that fertilizer application increased the growth traits measured compared to the control with no fertilizer application. This could be linked to sufficient nutrient released by the fertilizer that enhanced Saba growth. He further stressed that soil amendment with 50 t ha^{-1} of PM enhanced better growth of Saba seedling which indicates that this amount is sufficient for plant growth and development. Olajide [23] examined the effect of four PM rates (0, 10, 20 and 30 t ha^{-1}) on early growth of Saba in the field. He reported that soil amendment with PM at 20 t ha^{-1} positively influenced all the growth attributes measured. When nutrients are supplied optimally, high quality and better nutritious plants are produced [57]. Ndukwe and Baiyeri [13] found that application of PM at 20 t ha^{-1} was optimum for the production of yellow passion fruit in either the nursery or field.

4. Baobab (*Adansonia digitata*)

4.1 Description

Baobab (*A. digitata* Linn.) is a deciduous tree of the family Bombacaceae [87]. It grows up to 20 to 30 m tall with a diameter of 2–10 m at adult age. The trunk is soft, it has vast girth, reddish brown to gray smooth bark and possesses fibers used in making rope and fish net [88]. The plant produces numerous branches, the lateral root system can be up to 50 m from the trunk (see **Figures** 7 and **8**). The root end is tubular while the taproot of the tree is shallow which does not grow beyond 2 m depth making them susceptible to storms [88]. The mature tree begins each season with the production of simple leaves having 2 or 3 leaflets. The plant produces white, large, pendulous flowers, and they appeared singly or paired in the leaf axils. The plant bears flower towards the end of dry or prior to the commencement of rains usually after shedding of leaves [89]. The fruit possesses outer shell, pulp and seeds (see **Figure 9**). The life span of baobab trees ranges from 200 to 300 years and some can live beyond 1000 years [89]. *A. digitata* has numerous uses, hence the name 'tree of life and small

Figure 7.
Baobab plants at 4 months after transplanting.

Figure 8.
Baobab plants pruned after 6 months of transplanting.

pharmacy' as a results of its benefits including food, clothing, medicine, protection, fiber, seeds, leaves and roots [90].

4.2 Origin and distribution

Adansonia digitata is a tree originating from African savannah, Madagascar, Australia and Arabia, of the family Malvaceae [90]. It is distributed in arid regions of most countries of the Sahara. The trees are normally found in the thorn woodlands of

Figure 9.
Baobab seed extraction.

African savannahs having low altitudes and 4 to 10 dry months yearly [90]. The tree may grow alone, although it occurs in small groups, which depends on the nature of the soil. Wherever baobab is found, it is majorly in the arid or semi-arid regions of the world [91]. Baobab tree is seen in both settlements and in the wild. In Nigeria, the baobab trees are widely found in North central States (Kogi, Benue, Niger and Kwara) and Sudano-Sahelian parts of the country like Kano, Katsina, Sokoto, Zamfara, Kebi and Jigawa States (Northwestern) and in the Northeastern (Yobe, Borno, Gombe, Bauchi and Adamawa States) [91].

4.3 Food and economic importance

Baobab is an important tree for the African countries [92]. Traditionally, the plants have been used in many ways by people occupying the areas where they are available. The fruit pulp plays a vital role in contributing to the diets of the local populace, and it serves as seasoning material as well as appetizer [93]. When the pulp is soaked in water, the liquid derived from it can be used in making drinks, it can also serve as sauce for food, it can be fermented and used in local brewing [88, 93]. Recently, the pulp has gained popularity and used as ingredient in making ice and other products in urban centers [94–96], the pulp is made into juices and jams. The baobab fruit pods can be burnt and the potash-rich salt obtained can be used for making soap [91]. The European Commission has permitted the importation of baobab fruit pulp as new type food for human use [97] which was approved as a food ingredient by the Food and Drug Administration of the United States of America [98].

The seeds can be consumed fresh or dried, it can also be made into powder which can be used to thicken soup, or roasted and made into a paste, or boiled, fermented and then dried for use [87, 99]. The seeds can be pounded for the extraction of vegetable oil used in soup preparation and it can be fermented into seasoning [94]. The oil extracted can also be used as fuel, cosmetic, medicine and for treating muscle spasms, swollen veins, injuries and dandruff [100–102]. The seed is a potential source of protein, and the roasted seeds are used to substitute coffee in Sudan and North Africa [103].

Baobab leaves are important in traditional diets of the rural people as leafy vegetables are rich in iron and vitamins. Young leaves are harvested, dried, made into powder and used in making soup [87]. Fibers from the bark are used for weaving bags, hats and mats [87]. The wood is light and whitish when dried and used for fuel [104]. The tree provides shelter, clothing and material for hunting and fishing [94]. It is a good source of dye and fuel. The roots, leaves, seeds and pulp are consumed and it serves as a basic source of livelihood. Baobab trees provide shelter and it can store water [94], with capacities of 1000 to 9000 liters per tree [105]. The products were traded centuries ago being popular in Cairo markets in the sixteenth century [87].

4.4 Medicinal value

Baobab possesses a lot of substances used for treating various diseases in African traditional medicine [106]. In many medicinal uses the stem bark is ground for internal use and it is effective as a result of the presence of soluble and insoluble tannin [107]. The plant parts are used treating diseases and specific uses that were documented includes the treatment of microbial infections, tuberculosis, malaria, anemia, fever, diarrhea, toothache and dysentery [108]. The leaves and fruit pulp are used as febrifuge and boosts the immune system [97, 109]. It is reported that baobab pulp is used externally with buttermilk for relief from diarrhea and dysentery in India, and also the fresh leaves are crushed and used to treat painful bruises [89]. In some West African countries, the seeds, leaves, fruit pulp are major ingredients in beverages, sauces and porridges [97, 107, 110].

4.5 Nutritional qualities

Previous studies showed that baobab leaves are good sources of nutrients. Study [23] that evaluated the effect of poultry manure application rates on nutritional qualities of two accessions of baobab grown in Nsukka, Enugu State, Nigeria revealed that ash ranges between 7.68–8.44%, carbohydrate varied from 51.70–58.50%, fat ranged from 2.90–6.10%, fiber varied from 4.73–5.39%, moisture ranges from 7.40–14.10% and protein ranged from 14.25–18.29%. This author reported that cyanogenic glycosides content obtained was 0.02 mg/100 g while flavonoids ranged from 19.22–25.33 mg/100 g, oxalate varied from 36.70–66.70 mg/100 g, phenol ranges from 50.00–146.00 mg/100 g, phytate ranged from 1.56–2.54 mg/100 g, saponin varied from 0.09–0.10 mg/100 g and tannin ranged from 5.35–5.66 mg/100 g. The concentration of phytate was within the tolerable limit of 5.72–9.22% [111] but oxalate, phenol, saponin and tannin were above the tolerable limits of 5% [112], 2% [113], 0.2% [114] and 3.3% [115], respectively. Since the leaves are not consumed raw, the anti-nutrient contents may be significantly reduced by heat during the cooking process. When plant parts are boiled in water, effects of poisonous anti-nutrients are reduced, hence increasing their palatability [116]. Olajide [23] reported that calcium, iron, iodine and zinc contents ranged from 89.70–98.10, 7.78–8.00, 7.83–8.44 and 0.90–0.93 mg/100 g, respectively. Value for vitamin B_{12} was 0.04 mg/100 g while vitamin B_6 varied from 0.54–2.28 mg/100 g, vitamin E ranged from 8.89–12.33 mg/100 g and carotenoids ranges from 81.10–124.40 mg/100 g. The study also found that poultry manure application rates significantly influenced moisture, iron, iodine, zinc, vitamin B_6, carotenoids and phenol contents in Baobab leaves. Baobab leaves contain protein (13.6%), fat (2.71%), ash (4.08%), crude fiber (2.45%), (0.01%), moisture (78.2%) and vitamin C (14.98 mg/100 g) [117]. Osman [118] observed that the seeds possessed high quantities of fat, fiber, crude protein but low carbohydrate contents. Consuming 20 g can provide 15 to 34% recommended daily allowance of protein for children; 60 g can meet 27% of the recommended daily allowance for pregnant women. Also, consuming 100 g can supply about 22% recommended daily intake of the energy for pregnant women and 29.4% recommended daily allowance of energy for children [110]. Previous work of Arowora *et al.* [119] revealed 31.43 mg/100 g of tannins, 124.36 mg/100 g of phenolics, 9.35 mg/100 g of alkaloids, 63.43 mg/100 g of flavonoids and 14.63 mg/100 g of glycosides. Enoch *et al.* [120] reported that baobab leaf contains sodium (0.870 mg/l), magnesium (1.260 mg/l), potassium (4.118 mg/l), calcium (0.780 mg/l), iron (3.640 mg/l) molybdenum (0.409 mg/l) aluminum (0.006 mg/l), nitrogen (0.278 mg/l) and phosphorus (0.162 mg/l). Baobab fruit contains α-carotene (0.17 μg/g) and lutein (1.53 μg/g) in dry weight [121]. Becker [122] found riboflavin, thiamine and niacin content with respective values of 0.07, 0.04 and 2.16 mg/100 g dry weight.

4.6 Climatic requirement

A. digitata is an enormous evergreen tree distributed across subtropical regions of Africa such as South Africa, Botswana, Nigeria, Tanzania and Madagascar. The baobab is also considered to be one of the oldest forms of life in Africa, some estimated to be up to 3000 years old [123]. The tree is restricted to hot, dry regions but lives in various environments outside both the northern and southern edges of tropical regions of Africa, more specifically outside latitude lines 16° N and 26° S [124]. Its semi diverse stretch reaches biomes like scrub, woodlands, wooded savannah and even semi-arid/semi humid tropical regions. *A. digitata* tree is usually seen in regions with annual rainfall of 500 to 800 m [125].

4.7 Cultivation

Due to the medicinal, nutritional and cosmetic applications of baobab, it has gained popularity and attracted the interest of a lot of pharmaceutical companies and researchers in the past decade. As a result of the high demand for baobab products in European Union and United States of America, the tree ought to be conserved, treasured and domesticated in other parts of the world [126]. The plant is found in hot, semi-arid regions, dry woodland and stony areas with low rainfall of 1500 mm per annum [94], it thrives on marginal soils but does well on well-drained, clays to sandy soils, but not on deep sands, where it will be difficult for the plant to obtain sufficient moisture and support [89]. In Africa, baobab is found at latitude 16° N and 26° S, these areas do not have more than one day of frost in a year. It has slow growth which could be as a result of low rainfall and low soil fertility. Assogbadjo *et al.* [127] determined the perception and preferences of baobab products in Burkina Faso, Benin, Senegal and Ghana, the study included women and men of different ages. According to the survey, if the bark is easier to harvest, then the pulp and leaves will be tastier; slimier pulp are less tasty; when the fruit capsules are longitudinally marked, the tastier the pulp will be. The study indicated that farmers can use selected combinations of attributes as a guide in germplasm collection. This knowledge could be employed during the selection of a suitable planting material and a guide for a domestication. Commercialization of baobab seed oil and fruit pulp is on the rise, in addition, exportation worldwide has led to mounting pressure on this resource [87].

4.8 Fertilizer requirement

In order to increase productivity to meet the nutritional requirements of human population and to increase the household income, enhancement of soil health is a critical factor. Soil fertility can be improved using organic or inorganic fertilizers and may be combined [14]. Frequent utilization of inorganic fertilizers solely cannot increase crop yield on poor soils [128]. Therefore, the need for organic soil amendments to increase soil fertility and enhance the physicochemical and biological properties for continuous production of crops. It was noted that amending the soil with organic and inorganic fertilizers support the best crop performance [129, 130]. Olajide [23] who worked on the influence of three rates of PM (0, 15 and 30 t ha^{-1}) on early growth of baobab in Nsukka, Enugu State, Nigeria found that plots amended with poultry manure performed better in terms of the growth attributes measured compared to the control. The higher values of the morphological traits obtained with poultry manure application suggests that baobab plants are highly responsive to manure application. Poultry manure is the richest out of the animal manures, and it is a valuable source of nitrogen and potassium as well as organic matter [12]. Organic manure as soil amendment is highly important in order to sustain crop production systems since it is a reliable source of nitrogen and carbon [131, 132] and it also moderates soil pH [133]. Olajide [23] reported that application of 15 t ha^{-1} of PM increased all the growth parameters evaluated than other poultry manure rates. Adebayo *et al.* [58] reported that when manure is available in adequate quantity, plants tend to grow at their optimal potential.

5. Kapok (*Ceiba pentandra*)

5.1 Description

C. pentandra is a tree belonging to Malvaceae family. It is a plant that is found in the wild. The plant is called Kapok or white Silk-Cotton tree in English. It is known as "Araba" in Yoruba, "Akpu-ogwu" in Igbo and "Rimi" in Hausa tribes of Nigeria [134, 135]. *C. pentandra* is a fast-growing tree and can grow up to 24–70 m high, having a diameter of 100–300 cm. Universally, kapok is known to be among the largest trees. The stem and large branches are usually crowded with conical spines (**Figures 10** and **11**). The palmate leaves consist of 5 to 9 leaflets and can be up to 20 cm (7.9 in) in length. The tree produces hundreds of pods measuring 15 cm (5.9in) with seeds surrounded by a fluffy fiber which combines lignin and cellulose (**Figure 12**). In Miami, Florida, one of the oldest known trees lives at 200 years [136]. Kapok fiber is light, water resistant, but it is highly flammable. The harvesting, processing and separating the fiber is done manually and is labor-intensive. Although it is difficult to spin, it is used alternatively to down for filling in mattresses, pillows, upholstery, zafus and stuffed toys like teddy bears and for insulation. Earlier, it was used in making life jackets and similar devices until synthetic materials largely substituted the fiber. Oil extracted from the seeds is used locally in soap and fertilizer production [137].

Figure 10.
Kapok plants at 5 months after transplanting.

Figure 11.
Kapok plants at 11 months after transplanting.

5.2 Origin and distribution

Ceiba pentandra is native primarily in West Africa where it is found in rainforests typically at elevations of 900 to 1200 meters [138]. It is widely distributed in South America—Brazil, Venezuela, Ecuador, Peru, Bolivia, Colombia, the Guyanas; North through Central America to Mexico; Caribbean; West tropical Africa [75]. Kapok is a multipurpose tree with numerous uses for the local populace where it is grown. It is highly valued for its fiber, and it also provides food, medicines and other products.

Figure 12.
Kapok pods.

Kapok is widely grown in the tropics for its medicinal value and fiber, the plant exists naturally in many areas [139]. It has long been planted around buildings in villages for food, medicine, beautification purpose and other uses. However, the floss is harmful as it causes irritation to the eye and nose, so the tree is not suitable for town planting. Commercially, kapok is grown for the fiber from the pod in Java [75].

5.3 Food and economic importance

The succulent leaves of kapok are used in soup preparation, which is comparable to Okra, and it is used for eating starchy balls made from millet, cassava and yam [140, 141]. The leaves are dried and made into powder used to prepare delicious soup known as 'kuka' during dry periods [23]. Fresh and dry leaves made into powder are hawked in the villages, which contributes to the rural farmer's economy. Vegetable oil extracted from the seeds can be used for bio fuel, soap making, paint preparation and can also be used in manufacturing fertilizer [130]. The plant provides fiber and timber. The whitish cotton (floss) can be used for making mattresses, absorbent material, pillows and tinder [141]. The wood is widely used in plywood manufacturing and in making canoes. It is also used for musical instruments, mortars, carvings, lightweight furniture and other items [130]. The foliage can be used in feeding ruminant animals, trunk for plywood and wood pulp for paper. The fiber is used while dressing injuries; applying the oil can treat rheumatism [130]. *C. pentandra* is known in folktale, it is noted to be a sacred plant and its image is used as the national emblem of Guatemala, Puerto-Rico and Equatorial Guinea. It appears on the coat of arms and flag of Equatorial Guinea.

5.4 Medicinal value

Leafy vegetables play a vital role in maintaining health of the populations and diseases prevention. Large quantities of micro-minerals are obtained from dark green leafy vegetables which are essential in nutrient metabolism and slow down degenerative diseases [142]. Gropper *et al.* [143] emphasized the importance of consuming vegetable based meals to prevent colon cancer. Ball [144] found high vitamin, dietary fiber and mineral contents in vegetables and the role they play in keeping up alkalinity in the body. The high amount of fibers in green leafy vegetables assists in regulating the digestive system, improving bowel health and weight management [145, 146]. The leaves are recognized as having emollient and sedative contents. In Senegal, Kapok leaves are mashed in water, which is drunk for general fatigue, stiffness of the limbs, headache and bleeding of pregnant women. Young leaves are warmed, mixed with palm oil and eaten against heart diseases. Leaf sap is used in treating skin infections and mental illness. Leaf decoction is used by veterinary doctors in the treatment of trypanosomiasis among others [137]. The bark contains a blackish mucilaginous gum; it is astringent and is used in India and Malaya for bowel complaint and West Africa for diarrhea [66]. It is also used for treating skin infection and tooth troubles in Senegal. Bark macerations are credited with stimulant and antihelminthic properties. It is also a cure for heart trouble and hypertension [130]. The root forms part of preparations to treat leprosy. The flowers are eaten to treat constipation and fruits taken with water against intestinal parasites and stomach problem.

5.5 Nutritional qualities

Many of the local vegetables are underutilized due to inadequate information on their potentials to nourish the body with nutrients [147]. High utilization and consumption of vegetables is crucial in alleviating universal incidence of nutritional deficiencies [148]. Chemical analysis has shown that the leaves of *C. pentandra* contain anti-nutrient, proximate, mineral and vitamin contents. Olajide [23] reported that the succulent leaves of *C. pentandra* contain ash (8.59%), carbohydrate (55.60%), fat (1.50%), fiber (18.80%), moisture (3.34%) and protein (12.34). As reported by Osuntokun *et al.* [137], it has protein (16.25%), fat (5.34%), fiber (8.53%) ash (8.72%), moisture (7.32%) and carbohydrate (53.72%). Enechi *et al.* [149] found 47.37% for moisture, 16.81% for protein, 25.23% for carbohydrate, 4.47% for fiber, 2.23% for fats and 2.14% for ash. According to Olajide *et al.* [150], succulent leaf of *C. pentandra* possess calcium (9.93 mg/100 g), iron (19.05 mg/100 g), potassium (35.80 mg/100 g), magnesium (60.79 mg/100 g), phosphorus (78.50) and zinc (0.59 mg/100 g). Shahin *et al.* [151] reported that the leaf contains 177.0 mg/100 g of calcium, 153.66 mg/100 g of potassium, 48.15 mg/100 g of magnesium, 27.09 mg/100 g of zinc and 1.54 mg/100 g of iron. Olajide [23] obtained 0.012 mg/100 g for vitamin B_{12}, 0.59 mg/100 g for vitamin B_2 and 0.97 mg/100 g for vitamin C in succulent leaves of Kapok from Kogi State, Nigeria. Earlier report of Adepoju and Ugochukwu [141] found that Kapok leaves contain vitamin B_2 (0.19 mg/100 g) and B_{12} (0.24 mg/100 g). Friday *et al.* [140] found that it contains phenol (173.94 mg/100 g), oxalate (0.10 mg/100 g), tannin (0.48 mg/100 g) and saponin (1.55 mg/100 g). Olajide *et al.* [150] also reported 0.19 mg/100 g of saponin, 107.10 mg/100 g of oxalate, 18.20 mg/100 g of phenol, 2.60 mg/100 g of phytate, 4.55 mg/100 g of tannin and 3.96 mg/100 g cyanide in succulent leaves of *C. pentandra* sourced from Kogi State, Nigeria. The concentration of saponin content was within the tolerable limits of 0.2% reported by Codex [152], but oxalate, phenol, phytate and tannin contents of succulent leaves of *C. pentandra* were above the tolerable limits of 5% [112], 2% [113], 9.22–5.72% [114] and 3.3% [115], respectively. Since the leaves are not eaten raw, the anti-nutrient contents may be significantly reduced by heat during the cooking process. Boiling significantly reduced the poisonous effects of anti-nutrients and increased the leaf consumption [116, 153]. Olajide [23] conducted a study on the impact of integrated application of poultry manure and inorganic fertilizer on mineral and vitamin constituents of *C. pentandra* leaves grown in Nsukka, Enugu State, Nigeria. The results indicated that fertilizer application rates only influenced zinc and calcium with 20 t ha^{-1} of PM having the highest concentration of Zinc (0.45 mg/100 g) and integrated application of 5 t ha^{-1} of PM + 200 kg ha^{-1} of NPK recorded the highest value for calcium (145.00 mg/100 g). Protein consumption is necessary due to role carried out by its essential and non-essential amino acids as building blocks for protein biosynthesis not only for the growth of infants and children, but also for the steady replacement and turnover of body protein in adults [147]. Flavonoids have antiviral, antibacterial, antineoplastic, anti-inflammatory and anti-allergic properties [148]. Tannins possess antioxidant, antimicrobial and anti-inflammatory properties. Phenols are known as powerful antioxidants, preventing oxidative damage to biomolecules such as DNA, lipids and proteins which are active in chronic diseases such as cancer and cardiovascular diseases [137]. Ascorbic acid/vitamin C empowers the body's immunity against infection, helps in collagen and thyroxin synthesis and improves absorption of iron [147]. In living organisms, ascorbate (anion of ascorbic acid) is an antioxidant that protects the body against oxidative stress and is a co-factor in several vital enzymatic reactions [154]. These constituents give the leaves their protective, preventive and therapeutic properties thus improving the gains that can be obtained by consuming these leaves.

5.6 Climatic requirement

C. pentandra needs abundant rainfall but requires drier period for flower and fruit production. It grows at altitudes as high as 4000 m [155]. Night temperature less than 17°C decreases the germination of pollen grains. This reduces areas where the plant can be found. It thrives in latitudes of 20°N and 20°S and requires annual rainfall of 1500 mm. Where kapok is naturally distributed, the average rainfall is 750 to 3000 mm per annum. The dry season should not exceed 4 months. In drier regions, water requirement by the plant can be met by providing irrigation [23]. For optimum production, the tree is grown on deep, fertile well drained soils. Kapok is prone to heavy winds and cannot survive bush fire. The plant is usually found in rainforest and drier zones. Kapok is dominant in secondary forest and along the riverbanks and is hardly found in primary forest. It is a fast growing plant with canopy developing within few months if left undisturbed [156]. The tree may occur in large numbers in humid to semi-arid regions. Kapok can be grown or self-sown; the seedlings should be protected from fire and livestock [23].

5.7 Cultivation

Kapok is a tropical plant found at height of 1200 meters, however, productivity declines beyond 460 meters [144]. Optimal growth is achieved in locations with yearly daytime temperatures range of 17–38°C, and kapok can withstand 12–40°C. The plant could die at −1°C or less [72]. Fruit production could be delayed at nocturnal temperature of 20°C. Kapok enjoys a mean rainfall of 1500 to 2500 mm per annum, although it withstands 750 to 5700 mm [72]. It tolerates long dry period range of about 0–6 months [157]. Kapok is the tallest indigenous plant in Africa [158]. It thrives in a fertile, deep, moisture-retentive but well-drained loamy soil is preferred

[139, 159]. It does well at a pH range of 5.5–6.5 and it can also tolerate 5–7.5 [74]. Kapok is prone to wind; it prefers wind break for protection against strong winds [139]. The tree may start bearing fruit at 4–5 years, with increased production till 8 years. The economic or production life of kapok tree can last for almost 60 years [74]. Leaf and flower production season are stable in drier regions where the plant is distributed; in wet regions, production of leaf and flower are not regular. Anthesis occurs in the night and ended at midday. The flower releases strong scent and secretes nectar at the flower base, which is large and bisexual. Ripening of fruits occurs at 80–100 days after flowering, the dehiscent types splits with loosely fixed seeds released and dispersed [74]. The light seeds are spread widely and find ideal germination conditions in abandoned agricultural land [74]. A single tree can produce over 300 pods yearly with an output of 20 kg of fiber from 5 years to 50 years [139]. The tree responds positively to coppicing. It has vigorous root system causing damage to buildings and roads [160].

5.8 Fertilizer requirement

Soil degradation as a result of deforestation, nutrients lost through leaching and erosion has led to depleting fertility and caused decline in soil organic matter levels [161, 162]. Soil amendment using organic manure is vital in increasing crop yield. A fertile soil should possess an organic matter content of more than 3%. Soil amendment could be in form of organic or inorganic or combined [14]. Previous studies confirmed that combined application of organic and synthetic fertilizers supports the best crop performance [129, 130]. Olajide and Baiyeri [163] who worked on the effect of these rates (No fertilizer, 5 t ha^{-1} PM + 200 kg ha^{-1} NPK, 10 t ha^{-1} PM, 20 t ha^{-1} PM, 450 kg ha^{-1} NPK and 20 t ha^{-1} PM + 100 kg ha^{-1} NPK) on growth of kapok in the field reported that 20 t ha^{-1} of PM applied solely increased the performance of kapok plants. Olajide and Baiyeri [163] also found that soil amendment using 450 kg ha^{-1} of NPK and other treatments combined with NPK reduced the growth of kapok when compared with the plants in plots where no fertilizer was applied. Inability of NPK fertilizer in increasing kapok might be associated with acidification of the rhizosphere. It has been established that application of NPK reduces soil pH and boosts soil acidification but addition of organic manure improves soil acidification [164].

6. Conclusion

It is quite evident from this review that tropical plant species provide a lot of benefits to ensure food security, improve the health and socio-economic status of the populations. These crops have nutritional, economic, medicinal and industrial potential and can ensure healthy food system for the people. They can also play an important role in climate resilience for sustainable environment. Their full potential should be harnessed as it has been established that these crops are highly responsive to fertilizer. They can be brought under regular cultivation culture and the fruits and leaves accessed without the traditional search for them. This information could encourage the domestication of these indigenous plant species and guide the utility of these crops. Empirical studies [23] copiously quoted in this review supports the possibility of adapting these forest species to regular cultivation culture.

Furthermore, in harnessing the potential of these tropical plant species to the fullest, this review outlined some key factors that could unlock their vast potential.

- *Awareness*: It is crucial to supply the stakeholders with knowledge with respect to their nutritional, economic, industrial and medicinal values in order to increase their acceptability and domestication.

- *Domestication*: No effort has been made to specifically cultivate these crops for food. Most of these tropical plant species have been neglected for long; farmers should be encouraged to grow them commercially. Government and the scientific community should work alongside the farmers.

- *Intervention of government*: Production, harvesting, processing and marketing of these crops require the support of government. Incentives can be made available to farmers to boost their morale.

Conflict of interest

The authors declare no conflict of interest.

Author details

Kayode Paul Baiyeri[1]* and Kolawole Olajide[2]

1 Department of Crop Science, University of Nigeria, Nsukka, Nigeria

2 Division of Agricultural Colleges, College of Agriculture, Ahmadu Bello University, Kabba, Kogi State, Nigeria

*Address all correspondence to: paul.baiyeri@unn.edu.ng

References

[1] Houston AM. Biological Diversity: The Coexistence of species on Changing landscapes. Cambridge University Press; 1995. p. 1

[2] Osunsina OO. Ecological Interaction of soil, flora and faunal resources of Alabata University of Agriculture, Abeokuta Permanent Site. Abeokuta: University of Agriculture; 1999

[3] Mukhtar RB. Effect of light intensity on seedling growth of *Diospyros mespiliformis*. In: Proceedings of the Forestry Association of Nigeria. Sokoto; 2013. pp. 320-328

[4] Akinyele AO. Mint. Silvicultural requirement of seedlings of *Buchholzia coriacea* Engler. [thesis] submitted to the Department of Forest Resources Management, University of Ibadan; 2007

[5] Enzonga J, Jean P, Arielle M, Mvoula T. Mint: Biochemical characterization of four less exploited edible fruits in Congo-Brazzaville: *Passiflora edulisf. Flavicarpa, Aframomum alboviolaceum, Saba comorensis* and *Clitandra cymulosa*. African Journal of Agricultural Research. 2019;**14**(34):1913-1920

[6] FAO, IFAD, WFP. The State of Food Security in the world. [Internet]. 2015. Meeting the 2015 International Hunger Targets: Taking Stock of Uneven Progress. 2019. Available from: http://www.fao.org/3/a-i4646e.pdf

[7] Ahodegnon DK, Gnansounou M, Bogninou RGS, Kanfon ER, Chabi B, Agbangnan Dossa PC, et al. Biochemical profile and antioxidant activity of Parkiabiglobosa and *Tamarindus indica* fruits acclimated in Benin. International Journal of Advanced Research. 2018;**6**(11):702-711

[8] Cernansky R. Mint: The rise of Africa's super vegetables. Nature. 2015;**552**:146-148

[9] Ebert AW. Mint: Potential of underutilized traditional vegetables and legume crops to contribute to food and nutritional security, income and more sustainable production systems. Sustainability. 2014;**6**:319-335. DOI: 10.3390/su6010319

[10] Okeke EC, Eneobong HN, Uzuegbunam AO, Ozioko AO, Umeh SI, Kuhnlein H. Mint: Nutrient composition of traditional foods and their contribution to energy and nutrient intakes of children and women in rural households in igbo culture area. Pakistan Journal of Nutrition. 2009;**8**:304-312. DOI: 10.3923/pjn.2009.304.312

[11] Umakanta S, Shinya O, Moses AD. Nutrients, minerals, antioxidant pigments and phytochemicals, and antioxidant capacity of the leaves of stem amaranth. Scientific Reports. 2020;**10**:3892

[12] Ani JU, Baiyeri KP. Mint: Impact of poultry manure and harvest season on juice quality of yellow passion fruit (*Passiflora edulis var. Flavicarpa* Deg.) in the sub-humid zone of Nigeria. Fruits. 2008;**63**:239-247

[13] Ndukwe OO, Baiyeri KP. Mint: Agronomic evaluation of two passion fruit genotypes in containers (mega-pot) predicted corresponding performances under field conditions. Tropical and Subtropical Agroecosystems. 2020;**23**:1-15

[14] Aba SC, Baiyeri KP, Ortiz R. Mint: Effect of fertilizer treatments on fruit nutritional quality of plantain cultivars and derived hybrids. Fruits. 2020;**75**(6):281-287

[15] GRIN. *Plukenetia conophora* Müll. Arg.Germplasm Resources Information Network (GRIN) Taxonomy for Plants. Beltsville: United States Department of Agriculture (USDA) and Agricultural Research Services (ARS); 2010

[16] Oyekale KO, Odutayo OI, Esan EB, Ogunwemimo KO, Denton OA, Bolaji DT. Mint: Comparative studies on phytochemical and proximate composition of four morphological distinct segment of the conophor seedlings (*Tetracarpidum conophorum* Hutch and Dalziel). Brazilian Journal of Biological Sciences. 2015;**3**:91-100

[17] Nwachoko N, Jack IR. Mint: Phytochemical screening and anti-diarrhea activities of *Tetracarpidium conophorum* induced in albino rats. Sky Journal of Biochemistry Research. 2015;**4**(4):21-24

[18] Janick J, Paul RE. The Encyclopedia of Fruits and Nuts. England: Oxfordshire; 2008. p. 68

[19] Ekwe CC, Ihemeje A. Mint: Evaluation of physiochemical properties and preservation of African walnut (*Tetracarpidium conophorum*). Journal of Natural & Applied Sciences. 2013;**4**(6):501-512

[20] Edem CA, Dosunmu LM, Francesca LB. Mint: Determination of proximate composition, ascorbicacid and heavy metal content of African walnut (*Tetracarpidium conophorum*). Pakistan Journal of Nutrition. 2009;**8**(3):225-226

[21] Oluwole SO, Okusanya OT. Mint: Dormancy and seed germination in the African walnut (*Tetracarpidium conophorum* Mull Arg). Journal of Scientific Research and Development. 1993;**1**:9-14

[22] Onawumi OOE, Faboya OOP, Ayoola PB. Mint: Chemical evaluation and nutritive values of African walnut leaf (*Plukenetia conophora* Mull. arg.). International. Journal of Herbal Medicine. 2013;**1**(3):122-126

[23] Olajide K. Agronomic and biochemical evaluation of four underutilized indigenous plant species (*Plukenetia conophora, Saba senegalensis*, *Adansonia digitata* and *Ceiba pentandra*) under different fertilizer treatments. [Thesis] Faculty of Agriculture, University of Nigeria, Nsukka; 2021

[24] Udedi SC, Ani ON, Anajekwu BN, Igwilo IO, Ononamadu CJ, Adindu CS, et al. Mint: Comparative proximate analyses of raw and cooked *Tetracarpidium conophorum* (African Walnut) Found in Awka, Anambra State, Nigeria. The Bioscientist. 2013;**1**(2):114-113

[25] Oke OL. Leaf Protein Research in Nigeria. Ibadan, Nigeria: Ibadan University of Ibadan Press; 1995

[26] Obianime AW, Uche FI. Mint: The effects of aqueous extracts of *Tetracarpidium conophorum* seeds on the hormonal parameters of male guinea pigs. Asian Pacific Journal of Tropical Medicine. 2010;**3**(1):21-24. DOI: 10.1016/S1995-7645(10)60024-7

[27] Nwaichi EO, Osuoha JO, Monanu MO. Mint: Nutraceutical potential of *Tetracarpidium conophorum* and *Buccholzia coriacea* in diet-induced hyperlipidemia. Journal of Chemical Health Risks. 2017;**7**(3):157-170

[28] William RD. *Juglans nigra* L. Black walnut. In: Burns RN, Honkala BH, editors. Silvics of North America. Washington.: USDA for Serv Agric; 1990. pp. 386-390

[29] Cogliastro A, Gagnon D, Bouchard A. Mint: Experimental determination of soil characteristics optimal for the

growth of ten hardwood planted on abandoned farmland. Forest Ecology and Management. 1997;**96**(1-2):49-63. DOI: 10.1016/S0378-1127(97)00042-X

[30] Egharevba RKMI, Ikhatura KC. Mint: Influence of seed treatments and growing media on seedling growth and development of African walnut *Plukenetia conophorum*. African Journal of Biotechnology. 2005;**4**:808-811

[31] Oyenuga VA. Nigerian Food and Feeding Stuff. Ibadan: University Press Ibadan; 1997. pp. 11-12

[32] Babalola FD. Cultivation of African walnut *Tetracarpidium conophorum* Mull. [Arg] on agricultural plantation: An approach to Conservation Agriculture in Nigeria. Presentation made at World Conference on Conservation Agriculture and 3rd Farming System Design Conference [WCCA/FSD]; 26-29 September, Brisbane, Australia; 2011. pp. 180-181

[33] Chikezie UN. Mint: Phytochemical and proximate compositions of *Tetracarpidium conophorum* [African Walnut] seeds. International Journal of Research Studies in Biosciences. 2017;**5**(10):25-31

[34] Ayoola SO, Omoile LN. Mint: Effects of the walnut *Plukenetia conophora* shell in the diet on the growth performance and genotoxicity of the African catfish *Clarias gariepinus*. Aceh Journal of Animal Science. 2019;**4**(2):99-110

[35] Payal R, Mamshad A, Arti T. Mint: Study on antimicrobial and antioxidant properties of walnut oil. International Journal of Current Research in Chemistry and Pharmaceutical Sciences. 2014;**1**(7):51-55

[36] Urszula G, Agata D, Lukasz P, Iwona K. Mint: Nutraceutical potential of tinctures from fruits, green husks and leaves of *Juglans regia* L. The Scientific World Journal. 2014;**1-10**

[37] Ekhuosuehi A. Properties of Walnut plant in culture. [Internet] 2008. The Nigerian Observer. Available from: www.nigerianobservernews.com/19072010/..../features3.html

[38] Hogue P. Real black walnut stain. Primitive Enterology. 2000;**13**(23):3245-3248

[39] Nael AT, Mohammed AA. Mint: Utility and importance of walnut, *Juglans regia* Linn: A review. African Journal of Microbiology Research. 2011;**5**(32):5796-5805

[40] Wolters K. “Black walnut”. The Interne 2009. Walnut information from drugs.com.

[41] Faramarz Z, Leila GH, Arash D, Ahmad FS, Arsham D, Zahra LS. Mint: Antibacterial Effect of *Juglans Regia* Bark against Oral Pathologic Bacteria. International Journal of Dentistry. 2013;**5**:854765

[42] Olajide K, Baiyeri KP, Ndubuaku UM. Mint: Variability in Proximate Quality Traits of 10 Accessions of African Walnut (*Tetracarpidium conophorum*) from Southwestern Nigeria. Nigerian Journal of Horticultural Science. 2020a;**25**(2):84-90

[43] Agbo EA, Baiyeri KP. Mint: Location influenced proximate and mineral composition of fresh seeds of five accessions of African Walnut (*Plukenetia conophora Muell Arg*). Nigerian Journal of Crop Science. 2019;**6**(1):30-33

[44] Arinola SO, Adesina K. Mint: Effect of thermal processing on the nutritional, antinutritional, and antioxidant properties of *Tetracarpidium*

conophorum (African Walnut). Journal of Food Processing. 2014;**1-4**. DOI: 10.1155/2014/418380

[45] Suara KO, Azubuike CP, Okubanjo OO, Igwilo C. Mint: Neutraceuticals and antibacterial properties of methanol extract of *Plukenetia conophora* (Müll.-Arg.) family Euphorbiaceae leaves and physical properties of its cream formulations. Nigerian Journal of Pharmaceutical and Applied Science Research. 2016;**5**(1):91-98

[46] Olajide K, Baiyeri KP, Ndubuaku UM. Mint: Nutritional diversity in accessions of African walnut (*Plukenetia conophora* L.) sourced from southwestern Nigeria as influenced by collection center and processing. Agro-Science. 2020b;**19**(4):24-29. DOI: 10.4314/as.v19i4.5

[47] Enujiugha VN. Mint: Chemical and functional characteristics of Conophor nut. Pakistan Journal of Nutrition. 2003;**2**(6):335-338

[48] Enujiugha VN, Ayodele OO. Mint: Evaluation of nutrients and some anti-nutrient in less known underutilized seed. International Journal of Food Science and Technology. 2003;**38**:525-552. DOI: 10.1046/j.1365-2621.2003.00698.x

[49] Ayoola PB, Faboya OOP, Onawumi OOE. Mint: Comparative analysis of the phytochemical and nutrient evaluation of the seeds and leaves of *Plukenetia conophora* plant. Chemistry and Materials Research. 2013;**3**(9):91-99

[50] Negi AS, Luqma S, Srivastava V, Krishna N, Gupta MPD. Mint: Antiproliferative and antioxidant activities of African walnut (*Julgan regina* L.) fruit extracts. Pharmaceutical Biology. 2011;**49**(6):669-705. DOI: 10.3109/13880209.2010. 537666

[51] Amusa TO, Jimoh SO, Azeez IO, Awodoin RO, Kareem I. Mint: Stock density and fruit yield of African walnut, Plukenetia conophora Mull-Arg (Syn. *Tetracarpidium conophorum*) in tropical lowland rainforests of southwest Nigeria. Journal of Tropical Forestry and Environment. 2014;**4**(2):73-81

[52] Chijoke OC, Anosike CA, Ani CC. Mint: Studies on the phytochemical and nutritional properties of *Tetracarpidium conophorum* (black walnut) seeds. Journal of Global Bioscience. 2015;**4**(2):1366-1372

[53] Akpuaka MU, Nwankwor E. Mint: Extraction, analysis and utilization of a drying oil from *Tetracarpidium conophorum* (Nigerian walnut). Bioresource Technology. 2000;**73**(2):195-196

[54] Rob G. Effect of organic Fertilizer on seedling growth. [Internet]. 2000. Irrigation and green magazine. Available from: www.igin.com. Info@naturalenvironment.com

[55] Daramola DS, Adeyeye AAS, Lawal D. Effect of application of organic manures on the growth and dry matter yield of *Amaranthus cruentus*. Paper presented at 2nd National Conference; 2006; University of Ibadan, Nigeria, 11

[56] Rice RP, Rice LW, Tindal HD. Fruits and Vegetable Production in Warm Climates. Hong Kong: Macmillan Press Ltd; 1994. p. 486

[57] Baiyeri KP, Aba SC, Faturoti BO, Tenkouano A. Mint: Effects of poultry manure and bunch pruning management on fruit size, shelf life and pulp colour of 'PITA 24' and

'Mbi-Egome' plantains (*Musa sp.* AAB group). Journal of Animal and Plant Sciences. 2009;**3**(2):215-226

[58] Adebayo AG, Akintoye HA, Olufolaji AO, Aina OO, Olatunji MT, Shokalu AO. Mint: Assessment of organic amendments on vegetative development and nutrient uptake of *Moringa oleifera* Lam. in nursery. Asian Journal of Plant Science. 2011;**10**:74-79

[59] Sarr MG, Ndiaye ND, Ayessou NC, Faye PG, Cisse M, Sakho M, et al. Mint: *Saba senegalensis*: Key features and uses. Food and Nutrition Sciences. 2018;**9**:1099-1111. DOI: 10.4236/fns.2018.99080

[60] Arbonnier M. Trees, Shrubs and Creepers from the Dry Areas of West Africa. Cirad: Mauguio; 2000. p. 573

[61] Tiendrebeogo S, Ganou L, Compaore CS, Tapsoba FW, Dicko MH. Mint: Biochemical composition of *Saba senegalensis* fruits from Burkina Faso. African Journal of Food Science. 2020;**14**(10):322-329

[62] Nafan G, Jesus FI, Souleymane S, Lenifere SC, Emmanuel IA, Abdourahamane SA. Mint: Genetic variation of *Saba senegalensis* Pychon (Apocynaceae) and Few Nutritional Value. International Journal of Behavioral Accounting and Finance. 2013;**1**:121-135

[63] AFTD (Agro forestry tree Database). 2016. Available from: http://www.worldagroforestry.org/

[64] Burkill HM. The Useful Plants of West Tropical Africa. Vol. 2. Royal Botanical Garden: University Press of Virginia, Charlottesville; 2000. p. 636

[65] Berhaut J. Illustrated Flora of Senegal. Dakar: Clairafrique; 1971. p. 626

[66] Fatim DHM, Souleymane T, Mohamed C, Doudjo S, Kouakou B. Mint: Biochemical characterization and nutritional profile of the pulp of *Saba senegalensis* from Côte d'Ivoire Forest. American Journal of Food and Nutrition. 2019;**7**(1):19-25

[67] Tanor F. Study on the collection and analysis of data on non-timber forest products in Senegal Dakar. In: Partnership Program CE-FAO. 2001

[68] Leeuwenberg AJM, Van Dilst FJH. Mint: *Saba senegalensis* (Pichon), series of revisions of Apocynaceae. Bulletin du Jardin botanique National de Belgium. 1989;**59**(2):189-206

[69] Wild HC. Fruit from Senegal. Montpellier: University of Montpellier; 2000. p. 372

[70] Diop AG, Sakho M, Dornier M, Cisse M, Reynes M. Mint: Le baobab Africa in (*Adansonia digitata* L.): principales caractéristques et utilisations. Fruits. 2005;**61**:55-69

[71] Anon. The Useful Tropical Plants Database. The Interne 2016. Available from: http://tropical.theferns.info/

[72] Boamponsem G, Johnson FS, MahunuGK, AwiniboyaSF. Determination of biochemical Composition of *Saba senegalensis* (Saba fruit). Asian Journal of Plant Science and Research. 2013;**3**(1):31-36

[73] N'Diaye M, Fodé BK, Philippe M. Main harvest fruits consumed and marketed in Guinea. Fruits. 2003;**58**:99-116. DOI: 10.1051/fruits:2002040

[74] TPD. Ken Fern. tropical. [internet]. 2020. Available from: theferns.info.<tropical.theferns.info/viewtropical.php?id-Saba+senegalensis>

[75] Baiyeri KP, Ishieze PU, Kalu PC, Okpala CV, Olajide K. Mint: Preliminary investigation of nutritional quality of Saba (*Saba senegalensis*), a wild fruit species obtained from North Central Nigeria. Nigerian Journal of Crop Science. 2019;**6**(2):11-16

[76] Koné MW, Atindehou KK, Téré H, Traoré D. Mint: Quelques plantes médicinales utilisées en pédiatrie traditionnelle dans la région de Ferkessédougou (Côte-d'Ivoire). Bioterre, Revue Internationale Sciences de la Vie et de la Terre Numéro special. 2002:30-36

[77] Kini F, Saba A, Parkouda C, Ouedraogo S, Guissou P. Mint: Partial phytochemical characterization of the fruits of *Saba senegalensis* (Apocynacaea) and Landolphia heudolotii (Apocynaceae). Pharmacopée et Médecine Traditionnelle Africaines. 2012:1-11

[78] Nacoulma OG. Medicinal Plants and Traditional Medical Practices in Burkina-Faso: The Case of the Central Plateau. Burkina-Faso: University of Ouagadougou; 1999. p. 261

[79] Favier JC, Ireland-ripertj LC, Feinberg M. Composition Table of Exotic Fruits, African Picking Fruits. Orstom, INRA, Tec. and Doc. Lavoisier, Paris; 1993. 243

[80] Lauga SC. The Circle of Hedges, Landscapes of Fulani Agro-Pastoralists from Fouta Djallon (Plains of Timbi, Guinea) C. Blanc-Pamard (Coordinator), Paris; Vol. II, 1997

[81] Linda D. Nutritional Composition and Stability of *Saba senegalensis* Fruit Extract, Faculty of Agriculture University for Development Studies, Tamale, Ghana. International Conference on Neglected and Underutilized Species: For a Food-Secure Africa, Accra; 2013. p. 12

[82] Ndiour PA. Role and Importance of Forest Fruits from Fallow in the Formation of Household Incomes. Bignona: Department of Bignona; 1996

[83] Lamien A, Lamien CE, Compaoré MM, Meda RN, Kiendrebeogo M, Zeba B, et al. Mint: Polyphenol content and antioxidant activity of fourteen wild edible fruits from Burkina Faso. Molecules. 2008;**13**:581-594

[84] Diouf P, Diedhiou S, Goudiaby AOK, Ndoye I, Koita B. Mint: Morphological Variability of *Saba senegalensis* (A. DC.) Pichon en Casamance (Sénégal). European Scientific Journal. 2019;**15**:500-517

[85] Stefano P, Dris R, Rapparini F. Mint: Influence of growing condition and yield and quality of cherry. II. Fruit. Journal of Agricultural Functions. 2004;**2**:307-309

[86] Olajide K, Baiyeri KP, Ndubuaku UM. Early growth pattern of four accessions of Saba *(Saba senegalensis)* in response to seven fertilizer rates in the nursery. In: Orisajo SB, Ipinmoroti RR, Adedeji AR. Ed. Book of Proceedings of the 39th Annual Conference of Horticultural Society of Nigeria, Cocoa Research Institute of Nigeria; 14th–18th November; Ibadan; 2021. p. 1212

[87] Magashi AM, Abdulmalik U. Mint: Antibacterial activity and phytochemical screening of stem bark extracts of *Adansonia digitata* on some clinical isolates. Journal of. Microbiology Research. 2018;**3**(1):1-7

[88] Sidibe M, Williams JT. Baobab – Adansonia digitata. Fruits for the future 4, International Centre for Underutilised Crops; Southampton, UK; 2002. 96

[89] Jitin R, Manish KJ, Shshu PS, Pakesh KK, Anuradha AN, Anup K, et al.

Mint: *Adansonia digitata* L. (Baobab). A review of traditional information and taxonomic description. Asian Pacific Journal of Tropical Biomedicine. 2015;**1**(5):79-84

[90] Wickens GE, Lowe P. The Baobabs: Pachycauls of Africa, Madagascar and Australia. London: Springer; 2008. DOI: 10.1007/978-1-4020-6431-9

[91] Shuaibu R, Rabi B. Conservation of *Adansonia digitate* L. (Baoba tree) for sustainable Livelihoods in Sudano-Sahelian Region of Nigeria in selected Local Government Areas in Katsina State. Sudano-Sahelian Landscape and Renewable Natural Resources Development in Nigeria. A Proceedings of 37th Annual Conference of Forestry Association of Nigeria; 9th–14th November; Minna, Niger State Nigeria; 2014. pp. 723-727

[92] Venter SM, Witkowski ETF. Mint: Baobab (*Adansonia digitata* L.) density, size-class distribution and population trends between four land-use types in northern Venda, South Africa. Forest Ecology and Management. 2010;**259**:294-300

[93] Ajayi IA, Dawodi FA, Oderinde RA. Mint: Fatty acid composition and metal content of Adansonia digitata seeds and seed oil. La Rivista Italiana delle Sostanze Grasse. 2003;**80**:41-43

[94] Scheuring JF, Sidibé M, Frigg M. Mint: Malian agronomic research identifies local baobab tree as source of vitamin A and vitamin C. Sight and Life. 1999;**1**:21-24

[95] Gebauer J, El-Siddig K, Ebert G. Mint: Baobab (*Adansonia digitata* L.): A Review on a multipurpose tree with promising future in the Sudan. *Gartenbauwissenschaft*. 2002; **67**:155-160

[96] De Caluwé E, Halamová K, Van Damme P. *Adansonia digitata* L. – A review of traditional uses, phytochemistry and pharmacology. Afrika. Focus. 2010;**23**:11-51

[97] Buchmann C, Prehsler S, Hartl A, Vogl CR. Mint: The importance of baobab (*Adansonia digitata* L.) in rural West African subsistence—suggestion of a cautionary approach to international market export of baobab fruits. Ecology of Food and Nutrition. 2010;**49**:145-172

[98] Addy R. Baobab Fruit Approved as Food Ingredient in US. The Interne 2009. Available from: http://www.nutraingredients-usa.com/content/view/print/2595742009 [Accessed: April 16, 2011]

[99] Nnam NM, Obiakor PN. Mint: Effect of fermentation on the nutrient and antinutrient composition of baobab (*Adansonia digitata*) seeds and rice (*Oryza sativa*) grains. Ecology of Food and Nutrition. 2003;**42**:265-277

[100] Chivandi E, Davidson BC, Erlwanger KH. Mint: A comparison of the lipid and fatty profiles from the kernels of the fruit (nuts) of *Ximenia caffra* and *Ricinodendron rautanenii* from Zimbabwe. Industrial Crops and Products. 2008;**27**:29-32. DOI: 10.1016/j.indcrop.2007.06.002

[101] Kamatou GPP, Vermaak I, Viljoen AM. Mint: An updated review of *Adansonia digitata*: A commercially important African tree. South African Journal of Botany. 2011;77:908-919. DOI: 10.1016/j.sajb.2011.08.010

[102] Vermak I, Kamatou GPP, Komane-Mofokeng B, Viljoen AM, Beckett K. Mint: African seed oils of commercial Importance-Cosmetic applications. South African Journal of

Botany. 2011;77:920-933. DOI: 10.1016/j.sajb.2011.07.003

[103] Dirar HA. The Indigenous Fermented Foods of the Sudan. UK: A Study in African Food and Nutrition; 1993. pp. 85-97

[104] Venter F, Venter J. Making the Most of Indigenous Trees. Pretoria, South Africa: Briza Publications; 1996. pp. 26-27

[105] Craig GM. The Agriculture of the Sudan. USA: Oxford University Press; 1991. pp. 32-58

[106] Obizoba IC, Anyika JU. Mint: Nutritive value of baobab milk (gubdi) and mixtures of baobab (*Adansonia digitata* L.) and hungry rice, acha (*Digitaria exilis*) flours. Plant Foods for Human Nutrition. 1994;**46**:157-165

[107] Yusha M, Hamza MM, Abdullahi N. Mint: Antibacterial activity of *Adansonia digitata* stem bark extracts on some clinical bacterial isolates. International Journal of Biomedical Health Sciences. 2010;**6**:129-135

[108] Van Wyk BE, Gericke N. People's Plants: A Guide to Useful Plants of Southern Africa. 1st ed. Pretoria: Briza Publications; 2000

[109] El-Rawy EM, Gergis SM, Bazaid S, El-Mougy SA. Mint: The immuno stimulant effect of *Adansonia digitata* on the immune response of chicken vaccinated with avian cholera vaccine. Journal of Egypt Veterinary Medicine Association. 1997;**57**:959-970

[110] Chadare FJ, Linnemann AR, Hounhouigan JD, Nout MJR, Van Boekel MAJS. Mint: Baobab food products: A review on their composition and nutritional value. Critical Reviews in Food Science and Nutrition. 2009;**49**:254-274

[111] Okon EU, Akpanyung EO. Mint: Nutrition and anti-nutrition in selected brands of malt drinks produced in Nigeria. Pakistan Journal of Nutrition. 2005;**4**:352-355

[112] Caser M. Low Oxalate Diet. Pittsburgh, USA: University of Pittsburgh Medical Center Information for Patients; 2003

[113] Michałowicz J, Duda W. Mint: Phenols – Sources and toxicity. Polish Journal of Environmental Study. 2007;**16**(3):347-362

[114] AOAC. Official Methods of Analysis. 18th ed. Gaithersburg, MD, USA: Association of Official Analytical Chemists; 2005

[115] Elfadil AG, Sabahelkhier MK, Rayan MY, Daa MO, Nagla AH, Israa SB. Mint: Effect of Tannin and Plant Tannins on some Organs and Physic-Chemical Characters of Diabetic Wistar Rat. International Journal of Advanced Research. 2013;**1**(10):165-170

[116] Zafar IK, Kafeel A, Asma Z, Humayun B, Abrar H, Zile H, et al. Mint: Assessment of poisonous and anti-nutritional compounds in wild edible forages consumed by ruminant species. Journal of Environmental Science and Technology. 2015;**8**:91-101

[117] Compaore WR, Nikiema PA, Bassole HIN, Savadogo A, Mouecoucou J, Hounhouigan DJ, et al. Mint: Chemical composition and antioxidative properties of seeds of *Moringa oleifera,* and pulps of *Parkia biglobosa* and *Adansonia digitata* commonly used in food fortification in Burkina Faso. Current Research Biological Science. 2011;**3**:64-72

[118] Osman MA. Mint: Chemical and Nutrient Analysis of Baobab (*Adansonia digitata*) Fruit and Seed Protein

Solubility. Plant Foods for Human Nutrition. 2004;**59**:29-33

[119] Arowora KA, Ojochenemi EY, Christopher S, Terungwa JI, Kenneth CU. Mint: Chemical composition of baobab leaves and fractionation of its ethanolic extract using column chromatography. International Journal of Science and Research. 2018;**8**(7):812-821

[120] Enoch BB, Abubakar IM, Sakiyo DC, Bashiloni N. Mint: Comparative analysis of nutritional contents in the leaf, pulp and seed of *Adansonia digitata* L. consumed in Adamawa State. Nigeria. African Journal of Food Science. 2020;**14**(8):215-221

[121] Sena LP, VanderJagt DJ, Rivera C, Tin AC, Muhamadu I, Mahamadou O, et al. Glew RH Mint: Analysis of nutritional components of eight famine foods of the Republic of Niger. Plant Foods for Human Nutrition. 1998;**52**:17-30

[122] Becker B. Mint: The contribution of wild plants to human nutrition in the Ferlo (Northern Senegal). Agroforestry Systems. 1983;**1**:257-267

[123] FAO. Adansonia digitata. Corporate Document Repository. Forestry Department. Web. 26. [Internet]. 2014. Available from: http://www.fao.org/docrep/x5327e/x5327e0g.htm

[124] Watson R. The African Baobab. Cape Town: Struik Publishers; 2007. pp. 19-32

[125] Shivcharn SD, Gunnar G. Mint: Local management practices influence the viability of the baobab (*Adansonia digitata* Linn.) in different land use types, Cinzana, Mali. Agriculture, Ecosystems and Environment. 2004;**101**:85-103

[126] Sanchez AC, Osborne PE, Haq N. Mint: Identifying the global potential for baobab tree cultivation using ecological niche modeling. Agroforestry Systems. 2010;**80**:191-201

[127] Assogbadjo AE, Glèlè Kakaï R, Chadare FJ, Thomson L, Kyndt T, Sinsin B, et al. Mint: Folk classification, perception, and references of baobab products in West Africa: Consequences for species conservation and improvement. Economic Botany. 2008;**62**:74-84

[128] Kang BT, Balasubramanian V. Long-term fertilizer trials on alfisols in West Africa. In: Koshino M, Morooka M, Morishita T, Owa N, Harada Y, Yoneyama T, Ozaki Y, Eds. Transactions of the 14th International Congress of Soil Science; Kyoto, Japan; The International Society of Soil Science, 1990. pp. 20-25

[129] Shiyam JO, Oko BFD, Obiefuna JC, Ofoh MC. Mint: Optimizing the productivity of plantain/cocoyam mixture by mulching and fertilizer application. World Journal of Agricultural Science. 2011;**7**(5):633-637

[130] Osundare OT, Fajinmi AA, Okonji CJ. Mint: Effects of organic and inorganic soil amendments on growth performance of plantain *(Musa paradisiaca* L.). African Journal of Agricultural Research. 2015;**10**:154-160. DOI: 10.5897/AJAR2014.8645

[131] Liang B, Yang X, He X, Murphy DV, Zhou J. Mint: Long-term combined application of manure and NPK fertilizers influenced nitrogen retention and stabilization of organic C in loess soil. Plant and Soil. 2012;**353**:249-260

[132] Rinaldi S, De Lucia B, Salvati L, Rea E. Mint: Understanding complexity in the response of ornamental rosemary to different substrates: A Multivariate Analysis. Scientia Horticulturae. 2014;**176**:218-224

[133] Abubakari A, Atuah L, Banful B, Bayor H. Use of soil amendments for urban horticulture in the savannah and forest areas of Ghana. Journal of Soil Science and Environmental Management. 2015;**1**(6):51-57

[134] Chaiarrekij S, Apirakchaiskul A, Suvarnakich K, Kiatkamjornwong S. Kapok: Characteristics fibre as a potential pulp source for paper making. Bio Resources. 2011;**7**(1):475-488

[135] Mojica ERE, Merca FE, Micor JRL. Mint: Fiber of kapok (*Ceiba pentandra*) as component of a metal sensor for lead in water samples. Philippine Journal of Crop Science. 2002;**27**(2):37-42

[136] Adeniji IT, Olomola DB, Jegede OC. Mint: Seed germination and seedling growth of *Ceiba Pentandra* (L) as influenced by different soil types in Ibadan, Southwest Nigeria. Journal of Research in Forestry, Wildlife & Environment. 2019;**11**(3):316-230

[137] Osuntokun OT, Ayodele AO, Adeoye MI. Mint: Assessment of antimicrobial and phytochemical properties of crude leaf and bark extracts of *Ceiba pentandra* on selected clinical isolates found in Nigerian teaching hospital. Journal of Bacteriology and Mycology. 2017;**4**(1):17-23. DOI: 10.15406/jbmoa.2017.04.00079

[138] Beentje H, Smith S. Plant systematic and Phytogeography for the understanding of African Biodiversity. Systematics and Geography of Plants. 2001;**71**(2):234-286

[139] Bown D. Encyclopaedia of Herbs and Their Uses. London: Dorling Kindersley; 1995

[140] Friday ET, Omale J, Olupinyo O, Adah G. Mint: Investigations on the nutritional and medicinal potentials of *Ceiba pentandra* leaf: A common vegetable. International Journal of Plant Physiology and Biochemistry. 2011;**3**:95-101

[141] Adepoju OT, Ugochukwu IC. Mint: Improving vegetable diversity and micronutrient intake of Nigerians through Consumption of Lesser Known Silk Cotton (*Ceiba pentandra*) Leaf. International Journal of Nutrition. 2019;**4**(1):19-30

[142] Chu YF, Sun J, Wu X, Liu RH. Mint: Antioxidant and anti-proliferative activities of common vegetables. Revised Journal of Agricultural Food Chemistry. 2002;**50**:6910-6916

[143] Gropper SS, Smith JI, Groff JI. The Antioxidant Nutrients, Reactive Oxygen Species and Disease. Advanced Nutrition and Human Metabolism. 4th ed. Belmont: Thomson Wadsworth; 2005. pp. 368-377

[144] Ball GFM. Vitamins in Foods; Analysis, bioavailability and stability. CRC Taylor & Francis, Boca-Raton. Trichosanthes anguina (Snake tomato) Fruits from Nigeria. Pakistan Journal of Nutrition. 2006;7(1)

[145] McDougall GJ, Morrison IA, Stewart D, Hillman JR. Mint: Plant cell walls as dietary fibre: Range, Structure, Processing and Function. Journal of the Science of Food and Agriculture. 1996;**70**:133-150

[146] USDA. U.S. Department of Agricultural Research Service. Nutrient Data laboratory USDA Nutrient Database for Standard Reference vegetables. Biokemistri. 1998;**18**(2):121-125

[147] Awobajo FO, Olatunji-Bello II, Obilade TT, Odugbemi TO. Mint: Knowledge of the Nutritional and medicinal use of some vegetables among a cross section of market women in

two major food markets in Lagos State, South West Nigeria. Pakistan Journal of Nutrition. 2010;**9**(3):216-221

[148] Bamishaiye B. An Overview of the Role of Nutrition in Carcinogenesis, Nutrition, Growth and Cancer. London: Allan R, Liss Inc; 1998. p. 418

[149] Enechi OC, Peter CD, Ugwu OPC, Udeh SMC, Omeh YS. Mint: Evaluation of the nutritional potential of *Ceiba pentandra* leaves. Mintage Journal of Pharmaceutical and Medical Sciences. 2013:25-27

[150] Olajide K, Davidson GI, Baiyeri KP. Mint: Assessment of mineral and antinutritional qualities of *Ceiba pentandra* succulent and mature leaves. Nigerian Journal of Nutritional Sciences. 2021b;**42**(2):106-115

[151] Shahin A, Husna PN, Shahal A, Aminul A. Mint: Proximate and Mineral Compositions of Leaves and Seeds of Bangladeshi Bombax Ceiba Linn. World Journal of Pharmaceutical Research. 2016;**5**(7):1-13

[152] Codex Alimentarius. Request for comments at step 3 on the proposed draft standard for quinoa. Codex Alimentarius Commission, Joint FAO/WHO Food Standards Programme. Italy, Rome. 2017

[153] Dei HK, Rose SP, Mackenzie AM. Shea nut (*Vitellaria paradoxa*) meal as a feed ingredient for poultry. World's Poultry Science Journal. 2007;**63**:611-624

[154] Obahiagbon FI, Erhabor JO. Mint: The health implication of the dietary nutrients detected in the vegetable leaves intercropped with Raphia hooker Palms. African Journal of Food Science. 2010;**4**(7):440-443

[155] Andzouana M, Mombouli JB. Mint: Assessment of the chemical and phytochemical constituents of the leaves of a wild vegetable—Ochthocharis dicellandroides (Gilg). Pakistan Journal of Nutrition. 2012;**1**:94-99

[156] Bakare RI, Magbagbeola OA, Akinwande AI, Okunowo OW. Mint: Nutritional and chemical evaluation of Momordica charantia. Journal of Medicinal Plant Research. 2010;**4**(21):2189-2193

[157] Caunii A, Cuciureanu R, Zakar AM, Tonea E, Giuchici C. Chemical composition of common leafy vegetable. Studia Universitatis "Vasile Goldis", Seria Stiintele Vietii. 2010;**20**(2):45-48

[158] Duvall CS. Ceiba pentandra (L.) Record from PROTA4U. In: Brink M, Achigan-Dako EG, editors. PROTA (Plant Resources of Tropical Africa / Ressources végétales de l'Afrique tropicale). Wageningen, Netherlands; 2011

[159] Chinea-Rivera JD. *Ceiba pentandra* (L.) Gaertn. Ceiba, Kapok, silk cotton tree. Bombacaceae. Bombax family. USDA Forest Service, Southern Forest Experiment Station, Institute of Tropical Forestry; 1990; p. 4. Available online: http://www.treesearch.fs.fed.us/pubs/30402

[160] Huxley A. The New RHS Dictionary of Gardening. MacMillan Press; 1992

[161] Matson PA, Naylor R, Ortiz-Monasterio I. Integration of environmental, agronomic, and economic aspects of fertilizer management. Science. 1998;**280**(5360):112-115

[162] Kumar R, Mahajan G, Srivastava S, Sinha A. Green manuring: A Boon for Sustainable Agriculture and Pest Management – A Review. Agricultural Reviews. 2014;**35**(3):196-206

[163] Olajide K, Baiyeri KP. Mint: Early Growth Response of Ceiba Pentandra to Organic and Inorganic Fertilizers under Field Conditions in Southeastern Nigeria. Journal of Agricultural Research. 2021;**6**(3):1-9

[164] Qaswar M, Jing H, Ahmed W, Li DC, Liu SJ, Lu Z, et al. Mint: Yield sustainability, soil organic carbon sequestration and nutrients balance under long-term combined application of manure and inorganic fertilizers in acidic paddy soil. Soil and Tillage Research. 2020;**198**:104569

Chapter 5

Challenges and Advances in the Production of Export-Quality Macadamia and Its Integral Use with Green Technologies

Laura Graciela Mereles, Mario Smidt, Karen Patricia Martínez, Eva Eugenia Soledad Coronel, Edelira Velázquez and Laura Correa

Abstract

Macadamia nut is an alternative crop for agricultural production in tropical Latin American countries. Its cultivation in itself constitutes a challenge for countries with high relative humidity temperatures, especially in the postharvest period. Environmentally friendly technologies suggest a comprehensive nut in shell (NIS) and kernels treatment, taking advantage of the waste generated in the drying process, critical point. This chapter explores the methods of the literature and those applied in local research for the integral use, drying of macadamia nuts, and their processing until obtaining products of high nutritional quality (dried nut and oil) and with clean technologies applicable to small producers.

Keywords: *Macadamia integrifolia*, production, composition, quality, analysis, oil extraction, byproducts, tropical countries

1. Introduction

Since the first scientific evidence on the frequent consumption of nuts on health, macadamia nuts, produced in various tropical regions of the world since their origin in Australia, have gained great interest as part of a healthy diet. In vivo studies in rats, mice, and humans have shown that macadamia nut oil reduces total cholesterol, low-density lipoprotein (LDL) cholesterol, body weight, and body mass index, inhibits the development of sucrose–/fructose-induced hepatic steatosis, and attenuates adipocyte hypertrophy and inflammation in adipose tissue and macrophages [1]. In addition to the high concentration of monounsaturated, the main source of health nutrition and pharmacological properties of these oils comes from their minor components that are phytosterols, tocopherols, phenols, squalene, carotenoids, and others [2]. Its antioxidant potential is attributed to these compounds together. Dried macadamia nuts have shown good total antioxidant capacity and may be useful when consumed alone or in combination with traditional pharmacotherapy to reduce the risk of cardiovascular disease. Intervention studies consistently show that the consumption of macadamia nuts causes a decrease in plasma total cholesterol and LDL cholesterol, and despite its high fat content, the regular consumption of macadamia nuts has been shown to have no effect on the body weight [3]. These studies have been carried out using macadamia nuts which, due to their unique composition, are influenced by the quality of the kernel, so the production system has a major impact on their commercialization and health effects.

The great challenge of macadamia production for the world has been to maintain sensory and nutritional quality, due to its high composition of oils, which are susceptible to oxidation and require a careful system of cultivation, selection, drying, and packaging to reach the consumer with all their nutritional and bioactive properties [4]. Compositional differences observed in different studies on macadamia nut and oil could be explained by the variation in cultivars, growth conditions, harvest time, degree of maturity, and storage conditions [1]. Several studies have addressed this problem, and its composition has been studied in different production regions, which opened a field of research on its sustainable production and use of green technologies that generate new markets with which the macadamia nut has become a product of great value based on these experiences [5–8].

The macadamia producers par excellence are Australia (where it originates from), South Africa, and the United States [9]. In South America, the macadamia nut represents an alternative crop to those of great expansion such as soybean or corn; however, it is a noble crop that allows the use of the soil in the spaces left by the treetops, with the production of other foods such as pineapple or medicinal herbs [10]. This, added to the added value that can be obtained from its integral use, from the exocarp (green) and the mesocarp (brown) of the fruit and the dried and split nuts (kernel) that are used to obtain oil, is viable and attractive alternatives for small- and medium-sized producers [5, 11, 12]. The rich fatty acid composition of the oil obtained from macadamia nuts allows its diverse use in many industries, i.e., cosmetic and pharmaceutical [11]. Macadamia nuts are marketed in the export market mainly in two forms: shelled nuts (kernel) and nuts in shell (NISs) (**Figure 1**). The market by application

Figure 1.
Cross section of the NIS with the almond (kernel) inside and the woody shell outside (mesocarp).

can be divided into snacks, confectionery and bakery, cosmetics, and others. The international market by region is mainly distributed in North America (the United States, Canada, and Mexico), Europe (Germany, the United Kingdom, France, Italy, Russia, and Spain), Asia-Pacific (China, Japan, Korea, India, Australia, and Southeast Asia), South America (Brazil, Argentina, and Colombia), and the Middle East and Africa (South Africa, the United Arab Emirates, and Saudi Arabia). Many times, the internal challenges and opportunities of the smallest producers are not the same as for the largest producers.

In this chapter, we address the production of nutritional quality macadamia nuts, from the postharvest processing, drying, and packaging system. Its comprehensive use in tropical countries, based on the data established in the literature and the work team's own results, opens a series of accessible low-cost alternatives and green technologies based on the lessons learned for their production in tropical countries with often limited resources.

2. Macadamia nut and oil composition

The nut is rich in protein and has a high energy content, outstanding organoleptic and nutritional characteristics, and a high amount of oil (~ 74% w/w) [13, 14]. Its use for cooking food is limited, and it is mainly used as a flavoring oil [15]. In addition, dried macadamia nuts (kernel) contain other important dietary components, including protein (7.2–10.4%), good levels of dietary fiber (6.2–8.6%), fat-soluble vitamins (especially alpha-carotene), minerals (Mg, Ca, and K), and phytochemicals (phytosterols). They are low in carbohydrates (0.5–13.8%). A comparative study of the centesimal composition and lipid profile of 22 varieties of *M. integrifolia*, grown in Itapirá (Brazil), showed large variations in composition with respect to the composition of macadamia from other origins; however, the samples were taken from very young trees (7 years) [7]. After 8 years of cultivation, adult trees are considered with a more constant composition of nuts. In another study, it was observed that tocotrienols and squalene were largely affected by varieties and their content in seven macadamia cultivars produced in Hawaii ranged from 31 to 92 and 72 to 171 μg/g of oil, respectively [6]. With several studies on its chemical composition, it is now known that it can vary greatly due to the influence of the type of cultivar (genetics), the ripeness of the grain, the time of harvest, the geographical location, and the conditions of the crop [1]. Macadamia nut proteins have all the essential amino acids, and their limiting amino acids include tryptophan, lysine, and threonine [16]. The sugar content is mainly represented by fructose, glucose, maltose, and sucrose. The cultivars differ in the sucrose content of the nut, but not in the content of reducing sugars [17]. Regarding the mineral content, macadamia nuts are considered a source of magnesium, calcium, and potassium [13]. Macadamia nuts from Australia have been reported to contain 5.77 mg/100 g of iron and low levels of zinc and copper [18].

Regarding the oil, in most published studies, the content of monounsaturated fatty acids (MUFAs) predominates, among which oleic acid C18:1 and palmitoleic acid C16:1 stand out. Other monounsaturated fatty acids, such as gondoic acid C20:1 and erucic acid C22:1, are also reported by some authors among the components of macadamia oil [19, 20]. Macadamia nut oil has various food and nonfood applications including food fortification, development of skin, hair, and healthcare products. Rich in monounsaturated fatty acids (oleic and palmitoleic acid), macadamia oil also contains a significant concentration of bioactive phytochemicals including β-sitosterol, α-tocopherol, α-tocotrienols, ρ-hydroxybenzoic acid, and caffeic acid [1]. Macadamia oil contains significant concentrations of phytosterols (~165 mg/100 g), especially β-sitosterol in 82% with levels from 96.9 to 151 mg/100 g of oil. Other components identified in this fraction are campesterol (11.6 mg/100 g), stigmasterol (2.2 mg/100 g), and avenasterol (16 mg/100 g) [20–22]. A study carried out on different oils from nuts and avocado and sesame describes that macadamia nuts contain more phytosterols (184 mg/100 g oil) than oils from other nuts such as walnuts (165 mg/100 g oil), almonds (122 mg/100 g oil), and hazelnuts (89 mg/100 g oil); however, avocado and sesame oils contain much higher amounts (434 and 620 mg/100 g oil, respectively) where the majority component is always β-sitosterol [20]. In addition, the oil has good oxidative stability, which can vary with the harvest season and the crop [17]. There is a positive correlation between antioxidant activity and oxidative stability of the oil. Although there is considerable variation in the oxidative stability of macadamia oil among the studies found in the literature, it is always related to the composition of phytochemicals. In a study, the chemical composition and antioxidant capacity of macadamia oils obtained from 15 cultivars of *M. integrifolia* were comparatively analyzed, and the analysis strongly supported the positive contribution of polyphenols and squalene to the antioxidant capacity of macadamia oils [5].

All these bioactive compounds in macadamia nuts and its high levels of monounsaturated fatty acids make them beneficial for health with frequent consumption, and at an industrial level, they allow the nuts to be minimally processed or industrialized for the production of oil and defatted meal [11].

The comparison of tocopherols, tocotrienols, and squalene content in seven varieties of *M. integrifolia* revealed that macadamia oil has significant amounts of tocotrienols (46.5–91.6 μg/g oil) and squalene (72.4–171 μg/g oil), with variations in the total content of tocotrienols between harvests [6], suggesting a considerable environmental effect on the accumulation of these phytochemicals during nut development. Compared to other nuts, they tend to contain the lowest levels of tocopherols [23, 24] despite being the most stable oil compared to almond, hazelnut, and walnut oils. This seems to indicate that its stability is due in part to its mainly monounsaturated fatty acids (MUFAs) profile, with low percentages of polyunsaturated fatty acids (PUFAs), rather than to the composition of its antioxidant compounds. High oleic oils offer excellent oxidative stability and low-temperature flow properties for many applications. In the vegetable kingdom, oils with a high content of natural oleic acid stand out, such as avocado, macadamia, and olive oils. Macadamia oil has the highest monounsaturated oil content (80%) among common edible oils, followed by olive oil (74%) and avocado (65%) [2]. However, roasting nuts, a process normally used to improve sensory characteristics such as texture and flavor, can alter the fatty acid profile and minor components of the nuts depending on the roasting temperature [25], so recommendations indicate that its healthiest consumption is as dry nut, without roasting.

Macadamia nuts are one of the main sources of palmitoleic acid and can serve as the main dietary source of palmitoleic acid in the diet. Palmitoleic acid is an unusual omega-7 monounsaturated fatty acid found naturally in high levels in macadamia plants (17–20% of the oil) and shares the same structure as the endogenously synthesized form of palmitoleic acid in humans, for which it receives a lot of attention regarding metabolism and health. Recently, palmitoleic acid has been shown to be a lipokine with many beneficial health effects, including anti-inflammatory properties, reduction in body weight, blood glucose, and triglyceride levels, and improvement in insulin sensitivity [2]. The importance of regulating its content as an indicator of its nutritional quality and commercial value has been discussed, especially given its ability to increase insulin sensitivity and reduce the risk of diabetes. Other described effects on the pathogenesis of obesity, liver, and cardiovascular health remain unclear [26].

On the other hand, nuts such as peanuts or walnuts may present certain food intolerances; in the case of macadamia nuts, a systematic proteomic description has recently been presented. The most abundant proteins belong to the 11S globulins and the 7S vicilins. In silico analysis revealed homology and linear epitope similarities with known allergens from lupine, walnut, and peanut, among others. This opens a path in clinical diagnosis and food analysis toward possible protein candidates with allergenic and cross-reactive potential for further immunoglobulin E (IgE) characterization studies of allergenic foods [27].

3. Industrial processing of *M. integrifolia* nuts

In South America and mainly Paraguay, a country with a very hot tropical climate, the rational cultivation of macadamia began in the 1960s, with seedling of genetically improved macadamia species [14]. In the last 10 years, there has been a significant increase in the export volume of this nut (**Figure 2**), ~ 70 tons were exported in 2020, generating important profits to the productive sector.

Figure 3 shows the main stages of the production process, which begins with the harvest, carried out manually, where the ripe nuts naturally fall off and can be harvested directly from the ground or collected with nets. Harvesting of ripe fruits should be done at least once a week, and this frequency should be increased on rainy days [28]. In the processing plant, the first operation is dehulling, whose purpose is the removal of the green husk (pericarp) that protects the fruit within 24 h after

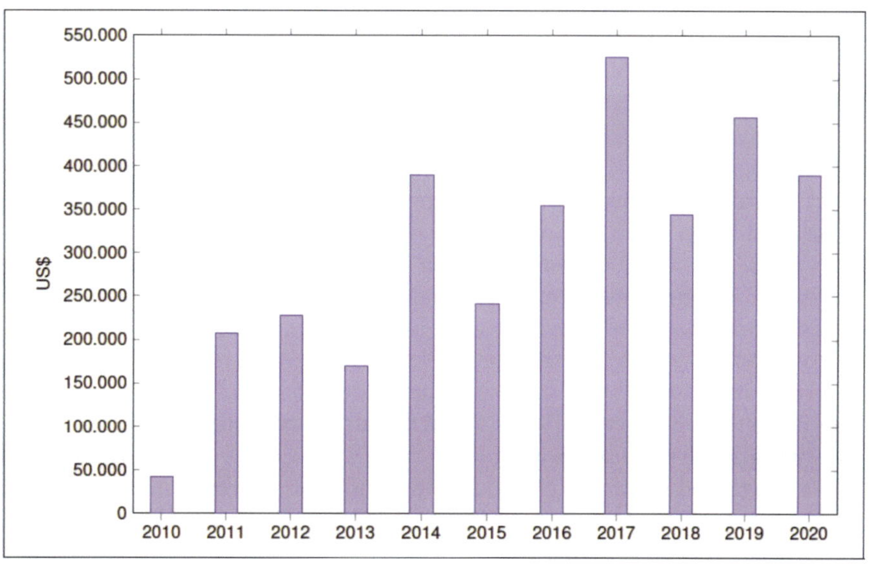

Figure 2.
Evolution of income derived from the export of M. integrifolia nuts in Paraguay. Source: Central Bank of Paraguay.

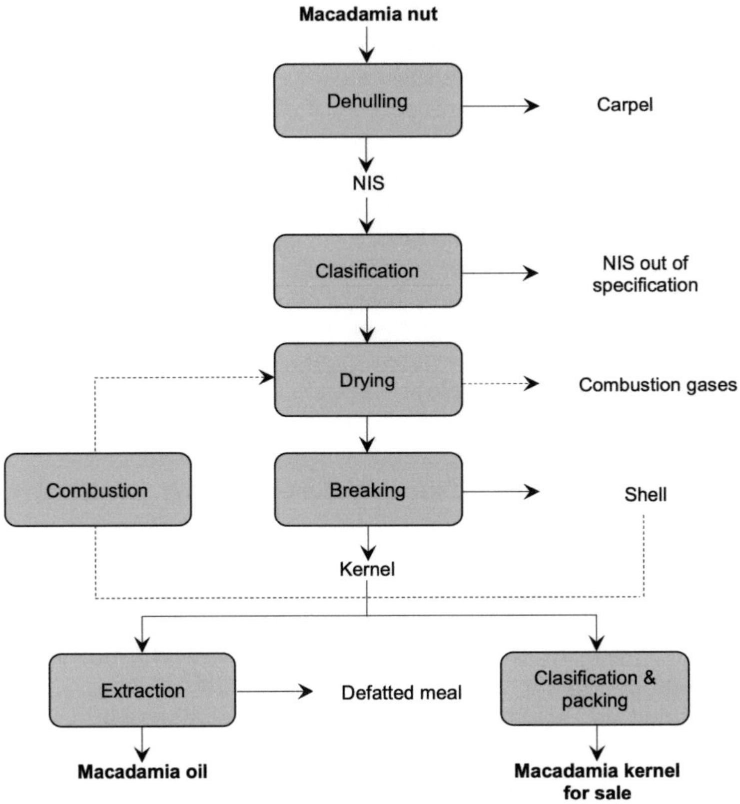

Figure 3.
Block diagram of the industrial processing of M. integrifolia nuts. Elaboration: Smidt, M.

harvest. This operation is important because its presence indicates to the seed that it must germinate, which leads to the translocation of sugars in the kernel and the beginning of the formation of a dark halo in the circumference of the macadamia. The carpel is a nutrient-rich material that is usually used as organic fertilizer for the crop itself. The NISs are then subjected to manual cleaning and classification by flotation, which allows the separation of unripe nuts because those with a high oil content sink. This operation also removes foreign particles and allows the product to be washed, but it is only effective when the moisture content of the NIS is greater than 17% [29].

The next operation is the most important phase in the processing of macadamia nuts, drying, which consists of removing water to a level that prevents the growth of fungi and bacteria, in order to allow the preservation of the nutritional quality of the grain as food or its viability as a seed. When the NIS falls from the tree, it has a moisture content of ~25% (wb) in the almond (kernel), which must be reduced to a percentage less than 1.8% (wb) [9], to avoid the attack of fungi mainly, as well as to avoid the oxidative deterioration of lipids.

Studies conducted in Paraguay indicate that moisture can be removed in two stages of drying at low temperature, which prevents kernel rancidity, evidenced by a high peroxide and acidity index. During the first stage, work is carried out at a temperature of 40°C for ~14 h, which allows a decrease from the natural moisture content of the NIS to ~8%, with a minimum air speed of 0.5 m/s. The next stage of drying allows the temperature to be increased to 65°C for ~15 h, a minimum air speed of 5.4 m/s, allowing the moisture content to decrease to 1.5%, value required almond packaging. The composition characteristics of macadamia nuts have led to an important assessment of their origin and nutritional and bioactive properties, which has allowed the regulation of the market based on well-established quality criteria. The physicochemical and microbiological quality specifications established and internationally accepted in most markets are moisture content less than 1.8%, peroxides 3–5 mEq/kg, free fatty acids 0.5% max, total aflatoxins 4–20 ppb, B1 aflatoxin 0–2 ppb, total plate count <30,000 cfu/g, molds and yeasts <20,000 cfu/g, *Escherichia coli* < 3/g, not detectable Salmonella 25 g, coliforms <300 cfu/g, no insect infestation, "normal" uniform cream color, oil-free appearance on surface, no foreign material, and no bad smells or flavors [9].

Laboratory-scale tests were conducted to evaluate the physicochemical and microbiological characteristics of the nuts subjected to drying, which were then transferred to the design and construction of an industrial dryer (**Figure 4**). The heat necessary to dry the nuts was produced by the combustion of the shell, obtained in the next stage of the macadamia processing (cracking); this material is hard and has a high calorific value (~ 19,600 J/g).

The air required for combustion enters through the ashtray, passes through a rack through its slots, and comes into contact with the fuel. The combustion gases circulate inside tubes with an extended surface (fins) to later be evacuated through the chimney. The drying air is driven by a centrifugal fan, circulates transversely to the finned tubes, and is heated by heat transfer from inside the tubes. Finally, the hot air

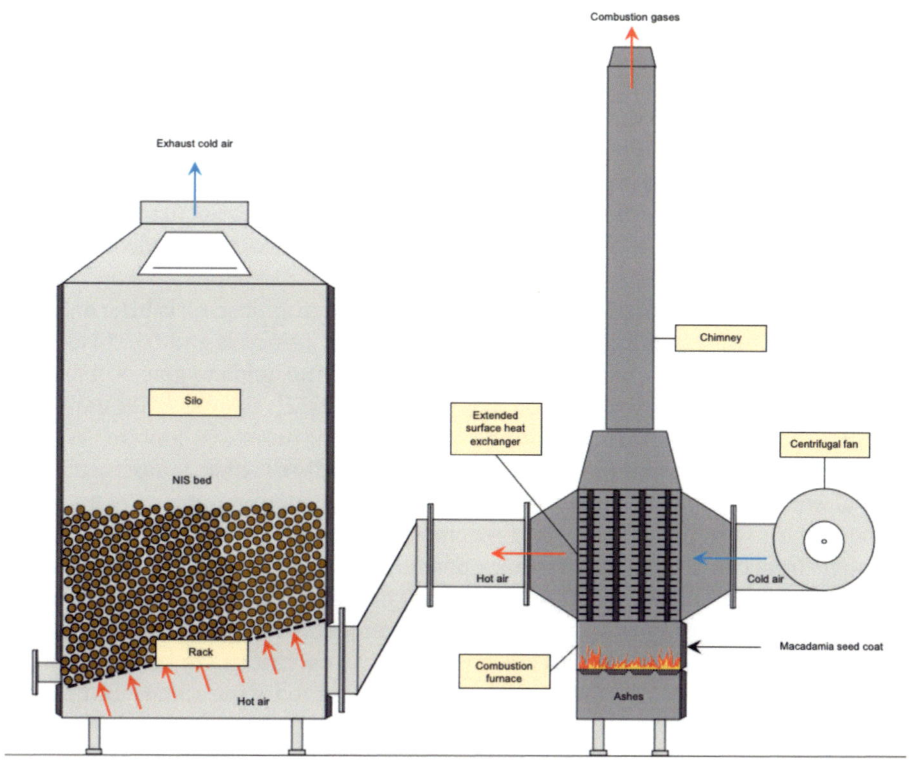

Figure 4.
Silo-type dryer with an air heating system. Source: Adapted from [30].

enters the silo from the bottom through an inclined rack. A detailed description of the design, construction, and operation of the equipment can be found in [30].

Continuing with the production process, the next operation, once the required moisture content (1.5%, wb) has been reached, is the breaking of the NIS carried out with a breaking machine. The separation of the kernel and the shell is done manually, obtaining a product with variable sizes, with whole nuts (~ 38%) being the most valued product, followed by whole nuts with small cuts (~ 12%). About 31% of the kernels obtained in this operation are halved in the breaking machine and 28% are smaller than half a kernel. Finally, the nuts are packaged to improve their shelf life and prevent spoilage.

Macadamia nuts from South America and other tropical countries have had to face certain challenges in their production and meet these quality criteria such as low levels of moisture and low levels of peroxides. The climatic and soil conditions of tropical countries were appropriate for the implantation of *M. integrifolia* crops and allowed the adaptation of varieties of the exotic species from Australia; however, it has become a nontraditional production item that has gained presence in the national and international market as an exportable resource. The process to reach production levels in South America has been slow and involves a great commitment to offer a product with characteristics that make it competitive in high-demand markets, which are characterized by demanding the satisfaction of rigorous quality criteria, sustainable over time.

The main factors affecting the production of macadamia nuts during postharvest are as follows:

- **Moisture:** The quality and storability of freshly harvested macadamia nuts are mainly limited by a high initial moisture content of approximately over 80% in the green carpel and 33% in the NIS. This high moisture content promotes microbial growth and lipolytic enzyme activity, leading to the development of rancidity and limited shelf life [31].

- **Rancidity:** This factor is closely related to moisture. Rancidity, caused by the high content of lipids in contact with moisture or air, constitutes the main quality defect of macadamia nuts that occurs in two main ways: oxidation and hydrolysis [25].

- **Browning:** A problem observed during the nut drying process is internal browning, caused by an accumulation of reducing sugars (glucose and fructose) in the center of the kernel, which, in turn, react with amino acids to give nonenzymatic browning products through Maillard reactions [17, 32]. On the other hand, macadamia nuts are temperature sensitive based on their moisture content. When fruits with a high initial moisture content are rapidly dried at temperatures above 38°C, sucrose (the predominant sugar in fresh macadamia nuts) may undergo hydrolysis to produce glucose and fructose, substrates for the Maillard reaction. Under these conditions, brown color develops in dried grains; therefore, browning may develop in the center of the kernel, or externally, depending on drying conditions. Physiological ripening also influences browning; the presence of unripe nuts (higher concentration of sugars) can contribute to kernel browning. Unripe grains have a higher content of sucrose and reducing sugars and more browning than ripe grains [17].

Therefore, the quality of the macadamia nut depends on the conditions of the initial composition of the nuts prior to the drying process and the methods by which they are processed, packaged, and stored. To combat these adverse factors for quality, adequate postharvest process, efficient drying systems, and the implementation of packaging systems that reduce susceptibility to lipid hydrolysis and oxidation are proposed in the industry.

Macadamia grain quality is more affected by slow drying at room temperature than by postharvest processes such as the type of huller used [33]. Grain drying consists of the removal of water to a level that prevents the growth of fungi and bacteria, so that the appearance and nutritional quality of the grain as food, or its viability as seed, is preserved [34]. The drying methods for the nuts can be summarized as follows:

- **Drying at low temperatures on the farm:** Those that use air at room temperature or heated 3°C ± 2°C above room temperature as a means of transporting moisture and energy. In these procedures, low specific air flows are used (2.0–5.0 m^3 / min.*ton*)[1]. The low air flows make the drying procedures at low temperatures typically slow, and the product obtained has good quality [34]. The first stage begins in the field when freshly harvested and shelled nuts, with a moisture content of 20–30% (db), are subjected to natural air convection drying, using aerated boxes, trays, and silos. Under these conditions, it takes 3–4 weeks to reduce the moisture content to 10% (db) [35].

- **Silo drying:** Only natural or slightly heated air is used. The procedure is relatively simple and inexpensive and maintains good grain quality. To apply this method, the moisture content of the grain must not be higher than 20%; if it is higher, the air temperature must be lower than 15°C and use higher air flows to avoid alterations in some layers of the grain mass [36]. It is recommended that the temperature does not exceed 40°C when the nuts have a moisture content greater than 8%; later, the temperature can be increased up to 70°C. The reason for this gradual increase in temperature to dry the nuts is to prevent the kernel from suffering color changes in the center; because the shell has a higher percentage of moisture than the kernel, the shell becomes saturated and has no space to release more moisture; then, the central part of the kernel is stained and changes its consistency with irreversible damage and impact on the quality of the kernel [30].

- **High-temperature drying:** They are characterized by the use of heated air, at least 10°C above room temperature. The specific air flow rates are higher and the drying time is shorter. The final stage of nut drying is carried out at a temperature between 50°C and 70°C, to reduce moisture to 1.5%. Conventional heating uses the application of energy on the surface, and this is transported to the interior of the material by conduction to perform the drying process [34]. The predried nuts that are transported to the industrial plant are placed inside a silo-type dryer, which operate with forced air flow heated by convection at a controlled temperature [35]. In places where the relative air humidity is high, as in many tropical countries, increasing the air flow rate is not enough to achieve drying, since this variable does not influence the drying potential of the air. In these cases, heating the air is favorable to increase the speed of movement of the drying front [34]. An alternative for kernel drying is microwave heating and

[1] Specific flow, which refers to the amount of air received by a cubic meter or ton of grain in a defined unit of time.

drying. Microwave heating works at frequencies between 300 MHz and 300 GHz [37]. Industrial ovens frequently combine both the conventional and microwave heat sources to obtain different degrees of browning and surface crispness, to accelerate moisture removal, and to reduce surface counts of microorganisms. Macadamia nut drying using microwaves, especially from the sensory point of view, allows obtaining a product with characteristics similar to those of the conventionally dried product. The advantage of this process is the lower impact on lipid rancidity, once the drying process is finished and after 6 months of storage [35]. The experiences carried out have been promising in terms of the greater drying speed, but have caused some problems of deterioration of the quality of the kernel, in terms of the characteristic cream color that it should present, without considering that this technology is even more expensive than conventional drying systems [30].

4. Macadamia nut packaging systems

Once dried, the macadamia nuts (kernels) contain a large amount of unsaturated lipids that can still be affected by oxygen, light, humidity, and heat during storage. For this reason, it is necessary that the relative humidity in the storage place does not exceed 70%. Similarly, an excessively dry environment can cause the nut to lose weight and lead to rancidity processes, so it is not convenient for the ambient humidity to be less than 40% [31]. In order to choose the appropriate preservation technique for the storage of dried kernels, it is necessary to know the estimated time and storage conditions of the packaged product.

Packaging fulfills the basic functions of containing and protecting the food, informing and attracting consumers. The use of packaging with new protection techniques has made it possible to extend the shelf life of these foods. The packages used in vacuum systems and protective atmospheres are made with coextruded materials, i.e., those that are made up of more than one polymer, generally low-density polyethylene and polyamides [38].

The appropriate packaging options have proven to be vacuum packaging in modified atmospheres or protective atmosphere.

- **Vacuum packaging:** It consists of the total extraction of the air that surrounds the product so that the packaging material folds around the food as a result of the decrease in internal pressure compared to atmospheric pressure. This material must have a very low permeability to gases, including water vapor.

- **Protective atmosphere packaging:** Mostly inert atmospheres of nitrogen and/or some noble gas such as argon and helium are used. Sometimes, carbon dioxide is also included for its antimicrobial action [38]. Organoleptic alterations are reduced when oxygen levels are below 0.2%. The characteristics of the most used gases for the packaging of the macadamia kernel are described below.

- **Carbon dioxide (CO_2):** CO_2 is a colorless gas with a slight pungent odor at very high concentrations. It dissolves easily in water (1.57 g/kg at 100 kPa, at 20°C), producing carbonic acid (H_2CO_3), which increases the acidity of the solution and reduces the pH. This gas is also soluble in lipids and other organic compounds. The solubility of CO_2 increases with decreasing temperature. Packaging with CO_2

registers a high diffusion through plastic materials; the high solubility of CO_2 can lead to the breakage of the package due to the reduction in headspace [39]. Also, CO_2 exerts a fungicidal and inhibitory effect on bacterial growth that reproduces rapidly in normal atmosphere. The absorption of CO_2 depends on the moisture and fat content of the products; therefore, most foods absorb this gas. In addition, high concentrations of CO_2 can cause discoloration and the development of pungent acid flavors [38].

- **Nitrogen (N_2):** N_2 is a gas that is not very reactive, odorless, tasteless, and colorless. It has a lower density than air, is nonflammable, and has low solubility in water (0.018 g/kg at 100 kPa, at 20°C) and other food components. The solubility of N_2 in foods prevents the breakage of packages, if enough gas is included in the atmosphere of the package to balance the reduction in volume due to the passage of CO_2 gas to the dissolved form [39]. Nitrogen is fundamentally used in a modified atmosphere to displace and eliminate the maximum amount of oxygen, avoiding oxidation of vitamins, aromas, color, and fats, inhibiting aerobic bacteria [40].

To take advantage of the benefits of the different gases, modified atmosphere packaging usually requires a mixture of at least two gases, with the optimum proportions varying from product to product. An example of commercially available products is the 50:50 gas ratio of N_2 and CO_2. Packaging in N_2, CO_2, and mixed N_2/CO_2 atmospheres can be a useful alternative to control the parameters that threaten the physicochemical and nutritional quality of macadamia nuts. In a practical case study on macadamia nuts produced in Paraguay, the influence of packaging the kernels in different atmospheres for 180 days was studied. The nuts were dried using a silo-type dryer in two stages:

The first consisted of predrying in a silo-type dryer at a temperature of 39°C and 0.932 m^3/min of air input, until reaching a moisture content of 8.0% ± 0.5%. In the second stage, the nuts were processed in a Sherwood 501 dryer from a moisture content of 8.0% ± 0.5% to 1.5% ± 0.3% at a temperature of 65°C and an inlet air flow of 2.00 m^3/min. Maintaining the height of the nuts bed, without exceeding one-third of the dryer capacity.

The dried kernels were packed in polyamide polyethylene bags using a vacuum packing machine up to 8 mbar and then injection of N_2, CO_2, and gas mixture

	Moisture (g/100 g)	**Lipids (g/100 g)**	**CH (g/100 g)**	**Proteins (g/100 g)**	**Ash (g/100 g)**	**DF (g/100 g)**
Postharvest	18.2 ± 1.25	66.26 ± 6.2	16.22 ± 0.78	6.05 ± 0.59	1.34 ± 0.05	15.71 ± 0.0
Dry raw material	1.05 ± 0.14	76.62 ± 1.7	8.36 ± 1.43	8.01 ± 1.67	1.59 ± 0.25	4.92 ± 0.0
CO_2/N_2	1.19 ± 0.21	77.06 ± 0.6	9.04 ± 0.85	7.30 ± 1.05	1.36 ± 0.03	7.19 ± 0.0
N_2	1.30 ± 0.04	76.18 ± 0.2	8.71 ± 0.32	8.09 ± 1.20	1.48 ± 0.01	7.76 ± 0.0
CO_2	1.01 ± 0.19	78.62 ± 1.3	7.29 ± 0.35	8.07 ± 1.53	1.35 ± 0.08	5.95 ± 0.0
Air	1.97 ± 0.17	77.37 ± 0.5	7.39 ± 1.37	8.28 ± 1.41	1.27 ± 0.01	5.16 ± 0.0
Vacuum	1.23 ± 0.14	77.55 ± 0.5	8.38 ± 0.65	9.18 ± 1.13	1.49 ± 0.04	7.46 ± 0.0

Table 1.
Centesimal composition of macadamia nuts in postharvest stages, drying and 180 days of storage in different protective atmospheres.

CO_2/N_2 (50,50) as a protective atmosphere up to a pressure of 300 mbar. Nuts were also vacuum packed and packed in polyethylene bags as a control for conventional packaging.

Values are presented as mean ± SD (n = 3). In each row, different lowercase letters indicate a statistically significant difference (ANOVA, post Tukey test, $p < 0.05$). Statistical analyses were performed separately at each stage, because the nuts analyzed were not the same in phase.

The results showed that under these drying conditions, the nuts maintain the quality required for export for up to 6 months, conserving the parameters of peroxide index, acidity, mesophilic aerobic count, fungi and yeasts, and coliforms at 45°C. This processing system used allowed to obtain a good quality product, in accordance with national and international standards, at the time of analysis, up to 180 days of storage and kept in a place with low relative humidity. Regarding the organoleptic characters through a sensory profile of taste and texture, there was a significant difference at 45 days; according to the tasters, a lower moisture content was perceived in the nuts packaged in an atmosphere of CO_2/N_2 (50:50). After 90 days of storage, a greater intensity of the bitter taste was perceived in the nuts packed in a CO_2 atmosphere, with a significant difference compared to the samples packed with other gases. At 180 days of storage, a significant difference was observed in the bitter taste of nuts packed under conventional conditions. The most important variation in the centesimal composition of the macadamia nuts was observed between the postharvest stage and the dry raw material; in the postharvest, the nuts presented 18% moisture and it was possible to reach 1% in the drying stage. The results of the centesimal composition of the macadamia nuts obtained in the different stages are shown in **Table 1**.

5. Drying versus browning: color analysis

Color is a very important quality parameter of macadamia nuts and varies from white to creamy [33]. Variations in ripeness, sugar composition, moisture content, and drying conditions of macadamia nuts contribute to internal browning [41] caused by an accumulation of reducing sugars in the center of the kernel, which, in turn, react with amino acids to give nonenzymatic browning products through Maillard reactions [42]. The roasting process in macadamia nuts is widely used because it improves the shelf life of the nuts, inactivating the oxidative enzyme system (lipoxygenic enzymes) and improves the flavor, aroma, texture, color, and appearance of the nuts through the reaction of Maillard and lipid peroxidation [43]. However, it is the main cause of intense color variations [44]. Conventional industrial roasting (120–160°C for about 10–20 min) must be controlled to achieve the desired color and sensory characteristics. Color variations can be detected by human vision, which is subjective, so detection by instrumental means is better [45]. Currently, color spaces and numerical values are used to create, represent, and visualize colors in a space of two or three dimensions [46]; in the case of food, as agreed by the International d'Eclairage Commission, it is adapted to a system by the CIE 1976 (ISO 11664–4, 2008) where the color space usually used is the L^*a^*b, which is the Euclidean distance between two different colors corresponding to the color difference perceived by the human eye. The L^* value represents lightness and darkness, its value is between zero for complete darkness and 100 for complete lightness, and the parameters a^* (from green to red) and b^* (from blue to yellow) are the two chromatic components, ranging from −120 to 120 [47]. It is necessary to know the color value in each pixel of the surface of the macadamia nut in order to perform a detailed characterization of it and thus evaluate its quality with greater precision. However, the colorimeters available in the market measure L*a*b in a few square centimeters (~ 2 cm^2), and therefore, their measurements are not very representative in heterogeneous materials, as is the case of macadamia nuts [48]. On the other hand, the high cost of colorimeters for small producers in tropical countries such as Paraguay has led to the exploration of other methods of color measurement under standard conditions through photographic images. An alternative to measure color is through computer vision techniques [46, 48–51]. With a digital camera, it is possible to record the color of any pixel in the image of the object using three color sensors per pixel. The most widely used color model is the RGB model in which each sensor captures the intensity of light in the red (R), green (G), or blue (B) spectrum, respectively, which can be analyzed and presented in a histogram; these results can later be converted to the L*a*b color model [50]. For this conversion, most of the computerized vision systems described in the literature use specialized equipment or algorithms that are not easily accessible to most researchers, a problem that could be overcome with the use of software already available on the market [48].

To solve the problem of color analysis, a simple, practical, and economical method for color measurement was developed, using a BYK byko basic ® light booth (Columbia, USA), illuminated with two 60-cm-long D65 "daylight" fluorescent lamps, GTI Color Matcher ® (USA), a Canon PowerShot SD1400 IS digital color camera, placed horizontally to the samples at a distance of 10 cm and Image J ® and Adobe Photoshop CC 4.0.1.192. software. Color variation was monitored throughout the nut industrialization process. The same parameters were measured on nuts at different conditions; fresh, dried at 65°C (5,36 m/s air flow), in a silo-type dryer (Sherwood model 501, fluid bed dryer), and nuts stored for 180 days vacuum packed in different modified atmospheres (air, vacuum, CO_2, N_2, and CO_2/N_2, 50:50; see **Figure 5**). On the other hand, the content of reducing sugars was measured by Clegg's anthrone method, in order to determine the packaging system that underwent the least Maillard reaction during the process from drying to packaging at different atmospheres [52, 53].

Regarding the color of fresh macadamia nuts, a light cream color was observed, with L* values on the external surface equal to 68.83 ± 1.85. In dry nuts, however, the values obtained for L* were higher (71.33 ± 1.07), with significant differences being observed (ANOVA and Tukey's a posteriori test, $p < 0.05$), which indicates a greater luminosity in the dry samples (**Figure 6**). Besides, in the internal surface, we

Figure 5.
Color of macadamia nuts at different packaging atmospheres.

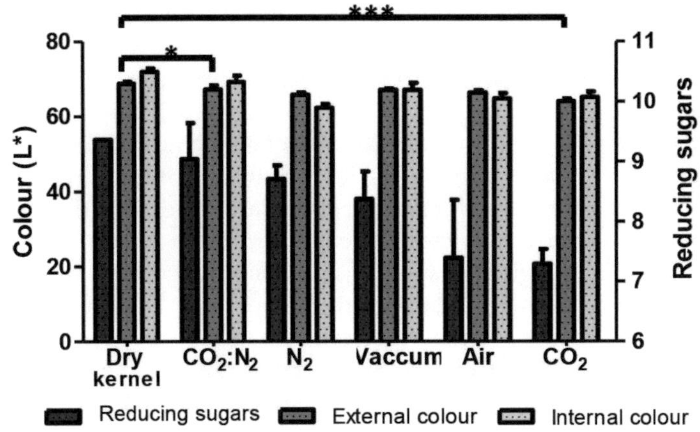

Figure 6.
*External and internal color and content of soluble sugars in dry macadamia nuts, prior to packaging and after 180 days of storage in modified atmospheres. The values are expressed as mean ± SD. The nuts analyzed in each condition are not the same. Same letters indicate that are no statistically significant differences (ANOVA and Tukey's a posteriori test, p < 0.05). *they express the degrees of statistical significance with respect to dried nuts prior to packaging.*

observed a value of $L^* = 72.00 \pm 2.92$ and in dry nuts 76.75 ± 3.82, observing statistically significant differences, which indicates a greater luminosity in the samples (ANOVA and Tukey's a posteriori test, $p < 0.05$).

After 180-day follow-up of packaged nuts, statistically significant differences were observed in both the external and internal surfaces, and nuts stored with CO_2/N_2 were observed to maintain a desirable cream color in external and internal surfaces that was stable for 180 days and good luminosity values (L^*), with respect to the other atmospheres used.

Studies on the influence of modified atmospheres with different gases on the Maillard reaction are limited in the literature; however, Birch et al. [54] evaluated the external color change and the sugar content in macadamia nuts when they were heated for 20 min at 135°C and observed that both L^* value and sugar content decrease as a function of time.

Once the issue of obtaining whole, dry nuts of high sensory and nutritional quality has been resolved, it is imperative to address the waste that this process entails. They are considered the main byproducts of this drying and packaging process; the cracked, broken nuts or powder of the broken and the shell of the NIS. From cracked nuts, one of the best alternatives is to obtain oil, considering a raw material with a high content of lipids, which require rapid processing due to the susceptibility to oxidation. The main alternatives used for the oil extraction with high added value are described below, as well as the use of biowaste such as the oil extraction cake and macadamia nut shells, which are found in frank development at the level of tropical countries in the productive sector of macadamia.

6. Advances in macadamia oil extraction with green technologies

The traditional extraction method is cold mechanical pressing, a technology that requires a large energy input and provides a product with fine suspended solids and low efficiencies (35–40% w/w) [55]. An alternative to increasing the efficiency and quality of the product would be extraction with organic solvents, but with enormous disadvantages from the energy, nutritional, and environmental point of view [56]. The growing trend in the consumption of healthy, safe, and functional foods has motivated studies on special cold-pressed oils, including macadamia oil. Consumers prefer cold-pressed macadamia nut oil (CPMO) over refined and solvent-extracted oil due to its exceptional quality and safety attributes [1]. In recent years, the technology of liquefied gases and supercritical fluids has gained relevance due to the use of solvents considered "green," since they are nontoxic, safe, and cheap [57]. The use of supercritical CO_2 (sc-CO_2) for the extraction of various products of interest has gained notoriety [58]; however, for the extraction of oils, it was reported that sc-CO_2 requires long extraction times, high temperatures that could degrade antioxidants present in macadamia, and high pressures that have been shown to negatively influence the profile of fatty acids since lower amounts of unsaturated fatty acids are obtained [11, 59–61]. Besides, subcritical CO_2 or liquefied CO_2 has numerous advantages, such as operating below the critical point, suitability for the separation of thermolabile compounds, avoidance of thermal degradation of components at the time of extraction, and high selectivity to flavor-representative and esterified components [62, 63]. Additionally, it is nonflammable, safe, cheap, odorless, nontoxic, highly available, and environmentally friendly [64]. However, CO_2 is a nonpolar molecule, which is why it is related only to compounds of the same nature. However, the addition of a suitable cosolvent can improve the solvent properties of CO_2, expanding the range of lipid extraction [59, 65]. Due to the ease of removing ethanol from the oil, its use as cosolvent is allowed in the food industry. It was shown that the lipid extraction range is extended if it is used with CO_2; it also decreases the viscosity and surface tension of the oil-CO_2 mixture and decreases the electrical permittivity of CO_2 and, therefore, the polarizability [59, 65]. As for the critical point of the CO_2-ethanol mixture, it increases with the alcohol fraction, a phenomenon that allows the subcritical work zone to be extended [11, 59–61].

Liquefied propane is another permitted cosolvent that has multiple advantages over solvents such as n-hexane (widely used for the extraction of edible oils), since it is cheap and does not leave a toxic residue [66]. The use of light hydrocarbons as cosolvents substantially improves the extraction kinetics due to the good solubility of triglycerides, making the operation faster and more efficient. Regarding the temperature and critical pressure of the CO_2-C_3H_8 mixture, it presents an antagonistic behavior, since as the fraction of C_3H_8 increases, the critical pressure decreases, while, for low fractions of C_3H_8, the critical temperature decreases below the critical temperature of pure CO_2, but then increases substantially. Low concentrations of C_3H_8 are said to push the critical temperature of the mixture [67].

A simple method to study the main operating variables that influence solid–liquid extraction with liquefied gases is the high-pressure Soxhlet, whose operation is similar to the conventional Soxhlet used to determine the fat content of a vegetable matrix (**Figure 7**). To create the necessary temperature gradient, the ends of the extractor must be subjected to a temperature difference such that, at the base, the evaporation of the solvent-cosolvent mixture is verified, while at the head, the condensation of these is allowed. This could be achieved by immersing the base of the extractor in a water bath at a controlled temperature and circulating cold water through the condenser, which must be conveniently located so that the film of condensate drips onto the sample contained in the extraction cartridge. The net effect would be the periodic recirculation of CO_2 in the extraction vessel, working under liquid–vapor equilibrium conditions. An extraction cycle is considered complete when the condensate in the

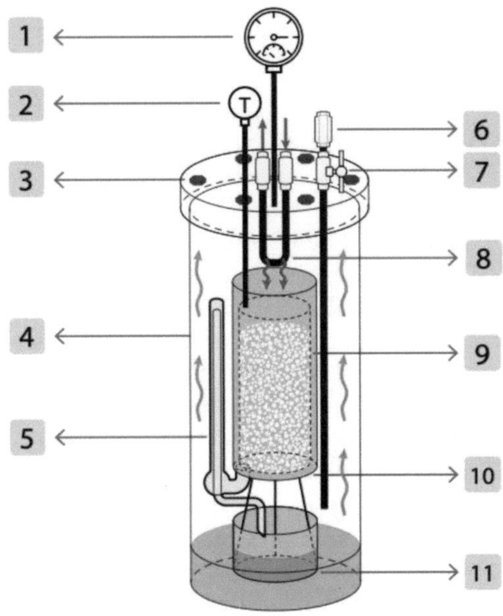

Figure 7.
High-pressure Soxhlet extraction system: (1) pressure gauge, (2) temperature sensor, (3) adjusting and closing nuts, (4) pressure vessel – Parr reactor, (5) siphon, (6) charge and decompression line, (7) shutoff valve, (8) condenser, (9) extraction cartridge, (10) solvent chamber, and (11) sample chamber. Elaboration: Smidt, M.

sample chamber is siphoned into the solvent chamber. The number of extraction cycles is determined by recording the temperature in the extraction chamber, where a temperature sensor is inserted at the same level as the siphon tube. The sensor will record a temperature gradient each time the solvent mixture is siphoned into the sample chamber. This process approximates a continuous multistage countercurrent extraction operation [68]. At the end of the extraction time, the container must be slowly depressurized; the solvent mixture will evaporate by sudden decompression (flashing) obtaining, on one side, an exhausted solid in the extraction cartridge and, on the other side, the extracted oil free of solvent in the solvent chamber.

The efficiency on any solid–liquid extraction operation with liquefied gases strongly depends on the proper selection of the solvent and cosolvent, as well as the nature of the solid matrix and the operating conditions [69]. The extraction temperature is an important variable, since it can significantly affect the quality of the product if it is very high and must be established together with the pressure, so that the working conditions are maintained in the subcritical region. Regarding the cosolvent fraction, it is important since the critical point of the mixture changes in relation to that of the pure solvents, so a fraction that allows the mixture to work in the subcritical region should be selected [59]. Respecting the granulometry, the efficiency is governed by this parameter since the components to be extracted must come into contact with the solvent, which will determine the extraction time and which, in turn, influences which components are extracted [70]. Initially, it is the solubility that controls the extraction process, and over time, it is the internal diffusion that governs the extraction process [71].

Experimental tests were conducted with subcritical mixtures of CO_2-ethanol and CO_2-C_3H_8, separately, by means of the high-pressure Soxhlet method, with the application of a multilevel factorial design, and the effect of extraction temperature, mass fraction of cosolvent, and average granulometry of the dry kernel on the extraction efficiency of *M. integrifolia* nut oil was evaluated. Twenty grams of kernels was used for each experimental condition, and these were subjected to the different conditions according to the experimental planning; previously, the moisture content of the macadamia was adjusted to <1.5%. For the CO_2-ethanol case, the extraction time was 6 h, while for the CO_2-C_3H_8 case, it was 1 h. At the end of the extraction time, the extractor was decompressed and the oil obtained was physicochemically characterized. The partially defatted solid was subjected to total lipid analysis to determine, indirectly, the extraction efficiency, as well as the characterization of some parameters of interest for its further use. It should be noted that for the CO_2-ethanol mixture, additional vacuum evaporation is required to strip the oil of the cosolvent.

In the case of ethanol-assisted extraction, the highest efficiency (66.5% w/w) was obtained at 45°C, average granulometry 4.05 mm, and 20% (w/w) of cosolvent fraction. On the other hand, the extraction assisted with propane had a maximum efficiency of 72.6% (w/w) at 38°C, average granulometry 4.05 mm, and 45% (w/w) of cosolvent fraction. Both efficiencies are higher than those reported for obtaining macadamia oil by cold mechanical pressing, which is the method mainly applied for the extraction of this oil [72]. **Figure 8** shows the behavior of the extraction efficiency with temperature (X1) and average granulometry (X2) with 20% (w/w) ethanol (**Figure 8a**) and 45% propane (**Figure 8b**). The lowest extraction efficiency with CO_2-ethanol required two cycles/h, while the highest required six cycles/h. On the other hand, the lowest extraction efficiency with the CO_2-C_3H_8 mixture required two cycles/h and the highest four cycles/h; these physical phenomena allow us to visually explain the differences between treatments that lead to higher extraction yields.

In edible oils, the recommended physicochemical quality control determinations include acidity index, peroxide index, and iodine value [4], which can be complemented with the saponification index, refractive index, and rancidity, in order to have a bigger picture of oil quality. Based on this, the aforementioned determinations were made for the oil obtained in the treatment with the highest efficiency for each solvent mixture (**Table 2**). The acidity index is low (< 4, **Table 2**), which implies a reduced

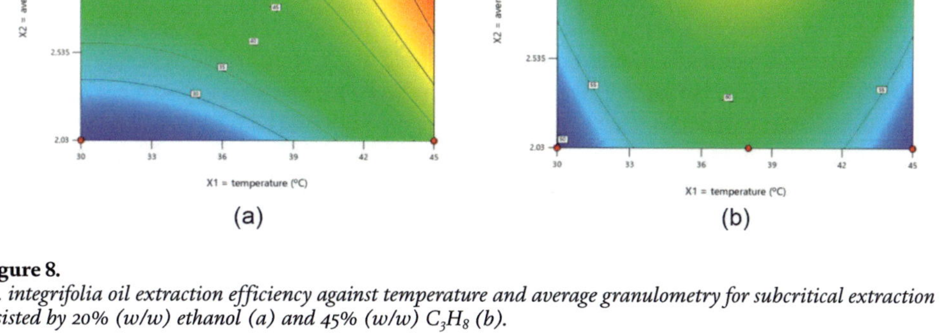

Figure 8.
M. integrifolia oil extraction efficiency against temperature and average granulometry for subcritical extraction assisted by 20% (w/w) ethanol (a) and 45% (w/w) C_3H_8 (b).

Determination	CO_2-ethanol	CO_2-C_3H_8	Reference	Source
Acidity index (mg KOH/g)	0.14 ± 0.03	0.29 ± 0.10	<4	[73]
Peroxide index (mEq O_2/kg)	0.95 ± 0.01	1.12 ± 0.17	<15	[73]
Iodine value (g I/100 g)	70.92 ± 2.93	71.75 ± 1.05	69–78	[74, 75]
Saponification index (mg KOH/g)	227.66 ± 1.10	218 ± 1.67	195–234	[12, 72]
Refractive index	1.46 ± 0.00	1.46 ± 0.00	1.46–1.47	[74, 76]
Rancidity	Negative	Negative	Negative	[76]

Table 2.
Physicochemical determinations of M. integrifolia oil quality carried out on the oil obtained in the best treatment of each solvent mixture.

amount of free fatty acids product of triglyceride degradation reactions that increase the quality of an oil, this complemented by a high oxidative stability verified by the low peroxide index (< 15, **Table 2**) [77], corroborated by a negative test for rancidity. The low iodine value classifies it as a nondrying oil (<100 g I/100 g), which is why a low content of polyunsaturated fatty acid is predicted [74]. Given the high saponification index, its refining is not recommended unless its final destination is the soap industry [72]. These results agree with what was reported by Mereles and Ferro [14], who have characterized freshly extracted macadamia oil from three consecutive harvests.

Regarding the fatty acids present in the extracted oils, these were contrasted with the fatty acid profile of virgin macadamia oil, extracted by the cold mechanical pressing of the raw material itself (**Table 3**). The values are close to each other, with a high and encouraging content of oleic acid as well as palmitoleic acid, both monounsaturated fatty acids of special interest in terms of nutraceutical and skin regenerative benefits [72]. A low content of polyunsaturated fatty acids was found, which is related to the low iodine value reported.

As for the defatted meal from both extraction processes, a high content of protein (> 18% w/w), fiber (>13% w/w), and carbohydrates (> 41% w/w) was verified, which makes it a candidate for the production of protein concentrates, isolated protein for the production of nutritional supplements, and partially defatted flours of interest for celiacs, among others [11, 78].

7. Activated carbon from macadamia nut shells

Several studies have addressed the use of large volumes of shell (mesocarp) that represents 40% of the dry weight of the fruit. In Paraguay, the potential of the shell of the *M. integrifolia* nut was studied as a precursor for obtaining activated carbon, whose volume in a macadamia processing industry is important. Activated carbon is the name applied to a series of porous carbon, which, after undergoing a carbonization and activation process, show great porosity and internal surface. It has been shown that activated is an extremely versatile adsorbent, with a crystalline structure similar to that of graphite, a characteristic that, together with the chemical nature of the carbon atoms that make it up, gives it surface adsorbent properties for a certain type of molecules [79]. The obtaining process basically consists of two stages: the pyrolysis of the carbonaceous material at temperatures below 800°C and the activation of carbon by physical or chemical methods.

Fatty acid	Formula	Virgin oil	CO_2-ethanol	CO_2-C_3H_8
Myristic	C14:0	0.61 ± 0.01	0.55 ± 0.01	0.62 ± 0.01
Palmitic	C16:0	8.14 ± 0.02	7.99 ± 0.05	8.28 ± 0.01
Palmitoleic	C16:1 ($\omega - 7$)	17.66 ± 0.05	15.89 ± 0.08	17.62 ± 0.02
Stearic	C18:0	3.93 ± 0.01	4.01 ± 0.01	3.92 ± 0.01
Oleic	C18:1 ($\omega - 9$)	62.28 ± 0.13	63.84 ± 0.10	62.58 ± 0.04
Linoleic	C18:2 ($\omega - 6$)	1.68 ± 0.01	1.66 ± 0.01	1.66 ± 0.00
Gondoic	C20:1 ($\omega - 9$)	2.29 ± 0.01	3.02 ± 0.02	2.84 ± 0.01
α – Linolenic	C18:3 ($\omega - 3$)	2.05 ± 0.01	2.21 ± 0.01	2.09 ± 0.00
Behenic	C22:0	0.75 ± 0.00	0.83 ± 0.01	0.76 ± 0.01

Table 3.
Fatty acid profile of M. integrifolia oil extracted by cold mechanical pressing, with CO_2-ethanol and CO_2-C_3H_8.

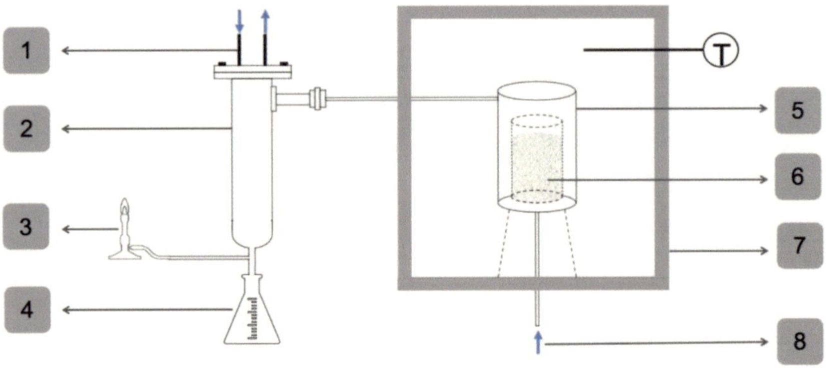

Figure 9.
Fixed bed reactor for obtaining activated carbon: (1) cooling water inlet/outlet, (2) condenser, (3) burner, (4) collector of liquid byproducts, (5) fixed bed reactor, (6) basket with macadamia shells, (7) pyrolysis oven, and (8) pipe for steam injection. Elaboration: Smidt, M.

The physical activation process was studied with a fixed bed reactor, consisting of a reaction chamber that contains a basket built with a metal mesh that supports the material to be carbonized (**Figure 9**). The reactor was placed in an electric furnace that allows the temperature to be controlled up to 1200°C; the steam necessary for activation was supplied by pumping liquid water with a peristaltic pump (not drawn) through a pipe inside the furnace, which allows the vaporization of this water. The gases produced are conducted out of the reactor, and the condensable fraction changes phase thanks to the removal of the heat in an indirect condenser through which cold cooling water circulates, while the incondensable gases are directed to a burner to burn the flammable fraction. About 750 g of macadamia nut shell (granulometry <1 mm) was introduced into the reactor and subjected to a heating rate of

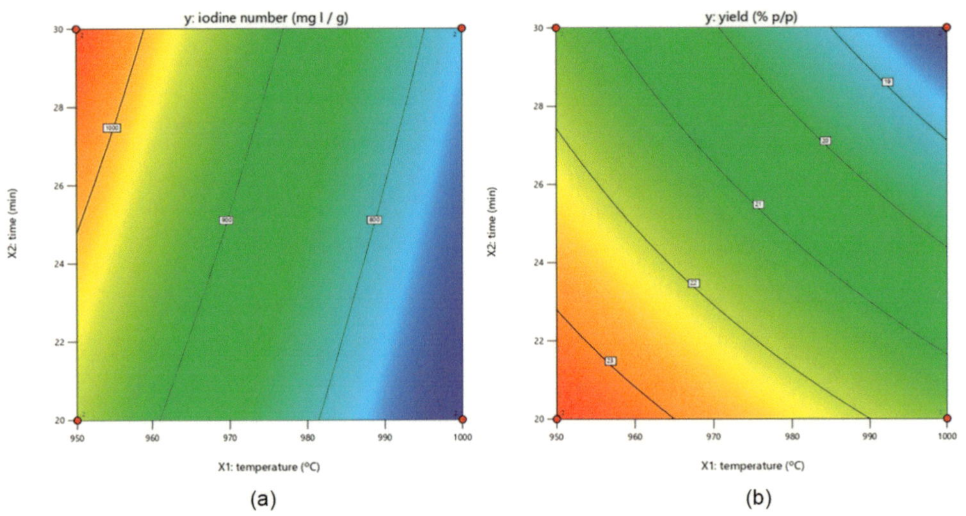

Figure 10.
Behavior of the iodine value (a) and yield of activated carbon (b) with temperature and activation time obtained from macadamia shells.

19°C/min up to 400°C and then 7°C/min until reaching the established temperature. The effect of temperature (950°C and 1000°C) and activation time (20 and 30 min) on the yield of activated carbon and the absorption capacity, evaluated by the iodine value, was studied using a 2^2 factorial design. The flow rate of water injected for activation was 42 mL/min. **Figure 10** shows the behavior of the absorption capacity and the yield obtained in the experimental region. For the case of **Figure 10a**, it is noted that the highest absorption capacity was obtained at 950°C and 30 min. However, higher yields are obtained at the same temperature, but with 20 min of reaction (**Figure 10b**) and with an average difference in the absorption capacity of only 96 mg I/g. The analysis of variance with a 95% confidence interval revealed that, for both the independent variables, the temperature and the activation time are significant. However, the interaction between these variables is only significant for the yield.

8. Use of the shell as nanosorbent in the chemical industry

Adsorption is a commonly used unitary operation for wastewater treatment due to its simplicity, cost-effectiveness, high removal capacity, and low energy consumption in large-scale processes [80]. Agricultural residues have the potential to be used as adsorbents due to their high availability and chemical composition [81]. Various byproducts of agricultural materials derived from adsorbents such as cactus leaves, hazelnut shells, banana peels, wool, almond shells, and coconut shells have been studied as potential adsorbents or contaminant removal from wastewater [82, 83]. Activated carbon is an adsorbent with a microporous structure, usually used for the elimination of polluting substances in wastewater, gases, etc. The microstructure of activated carbons depends not only on the natural texture of each raw material, but also on the activation process. Surface and porous areas are improved by carbonization and activation treatment. Improvements can be added by designing or modifying the activation process, such as carbonization at high temperature or in different atmospheres. Like other plant materials, the main component of the macadamia nut shell is cellulose (41.2%), which can be denatured to become activated [84]. Macadamia shell contains less inorganic content and high fixed carbon content compared to other biomass. Macadamia shells have a higher surface area than other nut shells [85, 86], and their ash content is very low [87]. Research has been conducted using macadamia nut shells to produce activated carbons for the removal of contaminants in water such as microcystin, aurocyanide, phenol, and methylene blue [88–91].

9. Conclusion

The fruits of macadamia nuts have wide advantages in their integral use, considering that the byproducts of the fruit are 100% industrializable, for which they present competitive advantages over other nontraditional crops. Several efforts have been made to improve the quality of nuts in countries such as Paraguay, with a tropical climate of high average temperatures and high relative humidity, which allow the macadamia nut to be a product of high nutritional and sensory quality, with a potential for sustainable and environmentally friendly production, compatible with current demands, within the framework of sustainable food for the future.

Acknowledgements

This work was financed by CONACYT through the PROCIENCIA, Program with resources from the Fund for Excellence in Education and Research—FEEI of FONACIDE. Projects 14-INV-001 "*Estudio del proceso de industrialización de nueces de macadamia, con calidad de exportación*" and PINV18-671 "*Evaluación del proceso de extracción de aceite de macadamia con CO_2 licuado + cosolventes y caracterización de subproductos.*" The authors especially thank the Faculty of Chemical Sciences, FUNDAQUIM, MACPAR, PLAPIQUI, Silvia Caballero, Juan Carlos Martínez, Patricia Piris, Lourdes Wiszovaty, Javier Michajluk, Vanessa Resquín, Antonella Elizaur, Andrea Paredes, Alejandro Satof, Blas Vázquez, Naomi Woodfin, Belén Díaz, Alicia Romero, and Pablo Hegel for their great collaboration for the execution of this work.

Author details

Laura Graciela Mereles*, Mario Smidt, Karen Patricia Martínez, Eva Eugenia Soledad Coronel, Edelira Velázquez and Laura Correa
Faculty of Chemical Sciences, National University of Asunción, San Lorenzo, Paraguay

*Address all correspondence to: laurameereles@qui.una.py

References

[1] Kaseke T, Fawole OA, Opara UL. Chemistry and functionality of cold-pressed macadamia nut oil. PRO. 2022;**10**:56. DOI: 10.3390/pr10010056

[2] Stolp LJ, Kodali DR. Naturally occurring high-oleic oils: Avocado, macadamia, and olive oils. In: Flider FJ, editor. High Oleic Oils. Champaign: AOCS Press; 2022. p. 7-52. DOI: 10.1016/B978-0-12-822912-5.00003-4

[3] Wood LG, Garg ML. Macadamia Nuts (Macadamia integrifolia and tetraphylla) and their Use in hypercholesterolemic Subjects. In: Preedy VR, Watson RR, Patel VB, editors. Nuts and Seeds in Health and Disease Prevention. Amsterdam: Academic Press; 2011. pp. 717-725. DOI: 10.1016/B978-0-12-375688-6.10085-4

[4] Mereles L, Martinez K. Manual de procedimientos analíticos para el control de calidad de nueces de macadamia secas. 1st ed. Asunción: Noemi Montiel; 2017. p. 43

[5] Shuai X, Dai T, Chen M, Liang R, Du L, Chen J, et al. Comparative study of chemical compositions and antioxidant capacities of oils obtained from 15 macadamia (macadamia Integrifolia) cultivars in China. Food. 2021;**10**:1031. DOI: 10.3390/foods10051031

[6] Wall MM. Functional lipid characteristics, oxidative stability, and antioxidant activity of macadamia nut (macadamia Integrifolia) cultivars. Food Chemistry. 2010;**121**:1103-1108. DOI: 10.1016/j.foodchem.2010.01.057

[7] dos ES P, Pio R, Rodrigues FA, LAC M, Costa FC. Análise de Frutos e Nozes de Cultivares de Nogueira-Macadâmia. Ciênc. Rural. 2011;**41**:2080-2083. DOI: 10.1590/S0103-84782011001200007

[8] Mereles L, Ferro E, Alvarenga N, Caballero S, Patricia P. Michajluk. Boris International Food Research Journal. 2017;**24**:2599-2608

[9] INC International Nut and Dried Fruit Council Macadamia

[10] Armadans A. Diferencias de Tamaño de Nuez de Tres Variedades de Macadamia (Macadamia Integrifolia Smith) de Seis Años de Edad. Investig. Agrar. 2013;**4**:5-7

[11] Navarro S, Rodriguez C. Macadamia oil extraction methods and uses for the defatted meal Byproduct. 2Trends. Journal of Food Science and Technology. 2016;**54**:148-154. DOI: 10.1016/j.tifs.2016.04.001

[12] Ribeiro A, Haddad F, Tavares T, Magalhães K, Pimenta C, Nunes C. Characterization of macadamia oil (macadamia Integrifolia) obtained under different extractions conditions. Emirates Journal of Food and Agriculture. 2020;**32**:295-302. DOI: 10.9755/ ejfa.2020.v32.i4.2095

[13] Segura R, Javierre C, Lizarraga M, Ros E. Other relevant components of nuts: Phytosterols folate and minerals. British Journal of Nutrition. 2006;**96**:36-44. DOI: 10.1017/BJN20061862

[14] Mereles L, Ferro E. Características Físicas, Composición Centesimal y de Minerales de Frutos de Macadamia Integrifolia Maiden & Betche, Cosechados En El Departamento de Cordillera. Vol. 14. Paraguay: Rojasiana; 2015. pp. 55-68

[15] Alasalvar C, Shahidi F, editors. Tree Nuts Composition, Phytochemicals, and Health Effects. Nutraceuticals Science

and Technology. 1st ed. CRC Press; 2008. 340 p

[16] Munro I, Garg M. Nutrient composition and health beneficial effects of macadamia nuts. In: Alasalvar C, Shahidi F, editors. Tree Nuts: Composition, Phytochemicals, and Health effects. Nutraceutical Science and Technology. Florida: CRC Press; 2008. pp. 249-258. DOI: 10.1201/9781420019391.ch15

[17] Wall MM, Gentry TS. Carbohydrate composition and color development during drying and roasting of macadamia Nuts (macadamia Integrifolia). LWT - Food Science and Technology. 2007;**40**:587-593. DOI: 10.1016/j.lwt.2006.03.015

[18] Chung KH, Shin KO, Hwang HJ, Choi K-S. Chemical composition of Nuts and seeds sold in Korea. Nutrition Research and Practice. 2013;7:82. DOI: 10.4162/nrp.2013.7.2.82

[19] Venkatachalam M, Sathe SK. Chemical composition of selected edible nut seeds. Journal of Agricultural and Food Chemistry. 2006;**54**:4705-4714. DOI: 10.1021/jf0606959

[20] SRP M, Kochhar SP, Dutta PC. Lipid components and oxidative status of selected specialty oils. Grasas y Aceites. 2012;**63**:143-151. DOI: 10.3989/gya.083811

[21] Derewiaka D, Szwed E, Wolosiak RP. Properties and composition of lipid fraction of selected edible Nuts. Pakistan Journal of Botany. 2014;**46**:337-343

[22] Maguire LS, O'Sullivan SM, Galvin K, O'Connor TP, O'Brien NM. Fatty acid profile, tocopherol, squalene and Phytosterol content of walnuts, almonds, peanuts, hazelnuts and the macadamia nut. International Journal of Food Sciences and Nutrition. 2004;**55**:171-178. DOI: 10.1080/09637480410001725175

[23] Abe LT, Lajolo FM, Genovese MI. Comparison of phenol content and antioxidant capacity of Nuts. Ciênc. E Tecnol. Aliment. 2010;**30**:254-259. DOI: 10.1590/S0101-20612010000500038

[24] Kornsteiner M, Wagner K-H, Elmadfa I. Tocopherols and Total Phenolics in 10 different nut types. Food Chemistry. 2006;**98**:381-387. DOI: 10.1016/j.foodchem.2005.07.033

[25] Buthelezi NMD, Tesfay SZ, Magwaza LS. Influence of roasting on antioxidants, fatty acids, sensory properties and oxidative stability of macadamia Nuts. Scientia Horticulturae. 2021;**278**:109850. DOI: 10.1016/j.scienta.2020.109850

[26] Hu W, Fitzgerald M, Topp B, Alam M, O'Hare TJ. A review of biological functions, health benefits, and possible de novo biosynthetic pathway of Palmitoleic acid in macadamia Nuts. Journal of Functional Foods. 2019;**62**:103520. DOI: 10.1016/j.jff.2019.103520

[27] Rost J, Muralidharan S, Lee NA. A label-free shotgun proteomics analysis of macadamia nut. Food Research International. 2020;**129**:108838. DOI: 10.1016/j.foodres.2019.108838

[28] Cadogan C. Manual de Cosecha y Post Cosecha. Asunción: REDIEX; 2013

[29] Quintas, G. Manual Técnico Para Productores de Nuez de Macadamia. 1st. Ed. México: Asoc. Mexicana de Productores, procesadores y exportadores de nuez de macadamia, A.C.; 2011. 264 p

[30] Martinez K, Martínez JC, Velázquez E. Guía de Secado y Envasado de Nueces de Macadamia. 1st ed. Asunción: Noemi Montiel; 2017. 39 p

[31] Borompichaichartkul C, Luengsode K, Chinprahast N, Devahastin S. Improving quality of macadamia nut (macadamia Integrifolia) through the use of hybrid drying process. Journal of Food Engineering. 2009;**93**:348-353. DOI: 10.1016/j.jfoodeng.2009.01.035

[32] Warangkana S. Effects of Postharvest Treatments on Macadamia Nut Quality [thesis]. Australia: The University of New South Wales; 2012

[33] Walton DA, Wallace HM. The effect of mechanical Dehuskers on the quality of macadamia kernels when Dehusking macadamia fruit at differing harvest moisture contents. Scientia Horticulturae. 2015;**182**:119-123. DOI: 10.1016/j.scienta.2014.10.053

[34] Valdecir Antoninho Dalpasquale, Marques Pereira DA, Sinicio R, Oliveira Filho D. Secado de granos a altas temperaturas – Contenido. Available from: https://www.fao.org/3/x5059s/x5059S00.htm. [Accessed: March 30, 2022]

[35] Silva FA, Marsaioli A, Maximo GJ, Silva MAAP, Gonçalves LAG. Microwave assisted drying of macadamia Nuts. Journal of Food Engineering. 2006;**77**:550-558. DOI: 10.1016/j.jfoodeng.2005.06.068

[36] Pereira JFA. Análisis de los factores que intervienen sobre la resistencia al flujo de aire en granos y semillas en operaciones de manejo poscosecha. Ingeniería. 2007;**17**:113-124. DOI: 10.15517/ring.v17i1.678

[37] Alvarado M. Variables influyentes en el secado discontinuo del carbón asistido con microondas. PRO. 2012;**10**:37-42. DOI: 10.15665/rp.v10i2.231

[38] Rodríguez Sauceda R, Rojo Martínez GE, Martínez Ruiz R, Piña Ruiz HH, Ramírez Valverde B, Vaquera Huerta H, et al. Envases inteligentes para la conservación de alimentos. Ra Ximhai. 2014;**10**:151-174. DOI: 10.35197/rx.10.03.e2.2014.12.rr

[39] Meneses SMO, Valenzuela JRC. La atmósfera modificada: una alternativa para la conservación de los alimentos. Revista Lasallista de Investigación. 2008;5:112-123

[40] Iglesias EG, Cabezas LG, Fernández JL. Tecnologías de Envasado en Atmósfera Protectora. Informe de Vigilancia Tecnológica. Fundación para el conocimiento madri+d CEIM. Madrid: 143 p

[41] Ozdemir M, Devres O. Kinetics of color changes of hazelnuts during roasting. Journal of Food Engineering. 2000;**44**:31-38. DOI: 10.1016/S0260-8774(99)00162-4

[42] Dominguez IL, Azuara E, Vernon-Carter EJ, Beristain CI. Thermodynamic analysis of the effect of water activity on the stability of macadamia nut. Journal of Food Engineering. 2007;**81**:566-571

[43] Srichamnong W, Srzednicki G. Internal discoloration of various varieties of macadamia Nuts as influenced by enzymatic Browning and Maillard reaction. Scientia Horticulturae. 2015;**192**:180-186. DOI: 10.1016/j.scienta.2015.06.012

[44] Duduzile Buthelezi NM, Samukelo Magwaza L, Zeray Tesfay S. Postharvest pre-storage processing improves antioxidants, nutritional and sensory quality of macadamia Nuts. Scientia Horticulturae. 2019;**251**:197-208. DOI: 10.1016/j.scienta.2019.03.026

[45] Pedreschi F, León J, Mery D, Moyano P. Development of a computer vision system to measure the color of potato chips. Food Research International. 2006;**39**:1092-1098. DOI: 10.1016/j.foodres.2006.03.009

[46] Cernadas E, Fernández-Delgado M, González-Rufino E, Carrión P. Influence of normalization and color space to color texture classification. Pattern Recognition. 2017;**61**:120-138. DOI: 10.1016/j.patcog.2016.07.002

[47] Wu D, Sun D-W. Colour measurements by computer vision for food quality control – A review. Trends in Food Science and Technology. 2013;**29**:5-20. DOI: 10.1016/j.tifs.2012.08.004

[48] Yam KL, Papadakis SE. A simple digital imaging method for measuring and Analyzing color of food surfaces. Journal of Food Engineering. 2004;**61**:137-142. DOI: 10.1016/S0260-8774(03)00195-X

[49] Colvin BC, Rowland DL, Ferrell JA, Faircloth WH. Development of a digital analysis system to evaluate Peanut maturity. Peanut Science. 2014;**41**:8-16. DOI: 10.3146/PS13-9.1

[50] León K, Mery D, Pedreschi F, León J. Color measurement in L*a*b* units from RGB digital images. Food Research International. 2006;**39**:1084-1091. DOI: 10.1016/j.foodres.2006.03.006

[51] Mendoza F, Aguilera JM. Application of image analysis for classification of ripening bananas. Journal of Food Science. 2006;**69**:E471-E477. DOI: 10.1111/j.1365-2621.2004.tb09932.x

[52] Coronel E, Martinez K, Elizaur A, Resquin V, Velazquez E, Mereles L. Influencia del envasado en atmósferas modificadas sobre parámetros fisicoquímicos y la reacción de Maillard en nueces de macadamia secas. In: Proceedings of the IV Simposio de Química Analítica y Fisicoquímica; 8-9 october 2018; San Lorenzo; 2018. p. 1-6. Available from: http://200.10.229.229/files/publicaciones/2018_QIAF/2018-_NUECES-DEMACADAMIA-SECAS.pdf

[53] Mereles L, Coronel E, Martinez K, Caballero S, Elizaur M, Resquin V, Velazquez E. Desarrollo de un método de análisis de imagen digital para medición de color L*, a*, b* En Nueces de Macadamia. In: Proceedings of the INNOVA; 27-29 september 2017; Montevideo. LATU; 2017. p. 91

[54] Birch J, Yap K, Silcock P. Compositional analysis and roasting behaviour of Gevuina and macadamia Nuts: Roasting of Gevuina and macadamia Nuts. International Journal of Food Science and Technology. 2009;**45**:81-86. DOI: 10.1111/j.1365-2621.2009.02106.x

[55] Ferro JM. Introducción a Las Variedades de Aceite. 1st ed. Michigan: Independently Published; 2020. 177p.

[56] Ibars A, Barbosa G. Operaciones Unitarias En La Ingeniería de Alimentos. 1st ed. Madrid: Mundi-Prensa; 2005

[57] Yen H, Chiang W, Sun C. Supercritical fluid extraction of lutein from Scenedesmus cultured in an Autotrophical Photobioreactor. Journal of the Taiwan Institute of Chemical Engineers. 2012;**43**:53-57. DOI: 10.1016/j.jtice.2011.07.010

[58] Rój E. Supercritical Fluid Applications. Puławy: New Chemical Syntheses Institute; 2016

[59] Adil J, Çetin H, Yener M, Bayindirli A. Subcritical (carbon dioxide + ethanol) extraction of polyphenols from apple

and peach pomaces, and determination of the antioxidant activities of the extracts. Journal of Supercritical Fluids. 2007;**43**:55-63. DOI: 10.1016/j.supflu.2007.04.012

[60] Silva CF, Mendes MF, Pessoa FLP, Queiroz EM. Supercritical carbon dioxide extraction of macadamia (macadamia Integrifolia) nut oil: Experiments and Modeling. Brazilian Journal of Chemical Engineering. 2008;**25**:175-181. DOI: 10.1590/S0104-66322008000100018

[61] Ungar Y, Osundahunsi OF, Shimoni E. Thermal stability of Genistein and Daidzein and its effect on their antioxidant activity. Journal of Agricultural and Food Chemistry. 2003;**51**:4394-4399. DOI: 10.1021/jf034021z

[62] Brunner G. Supercritical fluids: Technology and application to food processing. Journal of Food Engineering. 2005;**67**:21-33. DOI: 10.1016/j.jfoodeng.2004.05.060

[63] Tuan D, Ilangantileke S. Liquid CO2 of essential oil from star Anise fruits. Journal of Food Engineering. 1996;**31**:47-57

[64] Feng Y, Meier D. Comparison of supercritical CO2, liquid CO2, and solvent extraction of chemicals from a commercial slow pyrolysis liquid of beech Wood. Biomass and Bioenergy. 2016;**85**:346-354. DOI: 10.1016/j.biombioe.2015.12.027

[65] Al-Hamimi S, Abellan Mayoral A, Cunico L, Turner C. Carbon dioxide expanded ethanol extraction: Solubility and extraction kinetics of -Pinene and cis-Verbenol. Analytical Chemistry. 2016;**88**:4336-4345. DOI: 10.1021/acs.analchem.5b04534

[66] Jesus AA, Almeida LC, Silva EA, Filho LC, Egues SMS, Franceschi E, et al. Extraction of palm oil using propane, ethanol and its mixtures as compressed solvent. Journal of Supercritical Fluids. 2013;**81**:245-253. DOI: 10.1016/j.supflu.2013.06.011

[67] Martin A, Mothes S, Mannsfeld G. Application of fluid opacity for determining the phase behavior of binary mixtures near their critical loci - CO2 plus ethane and CO2 plus propane. Fresenius' Journal of Analytical Chemistry. 1999;**364**:638-640. DOI: 10.1007/s002160051401

[68] Cotabarren, N. Ingeniería Del Equilibrio Entre Fases En Biorrefinerías de Base Oleaginosa [thesis]. Argentina: Universidad Nacional del Sur; 2017

[69] Velasco R, Villada H, Carrera J. Aplicaciones de Los Fluidos Supercríticos En La Industria. Rev. Inf. Tecnológica. 2007;**18**:53-66. DOI: 10.4067/S0718-07642007000100009

[70] Luque-García J, Luque de Castro M. Ultrasound-assisted Soxhlet extraction: An Expeditive approach for solid sample treatment: Application to the extraction of Total fat from oleaginous seeds. Journal of Chromatography. 2004;**1034**:237-242. DOI: 10.1016/j.chroma.2004.02.020

[71] Cherchi G, Deidda D, Gioannis B, Marongiu B, Pompei R, Porcedda S. Extraction of Santolina Insularis essential oil by supercritical carbon dioxide: Influence of some process parameters and biological activity. Flavour and Fragrance Journal. 2001;**16**:35-43. DOI: 10.1002/1099-1026(200101/02)16:1<35:AID-FFJ942>3.0.CO;2-Y

[72] Rodriguez P, Silva A, Carrillo M. Caracterización Fisicoquímica Del Aceite de Nuez de Macadamia (Macadamia Integrifolia). Physicochemical Characterization of Macadamia

Nut (Macadamia Integrifolia) oil. CyTA-Journal of Food. 2011;**9**:58-64. DOI: 10.1080/19476331003597097

[73] Codex Alimentarius. CXS 210-199. Standard for Named Vegetable Oils [Internet]. Roma: FAO/WHO; 1999. [consultado el 10 de junio de 2022]. p. 14. Available from: https://www.fao.org/fao-who-codexalimentarius/shproxy/ar/?lnk=1&url=https%253A%252F%252Fworkspace.fao.org%252Fsites%252Fcodex%252FStandards%252FCXS%2B210-1999%252FCXS_210e.pdf

[74] Rengel A, Pérez E, Piombo G, Ricci J, Servent A, Tapia M, et al. Lipid profile and antioxidant activity of macadamia Nuts (macadamia Integrifolia) cultivated in Venezuela. Natural Science. 2015;7:535-457. DOI: 10.4236/ns.2015.712054

[75] O'Hare T, Trieu H, Topp B, Russel D, Pun S, Torrisi C, et al. Assessing fatty acid profiles of macadamia Nuts. Hort Science. 2019;**54**:633-637. DOI: 10.21273/HORTSCI13418-18

[76] Moodley R, Kindness A, Jonnalagadda S. Elemental composition and chemical characteristics of five edible Nuts (Almont, Brazil, Peca, macadamia and walnut) consumed in Sothern Africa. Journal of Environmental Science and Health. 2007;**42**:585-591. DOI: 10.1080/03601230701391591

[77] Badui Dergal S. Química de los Alimentos. 4th ed. Estado de México: Pearson Education, Addison Wesley; 2006. 736 p

[78] Bora P, Ribeiro D. Influence of PH on the extraction yield and functional properties of macadamia (macadamia Integrifolia) protein isolates. Food Science and Technology International. 2004;**10**:263-267. DOI: 10.1177/1082013204045779

[79] Velazquez E, Smidt M, Correa L, Barrios L. Obtención de Carbón Activado a Partir de Cáscara de Macadamia. San Lorenzo; 2014

[80] Feng Y, Gong JL, Zeng GM, Niu QY, Zhang HY, Niu CG, et al. Adsorption of Cd (II) and Zn (II) from aqueous solutions using magnetic hydroxyapatite nanoparticles as adsorbents. Chemical Engineering Journal. 2010;**162**:487-494. DOI: 10.1016/j.cej.2010.05.049

[81] Dubey SP, Gopal K. Adsorption of chromium(VI) on low cost adsorbents derived from agricultural waste material: A comparative study. Journal of Hazardous Materials. 2007;**145**:465-470. DOI: 10.1016/j.jhazmat.2006.11.041

[82] Singh CK, Sahu JN, Mahalik KK, Mohanty CR, Mohan BR, Meikap BC. Studies on the removal of Pb(II) from wastewater by activated carbon developed from tamarind Wood activated with sulphuric acid. Journal of Hazardous Materials. 2008;**153**:221-228. DOI: 10.1016/j.jhazmat.2007.08.043

[83] Acharya J, Sahu JN, Mohanty CR, Meikap BC. Removal of Lead(II) from wastewater by activated carbon developed from tamarind Wood by zinc chloride activation. Chemical Engineering Journal. 2009;**149**:249-262. DOI: 10.1016/j.cej.2008.10.029

[84] Kumar R, Ofosu O, Anandjiwala RD. Macadamia nutshell powder filled poly lactic acid composites with Triacetin as a plasticizer. Journal of Biobased Materials and Bioenergy. 2013;7:541-548. DOI: 10.1166/jbmb.2013.1387

[85] Antal MJ, Grønli M. The art, science, and Technology of Charcoal Production. Industrial and Engineering Chemistry Research. 2003;**42**:1619-1640. DOI: 10.1021/ie0207919

[86] Wartelle LH, Marshall WE. Nutshells as granular activated carbons: Physical, chemical and adsorptive properties. Journal of Chemical Technology and Biotechnology. 2001;**76**:451-455. DOI: 10.1002/jctb.408

[87] Strezov V, Patterson M, Zymla V, Fisher K, Evans TJ, Nelson PF. Fundamental aspects of biomass carbonisation. Journal of Analytical and Applied Pyrolysis. 2007;**79**:91-100. DOI: 10.1016/j.jaap.2006.10.014

[88] Correa L, Velázquez E. Utilización de carbón activado producido a partir de cáscara de Macadamia sp para la remoción de microcistina presente en agua destinada al consumo humano en condiciones de laboratorio. Invest. y Est. - UNA [Internet]. 30 de junio de 2020 [citado 16 de junio de 2022];11(1):02-9. Available from: https://revistascientificas.una.py/index.php/rdgic/article/view/798

[89] Poinern GEJ, Senanayake G, Shah N, Thi-Le XN, Parkinson GM, Fawcett D. Adsorption of the Aurocyanide, Au (CN)2- complex on granular activated carbons derived from macadamia nut shells - a preliminary study. Minerals Engineering. 2011;**24**:1694-1702. DOI: 10.1016/j.mineng.2011.09.011

[90] Rodrigues LA, De Sousa Ribeiro LA, Thim GP, Ferreira RR, Alvarez-Mendez MO, Coutinho ADR. Activated carbon derived from macadamia nut shells: An effective adsorbent for phenol removal. Journal of Porous Materials. 2013;**20**:619-627. DOI: 10.1007/s10934-012-9635-5

[91] Pezoti Junior O, Cazetta AL, Gomes RC, Barizão ÉO, Souza IPAF, Martins AC, et al. Synthesis of ZnCl2-activated carbon from macadamia nut endocarp (macadamia Integrifolia) by microwave-assisted pyrolysis: Optimization using RSM and methylene blue adsorption. Journal of Analytical and Applied Pyrolysis. 2014;**105**:166-176. DOI: 10.1016/j.jaap.2013.10.015

Chapter 6

Papaya: The Versatile Tropical Fruit

Parichart Burns, Pimpilai Saengmanee and Uthaiwan Doung-Ngern

Abstract

Papaya (*Carica papaya L*) is a versatile tropical fruit with its usage ranging from consumption, cosmetics, to pharmaceuticals. In 2020, it was the third most-produced tropical fruit crop in the world. Papaya is a trioecious herbaceous plant with distinct flower and fruit morphological appearances. The fruits from hermaphrodite papaya are favorable for both consumption and processing due to their superior quality. Papaya has a genome size of 372 Mb and chromosome of 2n =18. The male and hermaphrodite papaya have XY and XYh sex chromosomes, respectively, while the female has XX. Using omics and bioinformatics approaches, papaya cultivars with desired fruit quality can be selected and identified from germplasm for incorporation in breeding programs. Papaya production can be done either in open fields or under protected cultivation. Open field cultivation provides for large-scale production, but with the disadvantages of variability in fruit yield, quality, and limitations on growing and harvesting seasons. Under protected cultivation, papaya can be cultivated in all seasons, whilst delivering higher yields. Conversely, multidisciplinary approaches with selected papaya cultivars, good farm management, and suitable conditions provide high yields of quality fruit for both consumption and processing, whilst minimizing the adverse effects related to environmental conditions.

Keywords: Papaya, *Carica*, Y-chromosome, PRSV, tropical fruit, cultivation, nutraceuticals

1. Introduction

Papaya (*Carica papaya* L) is a tall herbaceous plant native to the Americas, specifically Mexico, Central America, and tropical areas of South America regions [1, 2]. Indigenous people have known of papaya and managed its cultivation since pre-Columbus times [3]. Papaya fruit, leaves, seeds, and sap have all been utilized widely as food, food additives including papain, and packaging for cosmetics. Although there are no remnants of papaya tissue in the archaeological record, analysis of 497 indigenous plant species databases confirm that papaya was one of the food sources for Mayans [4]. Following Spanish contact with central and south America, papaya was gradually introduced to Africa, South Pacific Islands, and the rest of the world as fruit [5, 6]. Presently, it is found in tropical and sub-tropical regions around the world. Green papaya fruit, young leaves, and shoots are used in many traditional Asian dishes, including those in India, Malaysia, Indonesia, and Thailand, and are consumed either fresh or cooked [7–9].

Countries	Plantation area (Ha)	Production (Tones)
India	142,000	6,011,000
Dominican Republic	12,395	1,271,303
Brazil	28,450	1,235,003
Mexico	18,983	1,117,437
Indonesia	11,404	1,016,388
Nigeria	92,338	877,120
Democratic Republic of the Congo	12,404	210,000
Colombia	7,309	194,332
Peru	12,359	186,508
Thailand	4,234	164,360

Table 1.
Top ten papaya production countries in 2020 [10].

Countries	Import value (US$)
United States	$134.2M
Germany	$33.3M
Portugal	$24.9M
Canada	$23.0M
Netherlands	$15.2M
Spain	$14.8M
France	$12.5M
United Kingdom	$11.6M
Singapore	$9.1M
Italy	$7.2M

Table 2.
Top ten papaya importers in 2020 by value [10].

In 2020, papaya was ranked the third most-produced tropical fruit crop in the world [10]. The major producing countries include India, the Dominican Republic, Brazil, Mexico, and Indonesia (**Table 1**). Papaya is also a highly traded fruit on international markets in fresh and processed form with the major importers being the USA, Germany, and Portugal (**Table 2**) [10].

2. Classification

Papaya is a member of the Family Caricaceae of which there are six genera *Carica*, *Jarilla*, *Horovitzia*, *Jacaratia*, *Vasconcella*, and *Cylicomorpha*. Papaya is a member of the Genus *Carica* of which there is only one species, *Carica papaya*. There are a total of 35 species in the Family Caricaceae; *Carica* (one), *Jarilla* (three), *Horovitzia* (one), *Jacaratia* (eight), *Vasconcella* (20), and *Cylicomorpha* (two). Although papaya is the only well-known edible fruit, other genera such as *Vasconcellea* (also called mountain papaya), *Jarilla* and *Jacaratia* are also consumed as fruit in Central America [3]. With the exception of the Genus *Cylicomorpha* which is native to Africa, all are native to the Americas.

2.1 Morphology and general characters

Plants in the family Caricaceae are stout-stemmed trees and exudate latex-like substances. Their leaves are palmately compound or lobed. The inflorescences are axillary and cymose and the flowers usually have fused petals. The fruits consist of numerous seeds surrounded by mucilage [11]. Of all the species in the family Caricaceae, papaya (*Carica papaya*) is the most well-known species ostensibly due to its fruit. Papaya is known by various names including pawpaw (Australia), Malagor (Thailand), tree melon (Brazil), and Fruitabomba (Cuba) [12]. Papaya is a herbaceous plant and, depending on the variety of which there are many, grows to a height of up to ten meters in height. The leaves are palmately-lobed with long and hollow petioles and the blades are divided into 5–9 segments. The flower buds are developed at the axils of the leaves. The fertilized fruit consists of up to 1000 seeds. The fruit skin is green at the unripened stage and turns yellow-orange when ripens. Generally, papaya plants have a life span of between five and ten years [13].

Most species in the family Caricaceae are dioecious. One is monoecious and two, *C. papaya* and *V. cundimarsensis* are trioecious [14]. Although the morphology of male, female and hermaphrodite papaya plants are very similar, their flowers and fruits were distinct. The male papaya plants produce small flowers in clusters with long peduncles and produce no or very small fruits. The female plants have large and round flowers while the fruits are round or ovule. The hermaphrodite flowers are cylindrical and produce cylindrical fruits. The fruits from hermaphrodite papaya have superior quality (size, shape, and flesh thickness) than those from female papaya plants. Based on fruit types, papaya cultivars can be divided into Solo and Formosa types. The fruits from the Solo group are small (500–700 g) with oval or pear shape, while those of the Formosa group are medium to large (≥1000 g) with cylindrical shape [15].

2.2 Genomics

Papaya (*Carica papaya*) has a relatively small genome of approximately 372 Mb across 18 chromosomes (2n = 2x = 18). Papaya chromosomes consist of autosomes and sex chromosomes. Male papaya has the sex chromosome XY and female XX, while the hermaphrodite papaya has XY^h. The nucleotide composition of the papaya genome is typical of dicot plants with a GC content of 36.51% GC and AT content of 63.49% [16]. The draft genomic sequence of a genetically modified variety of the female papaya, "SunUp", which was derived from the Hawaiian inbred cultivar "Sunset", was published in 2008 [17].

2.2.1 Sex determination

The papaya Y-chromosome deviated from the X-chromosome through deletions. Male-specific regions accounted for approximately 13% of Y-chromosome and share 99.6% of identity in male (MSY) and hermaphrodite (HSY) papaya [18, 19]. It consists of four knobs like heterochromatic structure and is heavily methylated

[18]. Expression of genes linked to X, Y and Yh chromosomes showed evidence of partial dosage compensation in X-link loci and a candidate gene associated with papaya sex determination and the transition to hermaphroditism, a homolog of the MADS-box protein short vegetative phase (SVG) [20, 21]. The dosage compensation of gene expression in papaya sex chromosomes was investigated further in female and male papaya and found to be at a gene by gene level. In addition, expression of most X-hemizygous genes was very low or none suggesting the role of gene silencing in controlling of transcriptional balance [22]. Recently, the landscapes of DNA methylation and transcriptomes were shown to be different in male and female papaya [23].

Using sequence information derived from papaya sex chromosomes, sex-specific primers were designed and used to screen plantlets/seedlings to identify fruit-bearing female and hermaphrodite types from males (MSY) [24]. More recently, a candidate gene, monodehydroascorbate reductase 4 (MDAR4), was identified from H-TSS No.7 line with X-chromosome mutant (3 bp deletion) resulting in all hermaphrodite progeny. MDAR4 is involved in a hydrogen peroxide scavenging pathway [25]. The marker developed from this gene has potential applications in papaya breeding, selection of potential lines for in vitro clonal propagation, and the production of high-quality commercial varieties of papaya seedlings.

2.3 Agronomic characteristics

Target genes related to papaya's important agronomic traits, including tolerance/resistance to abiotic and biotic stresses and fruit quality, were explored through omics and bioinformatics [26]. The papaya genome includes NBS genes which are disease-resistant genes with nucleotide-binding site motifs in the Toll/interleukin-1 receptor (TIR) and non-TIR subclasses [27]. Transcriptome profiles in young leaves of papaya ringspot virus (PRSV) resistant genetically modified variety "Sunup" showed high expression of several transcription factors (TFs) including MYB, ERF, WRKY, NAC, transporter proteins, and hormone-related proteins compared to susceptible "Sunset" papaya plants [28]. Under mild drought stress, stress-responsive genes were differentially expressed in papaya tissues with genes related to cell cycle and DNA repair processes. These stress-responsive genes were up-regulated in papaya leaves and sap while genes related to hormone signaling and sucrose metabolism were up-regulated in roots. Under severe drought stress genes related to oxidation-reduction, abiotic stress responses, and hormone signaling were also found to be up-regulated in all tissues [29]. Drought tolerant papaya had more photosynthetic II (PSII) efficiency than susceptible papaya. Drought susceptible plants displayed greater leaf abscission, less turgid shoots, and lower plant growth than those of tolerant papaya. Molecular analysis identified six transcription factors including *CpHSF*, *CpMYB*, *CpNAC*, *CpNFY-A*, *CpERF*, and *CpWRKY* that were highly expressed in tolerant papaya [30]. These genes were reportedly also involved in drought tolerance in rice and maize [31, 32]. Two transcription factors, RAP2.4 and DREB2 belonging to the ethylene response AP2/ERF family, have also been linked to extreme temperature responses in papaya. Overexpression of these genes in transformed tobaccos resulted in the cold (4°C) and heat tolerance (40°C) [33, 34]. In the regulation of fruit development and ripening, the papaya SQUAMOSA promoter binding protein Cp-SPL was found to be differentially expressed and cpmiRNA156 appears to play a critical role [35]. While in the carotenoid biosynthetic pathway, critical in the color development of papaya fruit, transcription factors HLH1 and HLH2 appear to regulate the transcription of lycopene β-cyclase genes [36].

3. Papaya responses to environment stresses

Environmental factors including soil, temperature, and water availability are external factors that significantly impact plant growth and development. The increased demand for papaya fruit as a source of food and plant-derived products and the marginalization of land available for cultivation due to increased housing needs has pushed papaya cultivation to less productive farms or in more developed countries, expensive protective housing cultivation. Further, climate change involving prolonged periods of adverse temperatures and extremes in weather patterns are contributing to the environmental stresses on papaya cultivation. These changes are further limiting not just the productivity, but also the zones of cultivation. Traditional cultivation areas are realizing significant reductions in yield or are forced to grow alternative crops. Papaya plants have optimal cultivation temperatures of between 25°C and 30°C. Temperature, moisture, light, and wind are also major environmental factors impacting papaya production [37, 38]. Temperatures lower than 16°C and higher than 36°C for extended periods negatively impact plant growth [39]. Under these climate extremes, different plant tissue types and organs, including roots, leaves, flowers, and fruit, exhibit variations in responses [40]. The use of traditional genotypes with desired plant fitness and plant developmental stages, fruit types, and yields are being negatively impacted pushing papaya breeding programs to develop varieties that are more adaptable and stable to seasonal changes [41].

In one study, the performance of a number of papaya cultivars including Solo, Formosa, and local commercial hybrids over two harvest seasons was compared. The results indicated that the summer harvest season with average temperatures of 24.9°C and maximums of 34.2°C were more productive than winter harvests where the average temperature was 21.9°C and the minimum 13.4°C [42]. Under low temperatures (<11°C), papaya plants produced fewer new leaves with no fruit set [43]. The optimal temperature for germination of papaya pollen was between 20 and 25°C at 72–80% humidity [44]. Further, extreme temperatures below 15°C and above 30°C negatively impacted germination with rates dropping to between 0 and 56%. Upon applying heat stress to papaya plants, it was shown that plants recovered from mild (37–41°C) and moderate (46°C), but not severe (49°C) heat stress. Photosynthesis was delayed while stress volatile production was induced [45].

Water stress has also been demonstrated to negatively impact papaya growth and development. For short periods of water stress, papaya leaves become droopy. If the water shortage continues, papaya plants will drop flower buds, and delay new fruit, flower, and leaf production. Water stress also results in leaf water potential and a reduction of stomatal opening. This, in turn, reduces carbon dioxide availability and consequently photosynthesis [46]. As a result of limited water availability, seed germination in many papaya cultivars is also delayed [47]. The genotype "Golden" papaya with less chlorophyll content outperformed high chlorophyll "Alianca" papaya under limited water availability [48]. Drought tolerant papaya was shown to have greater photosynthetic II (PSII) efficiency than susceptible ones. Susceptible plants displayed greater leaf abscission, less turgid shoots, and lower plant growth than those of more tolerant ones [30].

4. Papaya production

Papaya can be grown from seeds and vegetative tissues such as cuttings, grafting, and in vitro culture. The somatic embryos and somatic tissues can also be micro-propagated [49]. In many countries including India, Bangladesh, and Malawi papaya seeds are collected by growers from open-pollinated varieties of both female and hermaphrodite types and cultivated with little or no fertilization, irrigation, insect, or pathogen control. As a result, the yield and quality generally are variable as too the phenotype. Fruits are harvested and consumed within the household and are an important source of dietary fiber [50–52].

4.1 Open field

Commercial production of papaya is traditionally done under large open-field conditions. Selected cultivars, both inbred and hybrid varieties with the desired market characteristics of fruit color, weight, size, shape, and texture, are cultivated from seedlings. The plants are well maintained being fertigated, rouged to remove both off types as well as competing for vegetation, and sprayed with pesticides, fungicides, and other protective applications against insects and pathogens which negatively impact yield. Commercial papaya plantations are either rain-fed or irrigated by furrow, drip, sprinkler, or other mechanical means [53]. Integrated farm management has been practiced in some countries and has been shown to enhance papaya fruit yields and net return for growers even when compared to traditional management techniques [54]. Papaya fruit is harvested by hand using experienced pickers or, in some more developed countries, using mechanical harvesters. Fruit can be treated before packaging for long-distance transport. The use of cold room storage provides for the extended availability of fruit in the market and allows for farmers and wholesalers to also penetrate nontraditional markets and seasons offering significantly higher returns.

Large-scale open field conditions are the preferred cultivation for most papaya commercialization as it provides large-scale production with low to medium investment and operating costs. The drawbacks are less consistent fruit quality and yields as a result of seasonal variations and unexpected weather conditions such as flooding, in addition to physical damage to fruit as a result of environmental conditions, insects, and disease. Favorable cultivation conditions for papaya plants include cultivation temperatures of between 20°C and 30°C with a relative humidity of 66%, well-drained soil with a pH of between 6.0 and 6.5, low wind, adequate irrigation, and a balanced fertigation regime preferably via a drip irrigation system. Papaya plants are sensitive to frost with yields being negatively affected both through reduced temperatures and fruit quality. Generally, plants are productive after approximately nine months of transplanting and will yield for between two and four years, depending on the variety as well as weather conditions and inputs. Papaya is susceptible to a range of diseases and pests depending on the region. Papaya ringspot disease, caused by PRSV, is one of the most severe diseases and results in significant losses. Genetically modified PRSV resistant papaya varieties have been developed and have been found to be effective in controlling the disease [55]. Resistant varieties have been developed in a number of countries both on a research and commercial basis including the United States, Australia, Taiwan, China, India, Thailand, and The Philippines. In the United States two cultivars, Rainbow and SunUp, have been released on a commercial basis [56]. More recently, the use of gene mutation technologies, including CRISPR-Cas9, has allowed the mutation of cell receptors in papaya that facilitate cell infection by PRSV (as well in a range of potyvirus susceptible plants including species of *Capsicum*), thus rendering the mutated plant resistant.

4.2 Protected cultivation

Papaya production under protected conditions has been widely adopted using a variety of modifications applicable to local climatic conditions and cultivars used. The use of greenhouses with full climate control environments provides for year-round, high-quality fruit with maximum yields, albeit at very high capital input in addition to higher operating costs compared to open or protected field cultivation. In India, "Red Lady" papaya growing under greenhouse conditions performed well with reduced insect infestation and disease, and improved fruit quality [57, 58]. A combination of short stature cultivars and greenhouse conditions in Argentina and similar temperate regions allows for year-round papaya cultivation [59]. The use of closed plastic tunnels in the subtropical areas of Europe, including the Mediterranean and Canary Islands, has been successful in producing high yields. Similar methods have also been employed in Turkey where there is a widespread use of protected cultivation and greenhouses for a large range of crops [60]. A comparison of papaya cultivation and harvesting periods throughout the year indicated that greenhouse conditions result in the production of more uniform fruit quality in part due to a uniform season [61]. In southeastern Spain, five locally grown commercial papaya varieties were cultivated in multi-tunnel greenhouses covered with low-density polyethylene over fixed periods of 456 days and were considered a commercial success [62]. Under similar conditions and in the same region, five commercial cultivars of various geographic origins, have different plant and fruit types. Two varieties, "Siluet" and "Sensation", have been specially selected under greenhouse conditions and are both high yielding with the fruit of optimal quality for the European market including size, shape, weight, and importantly, total soluble solids (TSS), a major factor in determining sweetness. The greenhouses, with active climate control (ACC) incorporating both cooling and heating systems, enhanced "Siluet" papaya plant growth, flowering, fruit set, and yield resulting in doubling yields with both more and heavier fruit. Additionally, fruit quality factors including skin color, acidity, and TSS were not affected. Protected cultivation and the use of greenhouses offer an affordable and cost-effective strategy for papaya cultivation, especially in regions where open field cultivation, whether due to climate, soil, or other factors, is not feasible [63]. The compactness of the protected system producing large volumes of high quality, uniform fruit, aligned with readily available packaging and transport facilities lends itself to supplying both long and short distance high-value markets.

5. Nutraceuticals

Fruits, leaves, and seeds of papaya have a long history of human consumption and use. Fruit pulp is widely known for its nutritional value while the leaves and seeds are used in cooking in many cultures. Papaya pulp consists of macronutrients (protein, carbohydrate, lipid), fiber, minerals, carotenoids, and vitamins A, B, and C. Carotenoids from papaya are more bioavailable for human nutrition than those from tomato and carrot [64]. Other papaya tissues including leaves and seeds are of high nutrition, although with reduced levels of vitamins. Phytochemical analysis of papaya pulp reveals significant levels of phenols, terpenols, alkaloids, flavonoids, and saponins. Papaya leaves contain alkaloids, carpain, pseudocarpain and dehyrocarpaine, choline, carposide, saponins, pro-anthocyanin, benzyl isothiocyanate, while papaya seeds contain papaya oil, carpaine, benzyl isothiocyanate, benzyl glucosinolate, glucotropacolin, benzylthiourea, hentriacontane, and β-sitostrol. Papaya oil contains oleic acid (72.5%) and palmitic acid (12.5%) [49, 65]. Caffeic acid, myricetin, rutin, quercetin, α-tocopherol, benzyl isothiocyanate (BiTC), and kaempferol have been identified in papaya, all of which have antioxidant activities and in the plant either promote antioxidant enzyme expression or reduce reactive oxygen species (ROS) production [66]. Alkaloids, flavonoids, saponins, and oleic acid all poses anti-inflammatory activities [12, 67, 68]. Alkaloids, flavonoids, and saponins in papaya have been shown to inhibit the bacterial growth of *Enterobacter cloacae*, *Pseudomonas aeruginosa*, *Proteus vulgaris*, *Klebsiella pneumoniae*, and *Bacillus subtilis* [69–71]. Finally, papaya latex, often considered a nuisance and irritatant, has been shown to actively inhibits the growth of gram-negative bacteria [72].

6. Conclusions

Lifestyles in the twenty-first century (and beyond) are setting strikingly different needs and demands from those of earlier periods. Twenty-first century consumers have higher expectations with respect to fruit type and quality; more flavorful, of higher nutritional value with additional benefits to both health and to the environment. Growers are constantly under pressure in meeting not just the demand for quantity but, more importantly, quality and overall consumer satisfaction. With the ready availability of papaya genomics and transcriptome databases, desired agronomic characters associated with fruit quality, yield, and plant adaptation can be identified in germplasm and incorporated into breeding programs [73, 74]. Papaya can be produced "for all seasons" using open field cultivation under favorable conditions and in protected cultivation systems and climate-controlled greenhouses regardless of climate and region.

Papaya cultivation under open field conditions offers large-scale production with low to medium capital inputs and low operating expenses, offsetting the issues of inconsistent quality and quantity due to seasonal changes, disease, and abnormal weather events. Protected cultivation and greenhouses with controlled environments provide more expensive papaya fruit year-round, but in higher yields and into markets demanding higher quality and capable of absorbing the additional costs. These conditions provide flexibility for papaya production under varying conditions and locations.

Metabolites with nutritional values such as vitamin C and carotenoids in papaya can be consumed directly because of their high concentration in papaya flesh. Metabolites extracted from papaya fruit, including Papain, an enzyme widely known and used in the food industry and cosmetics, offer additional markets for the fruit. Other metabolites including antioxidants can be purified from papaya seed, flesh, and leaves and there are other active constituent molecules that are the subject of evaluation and many no doubt many others yet to be explored. With multiple applications and the potential for the increased demand for papaya fruit and products, papaya is undoubtedly the fruit for the future.

Conflict of interest

The authors declare no conflict of interest.

DOI: http://dx.doi.org/10.5772/intechopen.104624

Author details

Parichart Burns[1,2,3]*, Pimpilai Saengmanee[2,3] and Uthaiwan Doung-Ngern[4]

1 National Center for Genetic Engineering and Biotechnology, National Science, Technology and Development Agency (NSTDA), Thailand Science Park, Pathum Thani, Thailand

2 Center for Agriculture Biotechnology, Kasetsart University, Nakhon Pathom, Thailand

3 Center of Excellence on Agricultural Biotechnology: (AG-BIO/MHESI), Bangkok, Thailand

4 Faculty of Agriculture at Kamphaengsaen, Department of Horticulture, Kasetsart University Kamphaengsaen, Nakhon Pathom, Thailand

*Address all correspondence to: p.burns@biotec.or.th

References

[1] Mapes C, Basurto F. Biodiversity and edible plants of Mexico. In: Ethnobotany of Mexico. Springer; 2016. pp. 83-131

[2] Janick J. Development of new world crops by indigenous Americans. HortScience. 2013;**48**(4):406-412

[3] Badillo VM, Leal F. Taxonomy and Botany of the Caricaceae. Horticultural Review. 2020;**47**:289-323

[4] Fedick SL, Santiago LS. Large variation in availability of Maya food plant sources during ancient droughts. Proceedings of the National Academy of Sciences. 2022;**119**(1)

[5] Voeks R. Ethnobotany of Brazil's African diaspora: The role of floristic homogenization. In: African Ethnobotany in the Americas. Springer; 2013. pp. 395-416

[6] Galán Saúco V, Cubero JI. Contribution of Spain and Portugal to the exchange and acclimatization of new and old world crops. In: XXVIII International Horticultural Congress on Science and Horticulture for People (IHC2010). 2010. pp. 71-82

[7] Khanthapok P, Sukrong S. Anti-aging and health benefits from Thai food: Protective effects of bioactive compounds on the free radical theory of aging. Journal of Food Health and Bioenvironmental Science. 2019;**12**:54-67

[8] Nguanchoo V, Srisanga P, Swangpol S, Prathanturarug S, Jenjittikul T. Food plants in Hmong cuisine in Northern Thailand. Thailand Journal of Botany. 2014;**6**:131-145

[9] Harmayani E, Anal AK, Wichienchot S, Bhat R, Gardjito M, Santoso U, et al. Healthy food traditions of Asia: Exploratory case studies from Indonesia, Thailand, Malaysia, and Nepal. Journal of Ethnic Foods. 2019;**6**(1):1-18

[10] FAOSTAT. FAO Statistics [Internet]. 2022. Available from: https://www.fao.org/faostat/en/#data [Accessed: February 18, 2022]

[11] Olson ME. Intergeneric relationships within the Caricaceae-Moringaceae clade (Brassicales) and potential morphological synapomorphies of the clade and its families. International Journal of Plant Sciences. 2002;**163**(1):51-65

[12] Wadekar AB, Nimbalwar MG, Panchale WA, Gudalwar BR, Manwar JV, Bakal RL. Morphology, phytochemistry and pharmacological aspects of *Carica papaya*, an review. GSC Biological and Pharmaceutical Sciences. 2021;**14**(3):234-248

[13] Singh SP, Rao DVS. Papaya (*Carica papaya* L.). Postharvest Biology and Technology of Tropical and Subtropical Fruits. 2011:86-126e

[14] Aryal R, Ming R. Sex determination in flowering plants: Papaya as a model system. Plant Science. 2014;**217**:56-62

[15] de Oliveira EJ, Oliveira GAF, Costa JL, de Oliveira VJ, Dantas ACVL, Dantas JLL, et al. Genetic diversity and marker-assisted inbreeding in papaya. Science Horticulture (Amsterdam). 2012;**147**:20-28

[16] Araújo FS, Carvalho CR, Clarindo WR. Genome size, base composition and karyotype of *Carica papaya* L. Nucleonics. 2010;**53**(1):25-31

[17] Ming R, Hou S, Feng Y, Yu Q, Dionne-Laporte A, Saw JH, et al. The draft genome of the transgenic tropical fruit tree papaya (*Carica papaya* Linnaeus). Nature. 2008;**452**(7190): 991-996

[18] Zhang W, Wang X, Yu Q, Ming R, Jiang J. DNA methylation and heterochromatinization in the male-specific region of the primitive Y chromosome of papaya. Genome Research. 2008;**18**(12):1938-1943

[19] VanBuren R, Zeng F, Chen C, Zhang J, Wai CM, Han J, et al. Origin and domestication of papaya Yh chromosome. Genome Research. 2015;**25**(4):524-533

[20] Lee C-Y, Lin H-J, Viswanath KK, Lin C-P, Chang BC-H, Chiu P-H, et al. The development of functional mapping by three sex-related loci on the third whorl of different sex types of *Carica papaya* L. PLoS One. 2018;**13**(3):e0194605

[21] Chae T, Harkess A, Moore RC. Sex-linked gene expression and the reversion to hermaphroditism in *Carica papaya* L.(Caricaceae). bioRxiv. 2020

[22] Liu J, Han J, Sharma A, Wai CM, Ming R, Yu Q. Transcriptional regulation of dosage compensation in *Carica papaya*. Scientific Reports. 2021;**11**(1):1-9

[23] Zhou P, Zhang X, Fatima M, Ma X, Fang H, Yan H, et al. DNA methylome and transcriptome landscapes revealed differential characteristics of dioecious flowers in papaya. Horticultural Research. 2020;7

[24] Deputy J, Ming R, Ma H, Liu Z, Fitch M, Wang M, et al. Molecular markers for sex determination in papaya (*Carica papaya* L.). Theoretical and Applied Genetics. 2002;**106**(1):107-111

[25] Chen J-R, Ueno H, Matsumura H, Urasaki N, Lee C-Y, Chen F-C, et al. Genomic characterization of a rare *Carica papaya* X chromosome mutant reveals a candidate monodehydroascorbate reductase 4 gene involved in all-hermaphrodite phenomenon. Molecular Genetics and Genomics. 2021;**296**(6):1323-1335

[26] Zainal-Abidin R-A, Ruhaizat-Ooi I-H, Harun S. A review of omics technologies and bioinformatics to accelerate improvement of papaya traits. Agronomy. 2021;**11**(7):1356

[27] Porter BW, Paidi M, Ming R, Alam M, Nishijima WT, Zhu YJ. Genome-wide analysis of *Carica papaya* reveals a small NBS resistance gene family. Molecular Genetics and Genomics. 2009;**281**(6):609-626

[28] Fang J, Lin A, Qiu W, Cai H, Umar M, Chen R, et al. Transcriptome profiling revealed stress-induced and disease resistance genes up-regulated in PRSV resistant transgenic papaya. Frontiers in Plant Science. 2016;7:855

[29] Gamboa-Tuz SD, Pereira-Santana A, Zamora-Briseño JA, Castano E, Espadas-Gil F, Ayala-Sumuano JT, et al. Transcriptomics and co-expression networks reveal tissue-specific responses and regulatory hubs under mild and severe drought in papaya (*Carica papaya* L.). Scientific Reports. 2018;**8**(1):1-16

[30] Estrella-Maldonado H, Ramírez AG, Ortiz GF, Peraza-Echeverría S, Martinez-de La Vega O, Góngora-Castillo E, et al. Transcriptomic analysis reveals key transcription factors associated to drought tolerance in a wild papaya (*Carica papaya*) genotype. PLoS One. 2021;**16**(1):e0245855

[31] Song K, Kim HC, Shin S, Kim K-H, Moon J-C, Kim JY, et al. Transcriptome

analysis of flowering time genes under drought stress in maize leaves. Frontiers in Plant Science. 2017;**8**:267

[32] Mallikarjuna G, Mallikarjuna K, Reddy MK, Kaul T. Expression of OsDREB2A transcription factor confers enhanced dehydration and salt stress tolerance in rice (*Oryza sativa* L.). Biotechnology Letters. 2011;**33**(8):1689-1697

[33] Figueroa-Yañez L, Pereira-Santana A, Arroyo-Herrera A, Rodriguez-Corona U, Sanchez-Teyer F, Espadas-Alcocer J, et al. RAP2. 4a is transported through the phloem to regulate cold and heat tolerance in papaya tree (*Carica papaya* cv. Maradol): Implications for protection against abiotic stress. PLoS One. 2016; **11**(10):e0165030

[34] Arroyo-Herrera A, Figueroa-Yánez L, Castano E, Santamaría J, Pereira-Santana A, Espadas-Alcocer J, et al. A novel Dreb2-type gene from *Carica papaya* confers tolerance under abiotic stress. Plant Cell, Tissue and Organ Culture. 2016;**125**(1):119-133

[35] Xu Y, Xu H, Wall MM, Yang J. Roles of transcription factor SQUAMOSA promoter binding protein-like gene family in papaya (*Carica papaya*) development and ripening. Genomics. 2020;**112**(4):2734-2747

[36] Zhou D, Shen Y, Zhou P, Fatima M, Lin J, Yue J, et al. Papaya CpbHLH1/2 regulate carotenoid biosynthesis-related genes during papaya fruit ripening. Horticultural Research. 2019;**6**:80

[37] Nishina M, Zee F, Ebesu R, Arakaki A, Hamasaki R, Fukuda S, et al. Papaya production in Hawaii. 2000

[38] Kanatiwela-de Silva C, Damayanthi M, de Silva N, Wijesekera R, Dickinson M, Weerakoon D, et al. Immunological detection of the Weligama coconut leaf wilt disease associated phytoplasma: Development and validation of a polyclonal antibody based indirect ELISA. PLoS One. 2019;**14**(4):e0214983

[39] Saran PL, Choudhary R. Chapter 17. Advances in papaya cultivation. In: Achieving Sustainable Cultivation of Tropical Fruits. 1st ed. Cambridge: Burleigh Dodds Science Publishing Limited; 2020

[40] Gray SB, Brady SM. Plant developmental responses to climate change. Developmental Biology. 2016;**419**(1):64-77

[41] da Luz LN, Santa-Catarina R, Barros GB, Barros FR, Vettorazzi JCF, Pereira MG. Adaptability and stability of papaya hybrids affected by production seasonality. Crop Breeding and Applied Biotechnology. 2018;**18**:357-364

[42] de Barros FL, Kuhlcamp KT, Arantes SD, Moreira SO. Productivity and quality of Formosa and Solo papaya over two harvest seasons. Pesquisa Agropecuária Brasileira. 2017;**52**:599-606

[43] Allan P. *Carica papaya* responses under cool subtropical growth conditions. International Symposium on Tropical and Subtropical Fruits. 2000;**575**(26):757-763

[44] Tamaki M, Urasaki N, Sunakawa Y, Motomura K, Adaniya S. Seasonal variations in pollen germination ability, reproductive function of pistils, and seeds and fruit yield in papaya (*Carica papaya* L.) in Okinawa. Journal of the Japanese Society for Horticultural Science. 2011;**80**(2):156-163

[45] Okereke CN, Kaurilind E, Liu B, Kanagendran A, Pazouki L, Niinemets Ü. Impact of heat stress of varying severity

on papaya (*Carica papaya*) leaves: Major changes in stress volatile signatures, but surprisingly small enhancement of total emissions. Environmental and Experimental Botany. 2022;**195**:104777

[46] Osakabe Y, Osakabe K, Shinozaki K, Tran L-SP. Response of plants to water stress. Frontiers in Plant Science. 2014;**5**:86

[47] Mengarda LHG, Lopes JC, Alexandre RS, Zanotti RF. Germination performance of papaya genotypes subjected to salinity, water restriction and high irradiance stress. Australian Journal of Crop Science. 2016;**10**(1):7-15

[48] Ruas KF, Baroni DF, de Souza GAR, de Paula BW, Paixão JS, dos Santos GM, et al. A *Carica papaya* L. genotype with low leaf chlorophyll concentration copes successfully with soil water stress in the field. Science Horticulture (Amsterdam). 2022;**293**:110722

[49] Santana LF, Inada AC, Espirito Santo BLS, WFO F, Pott A, Alves FM, et al. Nutraceutical potential of *Carica papaya* in metabolic syndrome. Nutrients. 2019;**11**(7):1608

[50] Arya S, Prakash S, Joshi S, Tripathi KM, Singh V. Household food security through kitchen gardening in rural areas of western Uttar Pradesh, India. International Journal of Current Microbiology and Applied Sciences. 2018;**7**(2):468-474

[51] Ferdous Z, Datta A, Anal AK, Anwar M, Khan ASMMR. Development of home garden model for year round production and consumption for improving resource-poor household food security in Bangladesh. NJAS-Wageningen Journal of Life Sciences. 2016;**78**:103-110

[52] Jones AD, Shrinivas A, Bezner-Kerr R. Farm production diversity is associated with greater household dietary diversity in Malawi: Findings from nationally representative data. Food Policy. 2014;**46**:1-12

[53] Janthasri R, Chaiayaboon W. Present status of production and martketing of papaya for unripe consumption in Northeastern Thailand. International Journal of Agriculture Environment and Biotechnology. 2016;**9**(2):323-332

[54] Kulkarni S, Rathod A. Performance of papaya (*Carica papaya* L.) under integrated crop management practices. Journal of Pharmacognosy Phytochemistry. 2020;**9**(5):213-215

[55] Lius S, Manshardt RM, Fitch MMM, Slightom JL, Sanford JC, Gonsalves D. Pathogen-derived resistance provides papaya with effective protection against papaya ringspot virus. Molecular Breeding. 1997;**3**(3):161-168

[56] Lobato-Gómez M, Hewitt S, Capell T, Christou P, Dhingra A, Girón-Calva PS. Transgenic and genome-edited fruits: Background, constraints, benefits, and commercial opportunities. Horticultural Research. 2021;**8**

[57] Reddy PV, Gowda VN. Influence of greenhouse cultivation on fruit quality of 'Red Lady'papaya. International Symposium on Tropical and Subtropical Fruits. 2011;**1024**(19):109-114

[58] Kaur K, Kaur A. Papaya performance under various growing conditions cv. Red Lady 786. Agricultural Science Digest Research Journal. 2017;**37**(4):290-293

[59] Céccoli G, Panigo E, Gariglio NF, Favaro JC, Bouzo CA. Fruit yield and growth parameters of several *Carica papaya* L. genotypes in a temperate climate. 2013

[60] Gunes E, Gübbük H. Growth, yield and fruit quality of three papaya cultivars

grown under protected cultivation. Fruits. 2012;**67**(1):23-29

[61] Güneş E, Öziyci HR, Gübbük H. Quality of different papaya cultivars grown in the greenhouse throughout the year in subtropical regions. Food Science and Engineering. 2022:20-30

[62] Honoré MN, Belmonte-Ureña LJ, Navarro-Velasco A, Camacho-Ferre F. The production and quality of different varieties of papaya grown under greenhouse in short cycle in continental Europe. International Journal of Environmental Research and Public Health. 2019;**16**(10):1789

[63] Salinas I, Hueso JJ, Cuevas J. Active control of greenhouse climate enhances papaya growth and yield at an affordable cost. Agronomy. 2021;**11**(2):378

[64] Schweiggert RM, Kopec RE, Villalobos-Gutierrez MG, Högel J, Quesada S, Esquivel P, et al. Carotenoids are more bioavailable from papaya than from tomato and carrot in humans: A randomised cross-over study. The British Journal of Nutrition. 2014;**111**(3):490-498

[65] Kadiri O, Olawoye B, Fawale OS, Adalumo OA. Nutraceutical and antioxidant properties of the seeds, leaves and fruits of *Carica papaya*: Potential relevance to humans diet, the food industry and the pharmaceutical industry—A review. Turkish Journal of Agricultural Science and Technology. 2016;**4**(12):1039-1052

[66] Kong YR, Jong YX, Balakrishnan M, Bok ZK, Weng JKK, Tay KC, et al. Beneficial role of *Carica papaya* extracts and phytochemicals on oxidative stress and related diseases: A mini review. Biology (Basel). 2021;**10**(4):287

[67] Salas CE, Gomes MTR, Hernandez M, Lopes MTP. Plant cysteine proteinases: Evaluation of the pharmacological activity. Phytochemistry. 2008;**69**(12):2263-2269

[68] Dotto JM, Abihudi SA. Nutraceutical value of *Carica papaya*: A review. Science African. 2021;**13**:e00933

[69] Gunde MC, Amnerkar ND. Nutritional, medicinal and pharmacological properties of papaya (*Carica papaya* linn.): A review. Journal of Innovations in Pharmaceutical and Biological Sciences. 2016;**3**(1): 162-169

[70] Peter JK, Kumar Y, Pandey P, Masih H. Antibacterial activity of seed and leaf extract of *Carica papaya* var. Pusa dwarf Linn. Journal of Pharmacy & Bioallied Sciences. 2014;**9**(2):29-37

[71] Yogiraj V, Goyal PK, Chauhan CS, Goyal A, Vyas B. *Carica papaya* Linn: An overview. International Journal of Herbal Medicine. 2014;**2**(5):1-8

[72] Islam A, Al-Mamun MA, Parvin S, Sarker M, Zaman MK, Farhana P, et al. Evaluation of antibacterial activities of latex of Caricaceae (*Carica papaya* L.). Asian Journal of Pharmaceutical and Clinical Research. 2015;**8**(1):308-311

[73] Vidal NM, Grazziotin AL, Ramos HCC, Pereira MG, Venancio TM. Development of a gene-centered SSR atlas as a resource for papaya (*Carica papaya*) marker-assisted selection and population genetic studies. PLoS One. 2014;**9**(11):e112654

[74] Bohry D, Ramos HCC, Dos Santos PHD, Boechat MSB, Arêdes FAS, Pirovani AAV, et al. Discovery of SNPs and InDels in papaya genotypes and its potential for marker assisted selection of fruit quality traits. Scientific Reports. 2021;**11**(1):1-8

Chapter 7

Tamarillo (*Cyphomandra betacea* (Cav.)) Origin, Cultivation, Breeding and Management

Rafiq Ahmad Shah, Parshant Bakshi, Hamidullah Itoo and Gaganpreet Kour

Abstract

Tamarillo has a unique flavor and rich history. South American fruit is popular in New Zealand. Tamarillo is commercially grown in New Zealand and South America. It grows best under sub-tropical areas. It matures in 18 months. It's 2 m tall and has lifespan of about 7 years. For propagation, seeds or cuttings are employed, and plant trimming for effective output varies according to propagation method. Tamarillo plants are wind-sensitive and need cover or windbreaks. It's a beautiful fruit with smooth, shining skin. Yellow, red, and purple fruits are available. This fruit contains vitamins, minerals, fiber, and antioxidants. It has a very low-calorie count. Breeding focuses on fruit quality through selection, hybridization, and biotechnological treatments for plantation and post-harvest management. Diseases, pests, viruses, and physiological abnormalities can be treated with plant protection techniques. Like other fruits, it's edible after harvesting. Made into juices, concentrates, jams, gelatins, and sweets. If processing facilities and transport are available, it can be exported as pulp or concentrate. The tamarillo can diversify sub-tropical fruit production as a high-value cash crop, with excellent fruits commanding premium prices in Europe, North America, and Japan.

Keywords: botany, breeding, biotechnology, cultivation, composition, diseases, origin, production, processing, pests, trade, value addition

1. Introduction

The tree tomato (*Solanum betaceum* Cav., syn. *Cyphomandra betacea* (Cav. Sendt.)) is a lesser-known tiny shrub or semi-woody tree that grows at elevations of 500–2500 m. The tamarillo, sometimes known as a 'tree tomato' because of its flesh's resemblance to that of a tomato, is a member of the *solanaceae* (nightshade) family [1]. The tree tomato is related to a group of taxa that used to be classified under the *Cyphomandra* genus. The species of the genus *Cyphomandra* were reassigned to the genus *Solanum*, subgenus *Bassovia*, based on morphological and genetic data [2, 3]. The scientific community now uses the *Solanum* designations to refer to tree tomatoes and wild cousins. In New Zealand, the term "Tamarillo" has become the usual commercial identification for the fruit [4]. It can be found in subtropical and mild temperate climates all over the world. Commercially, it is grown in New Zealand and a few parts of South America. Although some high-performing uniform lines have been produced, commercial plantings are of seed propagated populations. Although it is a largely unexploited species, it presents a wonderful chance to diversify fruit production as a high-value income crop in many subtropical and mild temperate production locations. It comes in two colors red and yellow, with the red being the more popular and prevalent. It's a vivid red egg-shaped fruit with yellow-orange flesh and black seeds encased in purple gelatin. The red color comes from anthocyanin pigments, while the yellow-orange color comes from carotenoids. Tree tomatoes at the fruit-bearing stage require support to keep their branches from breaking off as they get packed with fruit. Because of its shallow root system, they are readily blown over by the wind. The plant is not affected by the duration of the day. *Cyphomandra* species such as *C. hartwegii*, *C. sibundoyensis*, and *C. cajanumensis* produce edible fruits in the wild. Other *Cyphomandra* species are employed as colors and in medical formulations. This category of plants is becoming more economically important, and it may have a lot of future promise.

2. Area and production

Tamarillo popularly known as arboreal tomato [5] is a fruit tree of Andean origin and is currently grown in California, Argentina, Colombia, Ecuador, Venezuela and New Zealand [6] for use in the fresh fruit market and for food processing industry. Tamarillo is also produced less widely in Zambia, Zimbabwe, Uganda, Sri Lanka and India [7, 8]. The fruit's distinctive flavor and nutritional benefits are causing an uptick in interest today [9], which is keeping prices high. New Zealand and the United States have developed extensive plant breeding programmes to develop new cultivars that are more appealing to customers [10]. In Colombia, tamarillo orchards cover a total cultivated area of 7646 ha distributed in 18 provinces. However, Antioquia and Cundinamarca alone account for two-thirds of the total production (50 and 14%, respectively) [7]. In New Zealand about 2000 tons are produced on 200 ha of land and exported to the United States, Japan and Europe. New Zealand's horticultural practices, handling, storage, and transportation methods have all been enhanced via research there. Some better cultivars and shipping containers and circumstances for overseas exports have resulted from these investigations. This crop is limited by factors such as a lack of clear differentiation among varieties, low fruit quality (heterogeneity and phytosanitary issues), the use of ineffective local and foreign variety substitutions, and the fact that the tree tomato is frequently a subsistence crop and not included in genetic conservation programmes.

3. Marketing and trade

New Zealand is one of only three countries to grow tamarillos commercially, the others being Colombia and Australia. New Zealand leads in the production and export followed by Colombia [11]. Tamarillos of New Zealand are exported to the United States, Japan and Europe (**Table 1**). For the export, the existing marketing channels developed for the kiwifruit are used [12] which have greatly benefited the farmers. Due to lack of international market and knowledge of the fruit, the export potential of this fruit crop has not been yet achieved to its level. There are relatively

Market	2014		2015		2016	
	Volume	Value	Volume	Value	Volume	Value
United States of America	12	100,036	11	74,318	8	60,475
Pacific Islands	0.01	98	0	0	0.01	52
Australia	42	182,239	0	0	0	0
Japan	0.05	753	0	0	0	0
Fiji	0	0	0.01	80	0	0
New Caledonia	0	0	0.02	19	0	0
Thailand	4	12,392	0	0	0	0
Total	59	$295,518	11	$74,417	8	$60,527
% change (yr/yr)	173%	62%	−82%	−75%	−26%	−19%

Source: Statistics New Zealand.

Table 1.
Tamarillo export markets 2014–2016 (year ending June, tonnes and $NZ FOB).

few barriers to trade and very few countries specifically refer to tamarillos in their tariff schedules. The United States of America continues to be the main export market for tamarillos. In Columbia majority of the produce is consumed locally, but some part is also exported to the Netherland, France, Canada, Germany and Spain. It is important to mention here that standards have been established among commercial growing country's that market tamarillo fruits (New Zealand, Ecuador and Colombia) and in other countries, there is no such regulation and therefore no commercialization capacity [13, 14]. Fresh tree tomatoes are in high demand in foreign markets, particularly in the United Kingdom, the Netherlands and Spain, where they are especially popular if they are grown without the use of pesticides. Fair Trade accreditation also helps to open up new markets for the product. Imported fruit is processed into juices, syrups and other beverages as well as gelatins and other treats. It might also be exported in the form of fruit pulp or concentrate if processing facilities and suitable transportation were available.

4. Composition and uses

This fruit is found to be a good source of vitamin A, C, B6, E and antioxidants [15–18]. Tamarillo has a relatively high content of vitamins A and C (**Table 2**). Levels of vitamin A are intermediate between those of tomato and carrot [19], while the ascorbic acid content is similar to that of citrus fruits. Tamarillo is rich in anthocyanins and carotenoids which are responsible for their color [20]. The presence of anthocyanins and carotenoids show its biological, therapeutic, and preventative properties [21]. Osorio et al. [22] using spectroscopic analyses revealed that tamarillo fruits are a rich source of natural pigments with potential antioxidant activity, giving them a remarkable added-value. Phenolics are the main antioxidants found in the tamarillo fruit pulp [23]. The seed of the fruit is consumed together with flesh [24]. Tamarillo seeds made up 1.0–1.5% wet basis of the fruit (50–80 g). The seeds of tamarillo were examined for their proximate components [25]. Protein content was

Constituents	Content per 100 g of edible portion
Moisture (g)	85.20
Ash (g)	1.30
Crude protein (g)	1.60
Crude fat (g)	0.00
Carbohydrate (g)	11.90
Total dietary fiber (g)	6.00
Calcium (mg)	11.20
Sodium (mg)	17.80
Magnesium (mg)	25.20
Potassium (mg)	410.60
Iron (mg)	0.30
Beta-carotene (vitamin A)	4.80 (mg/100 g DW)
Ascorbic acid (vitamin C)	55.90 (mg/100 g DW)

Table 2.
Nutritional composition of tamarillo fruits.

discovered to be 22.63%, fat content to be 21.13%, and ash to be 3.15%, with a total carbohydrate content of 43.87%, in the seeds. Tamarillo seed oil's fatty acid profile revealed that linoleic acid (70.47%), oleic acid (14.93%), palmitic acid (9.41%), stearic acid (2.23%) and linolenic acid were the most prevalent fatty acids (1.73%). Apart from these, the oil included arachidic acid (0.23%), phellonic acid (0.22%), and lignoceric acid (0.23%). The fruits are very low in calories (only about 40 calories per fruit). These fruits are available in both red and yellow varieties. However, the red varieties are more popular and more common [26]. Acosta-Quezada et al. [11] assessed fruits of purple and yellow/orange cultivars and did not find any relevant differences among them.

Tamarillo is grown mainly for their edible fruits and to a lesser extent as an outdoor ornamental. The tamarillo fruit has been described as "brazenly beautiful" and the aroma "unusual and attractive". They have several culinary uses and can be eaten raw in salads or as dessert but preferably cooked [27]. The flesh can be eaten fresh or made into a range of sweet and savory dishes and condiments. Tamarillos are tangy and usually sweet with a bold and complex flavor that differs by variety. The fruit can be stewed to use on cereal or as a pie or crumble filling, added to stews or made into a delicious chutney, which is especially good with chillies. In Jamaica and the West Indies, the fruits are considered to have beneficial effects in relieving disorders of the liver [27]. Tamarillo is an important component in Rwanda's exotic fruit industry and consumption of the fruit is traditionally recommended for people suffering from stomach ailments. It is processed into jams, juices and jellies or canned in syrup or prepared in combination with milk products like yogurts, milk shakes and ice-creams. The fruit has high level of pectins, which makes it especially suited for jams and preserves [28]. It is also used for canning in syrup and for producing pulp, chutney, sauce, baby food and in combination with milk products like yogurt, milk shakes and ice creams [29]. The fruit can be used much as a regular tomato, but it has less moisture, so more water, stock or gravy is needed for most cooked dishes.

5. Origin and distribution

The tamarillo *C. betacea* (Cav.) is native to the Andean region of South America where most *Cyphomandra* species are found in cultivated state. It is also grown in New Zealand, Brazil, Argentina and Colombia. The area where *C. betacea* originated is not known, but some wild or naturalized populations have been reported in southern Bolivia and northeastern Argentina and may give an indication of its area of origin [30, 31]. It is found that *S. betaceum* is closely related to *S. unilobum*, *S. roseum*, and in particular to *S. maternum*, all of which are found in Bolivia in wild status [2, 3, 32, 33]. Little information is available on the domestication of the tree tomato, and at present it is unknown when and where this process took place. In New Zealand it is cultivated as a minor fruit crop. It was introduced to New Zealand as early as the late 1800s and there have been a number of further introductions [34]. It is cultivated and naturalized in Venezuela and grown in the highlands of Costa Rica, Guatemala, Jamaica, Puerto Rico and Haiti. Tamarillo was adopted on January 31, 1963 by the growers of New Zealand as the official common trade name for *Cyphomandra betacea*. The tree tomato was named as tamarillo in New Zealand in 1967 [35]. Tamarillo is a Maori word that indicates leadership, and "rillo" comes from the Spanish word for yellow, "amarillo" which was the original type of tamarillo to be grown and only in the 1920s was a new red variety developed [9]. It is possible to grow it successfully in areas with Mediterranean climates, where it has good prospects as a developing new fruit crop [36–38]. In India it is found growing in Sikkim, Darjeeling hills of West Bengal, Tamil Nadu, Meghalaya and in other north-eastern states. In Nepal during survey, it was found that it is grown in home gardens for vegetable purposes. It has been grown in Queensland and Australia in home gardens for many years and is a practical crop in the highlands of the Australian part of New Guinea. In USA it is grown in California and occasionally in Florida in pots and indoors. The plants fruit satisfactorily in greenhouses. In Malaysia *C. betacea* is cultivated in Cameron Highland (Peninsular Malaysia), and Kundasang (Sabah) and is locally known as "Pokok Tomato" or "Tamarillo" in Peninsular Malaysia and as "Buah Cinta," "Moginiwang," or "Tamarillo" in Sabah.

6. Botany and taxonomy

The plant is a small shrub or half-woody, evergreen or partially deciduous, attractive fast growing, brittle tree, shallow-rooted reaching 3 m to 5.5 m in height rarely as much as 7.5 m. The dichasial branching is responsible for the shrubby habit of the tree, although seedling grown plants do go through a juvenile phase and the initial branching pattern may not occur until after 30 nodes growth [39]. The plant is not tolerant to drought stress, and can be damaged by strong winds because of shallow root system. The leaves are perennial, simple, 10–35 cm long and 4–12 cm broad, evergreen, alternate, more or less heart-shaped at the base, ovate and pointed at the apex, thin, softly hairy, with prominent veins and have musky smell and slightly tinged purple when the leaves are young. Flowers with fragrance are borne in small, loose clusters near the branch tips, 1.25–2 cm wide, have 5 pale-pink or lavender, pointed lobes with 5 prominent yellow stamens and green-purple calyx. The long-stalked, smooth, egg shaped are borne singly or in clusters of 3–12, pointed at both ends and capped with the persistent conical calyx. In length, it can be anywhere from 5 to 10 cm long, and in breadth, it can be anywhere from 4 to 5 cm wide. Red, orange, yellow, or red-and-yellow skin colors are all possible, as well as a variety of shades in between (**Figure 1**). Species-dependent

Figure 1.
Tamarillo origin and general aspects (A) Andean region from where tamarillo is native; (B) tamarillo tree with red fruits; (C) Tamarillo's flowers; (D) Tamarillo's fruits range of colors. (http://funnelandspade.blogspot.pt/2010/06/tamarillos-also-known-as-treetomatoes.html; http://www.fancyplants.de/en/exotichome/nwhspec/tamarillo/).

variations in flesh color include shades of orange-red, yellow, and cream-yellow. In the two longitudinal compartments of dark-purple and red fruits, as well as yellow and orange fruits, a soft, juicy, subacid to sweet pulp surrounds the seeds. The exterior layer of flesh is luscious and bland, whereas the skin is rough and unpleasant to eat. In spite of their diminutive size, the seeds are tough and bitter. A weak or under-ripe tomato with a subtle resinous aftertaste has a resinous scent and flavor. The plant bears in first or second year after planting, but peak production is reached after 4 years and has a life expectancy of only 5–12 years. In general, it forms a large spreading crown at a height of 1.5–2 m from base of a single upright woody trunk.

7. Varieties and cultivars

There are apparently no named cultivars, but there are local preferences according to fruit color. Tamarillo have been described as three strains yellow, red (red skin, yellow-orange flesh), and purple (red-purple skin, light orange flesh), with the red being more popular and more common but no named varieties have been analyzed. In Europe and in the USA, the red and purple cultivars are the preferred by consumers due to its attractive color, flavor and nutritional properties. In Malaysia, the red variety of tamarillo can be easily grown at Cameron Highlands, Pahang. It is an egg-shaped bright red fruit with yellow-orange flesh and black seeds that are surrounded by purple seed coat. In Kenya the main varieties grown are the Gold-mine, Inca red, Rothamer, Solid gold and Ruby red. Red fruits are chosen for the fresh fruit markets because of their appealing color. Different selections have been developed in New Zealand from time to time. Commercial plantings of New Zealand's current dominant "black" type were selected in the early 1920s as a variation on the yellow and purple variants that had been used in the past. A reselection process resulted in the creation of this massive, higher-quality red variety. New Black was chosen by William Bridge in 1927 for its huge fruit. Ruby Red has been a staple of the New Zealand economy for decades. Heart-shaped

DOI: http://dx.doi.org/10.5772/intechopen.106601

fruit with a deep crimson color and a savory flavor were selected in 1970. Inca Gold is a canning favorite because of its amber hue and oval shape. If you are looking for something a little more vibrant, Ecuadorian Orange is a great choice. In 1979, Kaitaia Yellow was chosen for its flavor and sweetness. New Zealand is testing a new cultivar named Goldmine. Rothemer is an 85-gram fruit from San Rafael, California, with a bright red exterior and a golden yellow pulp. Solid Gold is an orange, luscious fruit with a sweet flavor. Red Delight, Yellow, and Oratia Round are also made in New Zealand.

8. Breeding and crop improvement

Tamarillo's chromosomes are diploid (2n = 2x = 24). However, spontaneous tetraploids, aneuploids, and triploids have all been documented. The flowers are autogamous and self-compatible but flowers need to be shaken for pollination. Traditional plant breeding has largely ignored the tamarillo, and few crop improvement techniques have been applied. Most attempts to transfer this crop into new areas have failed since they have relied on a single cultivar. As a result, the ability to use variety for local adaptation has been limited. On the other hand, the loss of genetic diversity in this crop and allied wild species is a major worry [40, 41]. Plantation management, fruit quality, and postharvest management are all improved through research and breeding. Breaking seed dormancy, improving fruit taste, and increasing yields are all key breeding goals. Little "stones" of sodium and calcium that occasionally form in the fruit skin offer a difficulty for industrial applications and must be eradicated through breeding. The following are the most important breeding procedures in this crop.

8.1 Selection

Although the tamarillo is essentially autogamous, there is some variation among plants of the same accession, particularly in traditional growing methods. This variance is caused by spontaneous mutations or by the inadvertent introduction of genetic material from different populations through genotype crossover [42]. This variation could be used to find a favorable variant that could be used in a future breeding effort for crop improvement. As a result, screening valuable genetic variety in traditional growing areas for conservation, selection, and crop improvement for this underutilized fruit crop is critical.

8.2 Hybridization

Pollen tubes pierce the ovary and set fruits in most interspecific matings with *Cyphomandra*, according to [30], but seeds do not mature. Pringle and Murray [43] experimented with interspecific hybridization of *Cyphomandra* with nine different Cyphomandra species and discovered that the majority of the crosses failed after fertilization. However, a species hybrid between the two Brazilian species *C. corymbiflora* and *C. diploconos* was obtained very easily. Most species' combinations had reciprocal variances in compatibility. The results of these crosses indicate that the S locus is not involved in the regulation of interspecific incompatibility.

8.2.1 Intraspecific hybridization

There are no technical issues in crossing between different varieties of *C. betacea*. Tamarillo individuals have a high degree of homozygosity due to their autogamy, and crossing genetically dissimilar individuals produces homogeneous offspring. Commercial production of the hybrid tamarillo is simple because each fruit contains more than 300 seeds [44]. The agronomic behavior of F1 hybrids, on the other hand, is completely unknown. Most solanaceous horticulture crops have heterotic yield features [45], and the tamarillo is no exception. Obtaining segregant generations allows for recombination and segregation, allowing for the emergence of new superior genetic combinations. There are, however, a scarcity of investigations on the degree of diversity in segregant generations. Despite the fact that it takes several years to evaluate, tamarillo cannot be regarded a standard fruit crop, as individual genotype evaluation can take up to 20 years. Nonetheless, evaluating each individual takes 3–5 years, making yearly species breeding schemes impracticable, as it could take up to 10 generations before an improved cultivar is released. It will be impossible to determine the best effective breeding strategy for this crop until studies are conducted to determine the corresponding value of the additive and dominant components of genetic variation for each of the primary interest traits. Breeders are interested in developing pure lines or F1 hybrids that maximize heterozygosis's. Vegetative propagation, on the other hand, allows for the propagation of the most valuable genotypes that may arise in segregant generations, such as an F2 produced by complimentary or transgressive crossings. Cloning permits the entire genotype to be preserved, which might be useful in the production of novel cultivars. Micropropagation for tamarillo [46] provides for vegetative propagation without the risk of viral transmission.

8.2.2 Interspecific hybridization

Interspecific hybridization could be effective for transferring desirable traits from wild species to cultivated forms, such as disease and worm resistance. Crossings between *C. acuminata* Rusby and *C. uniloba* Rusby have proven successful. These species are morphologically quite similar to *C. betacea*. Fertility is low in Tamarillo hybrids with *C. acuminata*. Those obtained using *C. uniloba*, on the other hand, are vigorous and prolific [30]. Other Cyphomandra species, such as *C. hartwegii* (Miers) Sendt. Ex Walp. and *C. sibundoyensis* Bohs, are edible as well, and may have some future potential as stand-alone plants or sources of genetic variety for tamarillo breeding. There is no publicly available enhanced breeding material for tree tomatoes. Nonetheless, numerous hybrids between *S. betaceum* and *S. unilobum* are being tested in Colombian fields as a result of breeding operations [47].

8.3 Polyploidy breeding

Triploid and tetraploid seedlings are found at low frequency among commercial plantings of the diploid tamarillo and may arise from the union of unreduced gametes. Further, [48] reported successful induction of tetraploids in the tamarillo (*Cyphomandra*, *betacea* (Cav.) Sendt.) by the application of colchicine to the germinating seed. Triploids and aneuploids are produced from interploidy crosses. Most aneuploids produce primary trisomics ($2n = 2x + 1 = 25$), but some possesses 26 chromosomes and few may be hyperpolyploids. Aneuploidy with 25 chromosomes, instead of 24 as the diploid types, show good fertility and give a similar or even higher yield and fruit size than their diploid counterparts. The pollen fertility of these plants can vary from 10 to 90%. The morphological traits of aneuploids and diploids were the same. For commercial use, aneuploids have fruit and seed sizes comparable to those of diploids. Tamarillo polyploids have been proven to have low fertility and poor agronomic properties, whether they are spontaneous or manufactured [12].

8.4 Biotechnological methods

Traditional techniques have revealed themselves inadequate for improving cultivars because of the low success of cross-pollination, the high incidence of incompatibility and phytosanitary issues [49]. A valid alternative for this plant's breeding are biotechnological methods as *in vitro* cloning and genetic transformation [46].

8.4.1 In vitro regeneration systems

Tamarillo micropropagation methods have been described in various assays [49]. *In vitro* cloning can be achieved through (1) axillary shoot proliferation, which was the first method to be applied [10] (2) organogenesis, obtained on leaf explants [50] and (3) somatic embryogenesis, first obtained from mature zygotic embryos and hypocotyls cultures, and later from other explants [51, 52].

8.4.1.1 Somatic embryogenesis

Plant multiplication has the potential to be revolutionized by somatic embryogenesis, a powerful biotechnology tool. Auxin rich media is used to induce somatic embryogenesis in tamarillo, where embryogenic callus is first generated (induction phase) and subsequently developed into embryos after being switched to a medium free of auxin (development phase). Several explants of tamarillo have the potential to initiate embryogenic cultures including, mature zygotic embryos, young leaves, cotyledons and hypocotyls. Tamarillo mature zygotic embryos were one of the first explants tested for somatic embryogenesis. The formation of embryogenic tissue offers a great potential for large-scale production of plantlets [53] and is also useful in plant genetic transformation. Somatic embryos pass through different morphological phases similar to those occurring during zygotic embryogenesis [54]. The subculturing of somatic embryos, on auxin-free medium for a further 4–5-week period give rise to green normal plantlets.

8.4.2 Genetic transformation

A number of viral infections impact the health and vigor of tamarillo trees, as well as the appearance of the fruit. The tamarillo mosaic virus, the most important pathogenic virus, has no known resistance in Cyphomandra species (TaMV). There has been little success in using traditional breeding programmes to increase tamarillo's virus resistance. Plants can now be genetically modified to be resistant to a variety of viruses, thanks to recent advances in molecular biology, including the tobacco mosaic virus. *Agro bacterium* mediated transformation of tamarillo has been successfully achieved and protocols for transformation are now available [55]. The use of genetic transformation offers the great opportunity for the improvement of many characters for which there is not enough genetic variation in the local germplasm. Genetic transformation is also being used as a functional genomics tool, helping to better understand the process of somatic embryogenesis [6]. The most successful strategy so far has involved constitutive expression in the host plant of the coat protein gene of the target virus. Recently, the coat protein gene for TaMV has been cloned and sequenced which may allow TaMV resistance to be engineered into tamarillo. The application of the *Agro bacterium* mediated transformation method to obtain genetically modified tamarillo plants regenerated *via*, organogenesis was used to introduce the pKIWI110 binary vector into leaf disks by using a virulent LBA4404 [56] and some of them also expressed the b-D-glucuronidase (gusA) reporter gene and chlorsulfuron resistance. *Agrobacterium* mediated transformation was also used to obtain tamarillo plants resistant to tamarillo mosaic virus (TaMV) that were regenerated by shoot proliferation [6]. The NEP25 gene has been silenced in tamarillo using Agrobacterium-mediated genetic transformation procedures [10]. After undergoing somatic embryogenesis, these plants have been regenerated and are currently being evaluated to see if they have the same capacity for somatic embryogenesis as untransformed plants.

8.5 Breeding for yield

At present, the tamarillo is being introduced as a promising crop in a variety of environments, e.g., regions with a Mediterranean climate [37, 57]. But most attempts to introduce tamarillo culture have been based on a single cultivar and this has restricted the opportunity to exploit variation for local adaptation. For this purpose, germplasm screening is essential in order to select the most adapted types in local environments. Several evaluations of tamarillo germplasm have shown a high degree of genetic variation for yield and fruit weight. Differences among different plant genetic materials can be as high as two-fold for fruit weight [37, 57]. The fruit yield is highly affected environmental components and exploitation of most adapted material could maximize the selection for yield.

8.6 Breeding for quality

Among fruit quality parameters fruit shape is of interest because it is directly related to consumer acceptance because it affects fruit attractiveness and is important in terms of packaging and presentation. The fruit shape varies from round to elongated, with a ratio length/width higher than two among accessions. Round to oval shapes appear to be preferred. Fruit color varies from yellow types to purple and it may have stripes or not. Differences among cultivars have been found organoleptic characters like soluble solids, titratable acidity, ascorbic acid and other characters [37, 58–60]. Cultivars with a higher sugar/acid ratio are probably more suited for processing industry. Sugar/acid ratio and sweetness were found to be higher in one of the seven accessions studied by [38] that had reduced acidity (between 15% and 23% and similar levels of soluble solids). Recently developed 'Oratia Red' and 'Andys Sweet Red' have a high sugar/acid ratio (Boyes and Strubi, 1997). Breeding for nutritive value is also possible as lot of variation for ascorbic acid content and vitamin A have been found among genotypes [19, 38, 58, 60]. Aromatic compounds of tamarillo have been identified [61, 62] and selection of high aromatic cultivars is of interest. However, aroma is a complex character, in which many interactions among different compounds are involved and in which environmental influence is considerable.

8.7 Breeding for disease resistance

Few studies have been devoted to tamarillo breeding for disease resistance. Most work has dealt with TaMV resistance. Resistance to TaMV has not been achieved yet, even in wild species belonging to the genus *Cyphomandra*. Genetic transformation is being used to develop tamarillo cultivars resistant to TaMV. Strategies used up to now have mostly involved the use of genetic constructs which include sequences of the coat protein of this virus. The use of mutagenic agents, such as nitrous acid for the production of defective TaMV strains, which could be used for cross protection has not been successful [56]. Resistance to anthracnose is also being attempted by *in vitro* selection of cells capable of growing in the presence of crude filtrate of the fungus [63]. However, there are no reports of the efficacy of this strategy in developing mature plants resistant to this disease.

8.8 Breeding for early harvesting

Early harvesting is of significance since different cultivars differ in how quickly their fruit ripens [64]. This could be exploited to select the earliest cultivars. There is also quantitative variation in the response to postharvest applications of ethylene in some cultivars so it is possible to achieve successful postharvest ripening [65]. Despite being considered a non-climacteric fruit [65], ethylene applications stimulate ripening [66, 67]. Fruit can be harvested during the turning stage and allowed to ripen after harvest in materials where ethylene induces ripening. Fruits that are turning may be stored in this way and matured as needed.

8.9 Breeding for parthenocarpy

Occasionally, parthenocarpic (seedless) fruit-producing trees can be discovered in orchards. These trees require vegetative propagation because they are the result of spontaneous mutations. Fruits from parthenocarpy are ovoid-shaped, red to orange in color, and have stripes that range from green to coffee in hue. Orange is the color of the flesh. Fruit weighs normally only 20 g and is smaller than other fruit. Due to their lack of seeds, parthenocarpic tamarillos would be very intriguing; nevertheless, before this form of fruit can become well-known, low weight and poor yield issues must be resolved.

9. Soil and climate

Tamarillo plants grow best in light, deep, fertile and rich in organic matter soils. However, soils must be well drained, since the plants are not tolerant to water-logging. They grow naturally on soils with a pH of 5–8.5. The tamarillo requires full sun and freedom from competition with roots or shade from other plants. The tamarillo prefers subtropical climate, they grow in many parts of world with rainfall between 600 and 4000 millimeters and annual temperatures between 15°C and 20°C. Although species can also thrive in colder climates, in areas with temperatures not lower than 10°C and where extreme freezing does not occur [49], but it is intolerant to frost (below −2°C) and drought stress. Even though extreme cold could severely damage tamarillo plants, often the plant has the capacity of recovering. During first year of planting the cuttings and seedlings should be protected from frost as plants are highly susceptible to frost and can readily be killed. Frost kills the small branches and foliage of mature trees but not the largest branches and main stem. It is assumed that fruit set is affected by night temperatures. Areas where citrus is cultivated provide good conditions for tamarillos. Tree tomatoes cannot survive in areas with prolonged drought. They must have ample water during the dry season. The best way to retain moisture in a tree tomato plantation is to apply mulch, which also reduces weed growth. Branches of Tamarillo's are fragile, brittle and break easily when laden with fruit so wind breaks should be established before actual planting of the fruit plants in an area where wind may be a problem. Further these plants have a shallow root system and can be blown over by strong winds if not protected sufficiently. Hailstones can also damage the leaves and break the brittle branches. However, damage to fruit is not so severe as in other fruit crops due to their thick skin and strong attachment to the plant.

10. Propagation and rootstock

The plant is mostly multiplied through seeds, cuttings, or grafting [12] as well as more contemporary methods of in vitro clonal propagation, which include axillary branch proliferation, organogenesis, and somatic embryogenesis and are commonly referred to as micropropagation. Seed propagation is simple and best done in protected areas. Seeds produce a high-branched, erect tree, ideal for sheltered locations. Seeds for planting are first washed, dried in the shade and then placed in a freezer for 24 h to accelerate germination. They are then planted in boxes of rich soil by keeping 30 cm distance between seeds and 60 cm between rows and virtually 100% will germinate in 4–6 days. Seedlings should be kept in the nursery until they reach a height of 1–1.5 m for efficient growth. Cuttings should be taken from healthy plants that are free from pathogenic viruses from the basal or aerial suckers should be of 1- to 2-year-old wood with thickness ranging from 10 to 25 mm and length 45–75 cm. They are planted directly in the field until they reach a height of 1–1.5 m. Cuttings develop into a shorter, bushy plant with low-lying branches, suitable for exposed, windy sites. Tamarillo can be grafted on several closely-related rootstock species. In Java, *Cyphomandra costaricensis* is sometimes used as a rootstock to attain a longer-lived plant. The cuttings can also be grafted on wild tobacco trees (*Solanum mauritianum*). The Solanaceae family, which includes many tamarillo cousins, has strong and frequently undesirable alkaloids that can transmit to scions and into fruits grafted on such roots. This is important to remember. It is not advisable to integrate tamarillo on an unknown or untested rootstock.

Propagation through seeds is not recommended as this method is known to produce high degree of genetic variability that negatively affects fruit color resulting in rejection of fruits in the international market. Thus, methods of vegetative propagation are usually used to obtain uniform plants [49]. Vegetative propagation by cuttings has been found to transmit deadly viral diseases. Tissue culture offers a feasible solution to produce large numbers of disease-free planting materials. It is also known that *in vitro* propagated Tamarillo plants produce higher yields and shorter gestation period compared with traditional methods [68]. Another advantage offered by tissue culture methods is the large-scale availability of planting materials at any time of the year irrespective of the season.

Nodal explants are surface sterilized, added to Murashige and Skoog media containing Benzyl amino purine (BAP), and then incubated in a growth environment with a temperature of 25°C and a 16-h photoperiod. Rooting occurs without the use of an exogenous source of auxin 2 weeks after micro shoot development, and they grow to a height of 40 mm in 4 weeks. After that, the rooted plantlets are brought to the greenhouse to begin weaning. *In vitro* propagation of Tamarillo using axillary buds have been reported [69]. Somatic embryogenesis and direct organogenesis have also been reported by [49, 70]. Direct organogenesis has the potential to keep regenerated plants' genomes stable, whereas regeneration through an intermediary callus phase raises the probability of somaclonal changes [71]. A very effective and repeatable in vitro regeneration technique is moreover an absolute requirement for creating transgenic plants.

11. Layout and planting

The seedlings are transplanted in the field when they reach 5–7 cm in height, spaced 80 cm apart in rows and 2 m apart between rows. In New Zealand, the trees are kept 2.5–3 m apart in paired rows 2.5 m apart with 4.25 m between each pair. In mechanized production, single row planting distances of 1–1.5 m between plants and 4.5–5 m between rows is recommended. In unprotected areas where winds are strong closer planting is recommended by keeping 1.5–1.8 m between the plants and 2.5–3 m between the rows and the trees may be staked to prevent swaying and disturbing of roots. In India, the trees are planted in pits 1.2–1.5 m apart. On poorly drained soils, plants should be planted on ridges. In poor soils or soils where nematode infestations or virus infections are possible prevalent, replanting with clean stock will be necessary and is recommended after a period of 3 or 4 years.

12. Irrigation

To maximize and stabilize production, water and nutrient inputs should be provided when needed. The plants need continuous supply of water due to their shallow root system. Drought stress results in decrease of plant growth, fruit size and productivity. The tree tomato cannot tolerate prolonged drought and must have an ample water supply during extremely dry periods. A mulch may be very beneficial in conserving moisture at such times.

13. Nutrient management

The suggested fertilizer rates per hectare for intensive orchard production techniques are 170 kg of nitrogen, 45 kg of phosphorus, and 130–190 kg of potassium. While potassium and phosphorus are sprayed at the beginning of the growth season, nitrogen is applied all year round. It is recommended to apply 0.25–1.0 kg of NPK fertilizer per tree evenly between early spring and summertime. The producer should give the plants in their fifth or sixth year a special feeding of 2 parts superphosphate, 1 1/2 parts nitrate of soda, and 1 part sulfite of potash in late winter or early spring at a rate of 1–1.5 kg per plant or 100 kg per hectare.

14. Training and pruning

The plant should be trained by pruning back long shoots and pinching shoot tips to induce a compact growth and the production of the fruit clusters near the centre of the tree. Seedling trees are pruned back the first year after planting to a height of 0.9–1.2 m to encourage branching. Annual pruning thereafter is advisable to eliminate branches that have already fruited and induce ample new shoots close to the main branches since branches that have already carried fruits will produce smaller fruits with lower quality the next time. Annual pruning is also must otherwise, the tree will develop a broad top with fruits only on the outer fringe and wide-spreading branches are also subjected to wind damage. Pruning also facilitates harvesting and if done timely and appropriately can extend the total fruiting period. Light pruning facilitates the production of medium sized fruits and heavy pruning to large sized fruits. Under protected cultivation pruning can be helpful to prevent the excessive vegetative growth. Early spring pruning of in some trees brings about early maturity and fall pruning of other trees delays fruit maturity to the following fall. It is best to cut the roots on one side of the tree and lean it to the other when it is between one and one and half metres tall (in the direction of the midday sun at about 30–45 degrees). As a result, fruiting branches can develop along the entire trunk as opposed to simply at the top.

15. Intercropping and interculture

In plantations, tamarillo is frequently intercropped with citrus so that, for the first 4–6 years of its rapid growth to a productive size, it can be produced economically. Thereafter, the tamarillo plants are removed [72]. Mulching can aid in preserving soil moisture because plants are susceptible to drought stress. As traditional soil management techniques like plowing are impractical due to the thin and delicate root system, it can also be used as a tactic to control weeds. Because of the shallow root system, deep cultivation is not possible, but light cultivation is desirable to eliminate weeds until there is sufficient vegetative growth to shade them out. The plants have to be protected from wind. Their shallow root system does not provide enough stability and the lateral branches are fragile and break easily when carrying fruits. The tree has to be protected from the wind because its roots are weak and it can easily be blown over. It is also fragile, and strong winds can easily break its branches, especially when they are heavy with fruit. Before laying out the plantation, it is advised that windbreaks be built to protect the young plants for each 1/2 acre (1/5 hectare). In the New Zealand region of North Auckland, *Albizia lophantha* Benth. and *Hakea saligna* hedges are maintained clipped, narrow, and well-liked.

16. Flowering and fruit set

Tamarillo shows sympodial growth habit and each sympodial unit consists of 3–4 leaves and a terminal inflorescence. New season's growth begins in October with shoots that originate in the axils of the previous season's leaf scars. Before inflorescence commencement and branching, shoot extension advances up to 15 nodes at the start of the growth season. At the tip of the stalk, inflorescences develop, and later, continuation shoots develop, typically in the axils of the two most recent leaves. Monochasial and trichasial branches are also seen, despite dichasial branching being prominent. Although they begin at the shoot apex, inflorescences can eventually be found 30–40 mm above the branch fork. At maturity, the flowers and inflorescences hang pendulously, with the anthers and stigma facing downward. Typically, the inflorescence is a mono- or di-chasium with flowers arranged alternately along the rachises. Up to 15 flowers per rachis and 50 in total could be seen in each inflorescence. At 2–3-day intervals, flowers open in an acropetal succession. As a result, the same inflorescence may contain fruitlets, flowers, and flower buds. Particularly if a fruit had already set inside the inflorescence, apical flower buds on each rachis frequently abscise before flower opening. Inflorescences formed late in the season also lose their immature flower buds, and occasionally just one or two blooms fully opened.

In climates with little annual variation, tamarillo trees can flower and set fruit throughout the year. In climates with pronounced seasons (such as New Zealand), fruits ripen in autumn. The flower is pentamerous, radially symmetrical, stellate and hermaphrodite and 24 mm in diameter when the petals are fully reflexed. The petals have a variety of colors, ranging from white to pale pink to white with purple flecks, and are relatively lengthy and meaty. Each stamen is united to a petal and forms a cone around the style. The stamens have two anther sacs on either side of a wide, yellow connective tissue, as well as a short filament and an anther. The style extends 2–4 mm beyond the anther cone and is long and thin. Small, flat, and papillate describe the stigma. Pollen is formed in large quantities in anthers (800,000 pollen). Pollen grains in their desiccated state are oval and trilobed with a pitted exine bearing numerous fine spines. Individual flowers generally open before midday and the petals close around the anthers and style again in the evening. The following morning, each flower bloomed once more, and this pattern persisted for another 2–3 days before the petals fully closed. The number of days a flower is open has an impact on pollination. Nearly a day earlier than unpollinated blooms, pollinated flowers closed for the last time on average after 2.25 days (3.05 days). Either at flower opening or right before it, an anther opens. Pollen is not released spontaneously but anthers release a cloud of pollen via the apical pore if touched or squeezed. If the anthers are not disturbed the pollens remain in the anthers beyond final flower closure. All flowers bear stigmatic exudate on the day of flower opening and remain present on the stigma until Day 5, even though the petals had closed. On the stigma, pollen grains and tubes can be seen from days three to four, but their abundance is greatest on days zero and two. On Day 0, pollen grain germination is at its highest, and it gradually decreases over the following days. On any given day, there are observed to be comparable numbers of pollen tubes in the style and pollen grains on the stigma. After 24 h, pollen tubes are about two-thirds of the way down the style, and after 48 h, they are present surrounding the ovules, according to observations of the pollen tube growth rate.

It is self-compatible and usually autogamous, but the flowers need to be shaken by the wind or visited by insects for pollination to take place [30, 43]. If grown in conditions where flower vibration is limited, such as in a greenhouse, fruit set can be very low. Flowers pollinated on Days −3-1 had the highest probability of fruit set. Fruit set then decline sharply until no fruit set on flowers pollinated on Day 4. Fruit set is strongly influenced by treatments designed to modify the pollen source and the vector. On average, two fruit set in open pollinated inflorescences and 0.6 in inflorescences bagged to exclude external sources of pollen has been found. The flower pollination is carried out by honey bees (*Apis melliferd*) and bumblebees (*Bombus terrestris* or *B. hortorum*). Although fruit set is lower in both cases than in open-pollinated flowers, the quantity of fruits set per inflorescence is unaffected by the pollen source, whether "self" or "cross" (by hand pollination). Self-pollinated or cross-pollinated flowers produce between 0.7 and 0.9 fruit sets per inflorescence on average, as opposed to 2.0 from open-pollinated blooms (**Figure 2**).

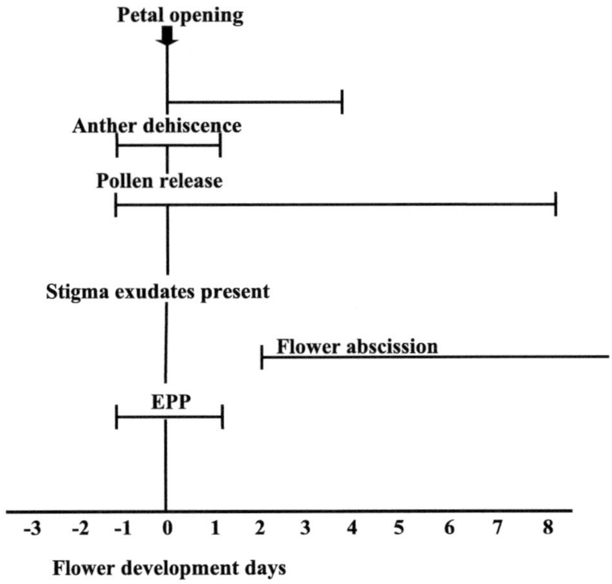

Figure 2.
Floral phenology of the tamarillo flower based on observations from 20 flowers grown under glasshouse conditions (EPP—effective pollination period) [73].

17. Fruit growth development and ripening

Fruit growth and development involves changes in its morphology, anatomy and physiology whilst fruit ripening is associated with dramatic changes in rind texture, color, juice composition, increase in softness due to changes in the cell walls, the metabolism of organic acids and the development of compounds involved in flavor and taste. During development it follows a simple sigmoid growth curve and during ripening it behaves as a non-climacteric fruit. Fruit growth shows an increase of fresh weight (and volume) fast and linear between the sixth and sixteenth week after this period the growth ceases [65, 73]. However, the weight dryness of the fruit continues to increase until reach a maximum in the twentieth week after anthesis. Tamarillos are commercially mature at 21–24 weeks after anthesis. Maturity is indicated by color, firmness, juice content and soluble solids content. Heatherbell et al. [74] reported that tamarillo fruits grew rapidly and reached full size within 16 weeks after anthesis and maturity is attained 11 weeks later i.e., at about 27 weeks. Marked changes in the skin and pulp color occur during development. After 15 weeks the first purple coloration of the skin appears, and thereafter increase in intensity. The green chlorophyll underlay disappears from the fruits between 18 and 22 weeks. Tamarillo is a non-climacteric fruit; therefore, it does not exhibit adequate self-stimulated increase in ethylene production and a consequent respiratory increase as part of its ripening behavior [65]. Pratt and Reid [65] harvested red and yellow tamarillos from 5 to 23 weeks of age and they monitored their respiration rate, demonstrating that the respiratory rate decreases as the age of the fruit increases. The ripe fruits presented a relatively high respiratory rate immediately after harvest (35 mg CO_2 kg^{-1} h^{-1} at 20°C), which then decreased gently until the beginning of the senescence. In this study the ethylene production of the fruits harvested was measured and it was shown to be negligible length of fruit development (less than 0.10 μl kg^{-1} h^{-1} at 20°C) until the beginning of the senescence, when it increases abruptly together with the respiratory rate.

Tamarillos of red cultivars remain green until the fruit growth around the sixteenth week after anthesis. From that moment, it begins to appear a violet color at the apical end subsequently, it will spread over the whole fruit. Around the nineteenth week after anthesis, the green base color begins to turn yellow and the apparent color purple reveals to be red. Violaceous coloring of the tissue surrounding the seeds is evident around the 12th week after anthesis, and intensifies until the Twenty-first week, when it begins decrease in intensity again. The fruits of yellow cultivars present a pattern of similar color change, except that the red coloration is very slight, resulting finally in a light orange color, and without that red pigmentation appears around the seeds. Changes in skin color are due to the degradation of chlorophyll and increase in the concentration of anthocyanins and carotenes. However, the increase in concentration of anthocyanins in the skin is very small in the 4–6 weeks before the commercial maturity. The concentration of anthocyanins, carotenes and chlorophylls from the tissue that surrounds the seeds is maximum when the fruit is in the violet state and then decreases.

The immature fruits have a high Starch content (14% of fresh weight) that, before the end of the stage of active growth, begins to decrease until reach less than 1% in ripe fruits. The decrease in starch content is accompanied by an increase in concentration of soluble solids and sugars. The sucrose is the predominant sugar in tamarillos. The citric and the malic are the organic acids predominant in tamarillo throughout the fruit development. The concentration of citric acid increases rapidly during the period of active fruit growth, reaching more than 2% of the fresh weight; the decreases up to 1.4% in mature fruits. Malic acid concentration is relatively low during the entire period of fruit development (0.2% of fresh weight in mature fruits). The pectin content decreases throughout the fruit development. The lysis of unions between pectin's and hemicelluloses results in increase in the solubilization of pectin's during ripening. Tamarillo's soluble solid content appears to rise to 10–12°B during ripening (normally, the values range between 10.0 and 13.5°B), while the titratable acidity gradually declines (typically, the values range between 1.0 and 2.4%), increasing the SSC/TA ratio and, as a result, the sensory flavor rating. The stems also undergo changes when the fruit ripens, turning yellow instead of green as a result of increased water loss and chlorophyll degradation [75]. Further, according to [76] an SSC value over 12% can qualify tamarillo either for raw consumption or industrial processing.

18. Fruit retention and fruit drop

Self-pollination occurs naturally in tree tomato blossoms. Unless there are bees to transmit the pollen, pollination may suffer if wind is fully stopped so as not to move the branches. Flowers that have not been pollinated will wilt quickly.

19. Harvesting and yield

The tree typically starts bearing when it is 1 1/2 to 2 years old and keeps producing for another 5 or 6 years. It may continue to bear fruit for 11–12 years if given appropriate nutrition. The crop does not ripen simultaneously and several pickings are necessary. Tamarillos are picked when fully colored. The optimum time to harvest the tamarillos of red cultivars is when they are violet color. Other recommended indices to determine the maturity of the fruit with greater accuracy are the firmness, the content of juice and the content of soluble solids [66]. The color of the pedicel in combination with the skin color was also suggested as Maturity index. Fruit can also be picked when they are at the turning stage (when the green color of the skin begins to change and the characteristic skin color begins to show) and then treated with ethylene to stimulate ripening [66, 69]. This early picking and subsequent postharvest ripening reduce the risk of crop failure, increases earliness and concentrates harvesting as it allows harvesting to be advanced by up to 1 month [66]. The fruits are clipped, leaving about 12.5 cm of stem attached. They are collected in bags worn by the harvesters. A single tree can produce more than 30 kg fruits per year. A regular orchard yields 15–17 tons per hectare.

20. Protected/high tech cultivation

Tamarillos are suitable for growing as indoor container plants, though their swift growth, their light, water and humidity requirements and their large leaves can pose a challenge within a limited space. Plants grown in the greenhouse are heavily pruned to prevent excessive vegetative growth. Greenhouse tamarillo production can be somewhat inconvenient due to its long life and extended growing cycle [77]. If grown in conditions where flower vibration is limited, such as in a greenhouse, fruit set can be very low [78].

21. Disease, pest and physiological disorders

21.1 Diseases

21.1.1 Powdery mildew and leaf spot

If left unchecked, this fungus can produce a significant amount of defoliation. It is possible to remove powdery mildew by using commercial pesticide washes and neem oil sprays (Oidium sp.).

21.1.2 Bacterial blast

It is caused by *Pseudomonas syringae* and *P. solanacearum*. It affects the shoot and leaves. *Anthracnose (Colletotrichum gloeosporioides).*

21.1.3 Anthracnose (Colletotrichum gloeosporioides)

The damage it inflicts is immense. Good orchard management can help lower the risk of pest and disease control in markets that demand immaculate fruit. As long as the nursery is kept clean, predator insects can reduce the need for chemical pesticides.

21.2 Viral diseases

21.2.1 Tamarillo mosaic virus (TaMV)

TaMV symptoms include a mosaic mottling on the leaf and ugly irregular spots on the fruit skin that are a darker red than the cultivar's typical skin color. Inside the fruit, there are no visible symptoms, and the eating quality is unaffected. Due to the milder background color typical of this variety of fruit, the darker red splotch on golden tamarillos is particularly ugly. The plants grow stunted as a result. TaMV can only be controlled after symptoms occur by removing trees that are seriously infested [79].

21.2.2 Cucumber mosaic virus and potato virus Y

These not only diminish the plant's strength and health but also distort the leaves, produce skin blemishes, and lower the plant's marketability. They result in fruit mottling and yield loss. The severity of the symptoms will be greater if multiple viruses have simultaneously attacked the plant and will be worse on young or sickly plants. In the valleys of Ecuador's Pichincha region, PLRV, ToRSV, PVY, and AMV were the most often discovered viruses linked to symptoms of viral diseases [80]. Alfalfa mosaic virus (AIMV), tomato spotted wilt virus (TSWV), arabis mosaic virus (ArMV), tobacco streak virus (TSV), and tomato aspermy virus (TAV) are other viruses that have affected the tamarillo, albeit the losses they cause are not as severe as those produced by TaMV [56]. The Colombian tamarillo crop's declining yield is caused by a virus complex [81]. Phytosanitary problem, among which anthracnose (*Colletotrichum acutatum*) and viruses (PLCR, CMV, Potyvirus and ToMV) and others, apparently less influential (AMV, ToRSV and TSWV) has major effect on expansion plans of the tamarillo cultivation in Colombia [82]. Good orchard hygiene, pruning and burring infected plants and a good pest management programme will help to reduce pests and diseases in crop. A suitable spraying schedule can also be helpful. However, there is no cure for a virus once it has infected a plant; the only option is prevention. Since aphids are the primary transmitters of viruses, it is crucial to have effective control over them. Since there is no proof that viruses affecting the tamarillo may be propagated by seed, seed-based reproduction is a strategy for restricting their spread.

21.3 Nematodes

The plant is also harmed by *Pseudomonas solanacearum* wilt, root knot (Meloidogyne sp.), root rot, crown rot, and other diseases. Good cultural norms ought to help prevent these issues.

21.4 Physiological disorders

21.4.1 Abnormality

In order to make jam, small, hard, irregular, semi-transparent stones that are present in the flesh of tree tomatoes must be strained out. If these resemble the two grit-filled bumps in the fruit's wall, that is unknown. Probably in the form of silicates, borates, aluminum-magnesium-oxygen complexes, aluminates, or magnesium oxides, these stones are highly concentrated sources of sodium and calcium.

21.4.2 Fruit scarring

Tamarillo fruit scarring is a cosmetic condition that costs New Zealand's tamarillo growers significantly lost revenue [83]. According to estimates, the condition affects 10–20% of the fruit, which will result in considerable economic losses. When the scars are about 3 cm long, they first appear as little, dark lines on the skin of young fruit. As the fruit grows, the scars turn into corky lesions. Phillips et al. [84] reported that a physical injury caused the scarring. A dark, uneven, and crazy corky scar was routinely produced when the epidermal layer was scratched with a toothbrush. Physical injury such as wind rub has been reported to cause corky scarring in other fruits also, such as avocado [72]. Wounding fruitlets by scratching or removing a patch of epidermis resulted in the characteristic scarring, which suggests that any type of physical epidermal damage incurred early in fruit development may result in scarring.

22. Package and transport

Fruits are sorted by size as small, medium, large and packed in paper-lined wooden boxes for marketing. Polythene films can also be used to reduce water loss and maintain fruit quality [85]. Packaging usually consists of wooden or cardboard boxes containing one tray of tamarillos or preformed plastic trays with 3–8 fruits. Because of its firm flesh and tough skin, the fruit can be transported to long distances without bruising. However, it deteriorates rather rapidly under ordinary storage conditions.

23. Postharvest handling and storage

Different studies carried out to improve post harvesting ripeness showed that application of ethylene or ethephon is helpful to decrease the risk of crop failure and an earlier delivery to the consumer, thereby enhancing the marketability of tamarillo [67]. The fragile lateral branches can break easily when loaded with fruits, so premature harvest helps to reduce this risk and allows storage of fruits up to 20 days at room temperature. Tamarillos can be stored for about 12–14 weeks at 3.5–4.5°C. A cold water dipping process, developed by the New Zealand Department of Scientific and Industrial Research also allows further storage of 6–10 weeks. At higher temperatures, postharvest diseases multiply rapidly. One of the main causes of postharvest loss is bitter rot caused by a *Colletotrichum* sp. Applications of postharvest fungicides greatly reduce the number of fruits affected. Another alternative for postharvest disease control is to dip the fruit in water at 5°C for 10 min [86]. Temperatures below 7°C will reduce softening, weight loss, TA reduction, and color change, among other postharvest handling characteristics that might impact quality. On the other hand, extremely low temperatures (between 0 and 2°C) raise the danger of chilling injury and worsen stem and calyx discolouration.

24. Processing and value addition

Pomegranate can be used in both savory and sweet dishes, and it can even be eaten raw. Fruits can be used to make preserves because of their high pectin content. However, if they are left untreated, they will oxidize and eventually fade. Yellow fruit is preferred for industrial production. As it turns out, it can be utilized in the same way as tomatoes are used in many other cuisines. Simply slice the tomatoes in half, sprinkle them with sugar, then consume the pulp and flesh by scooping it out. Chopped bacon and cream cheese go on top of sandwiches. Diced fruits, bread crumbs, butter, and seasonings are used as a stuffing for roast lamb. Dessert in the form of a pie sliced tree tomatoes, either on their own or accompanied by an apple, are available. They may be filled with water or sugar syrup, then packed into plastic containers with 50% sugar syrup and quickly frozen for later use as pie fillings or puddings. In a blender or on the stove, puree the peeled fruits, remove the seeds, and freeze in containers. Before blending, add some lemon juice into the purée to enhance the flavor even further. Several slices of fresh tree tomato are placed on top of the gelatin, milk, sugar, and lemon juice-cooked fruits. Cooking tree tomatoes with sugar, lemon zest, and juice results in a jam or chutney; the tomatoes can also be cooked with onions and apples for chutney, depending on the desired consistency. Chutney for the commercial market is made in an Auckland, New Zealand, facility. Pectin in the fruit makes it simple to turn into jelly, but the fruit quickly oxidizes and discolors if not treated further. Using whole, peeled fruits and sugar, a thick sauce is cooked to serve with ice cream. The peeled fruits can be used for tomatoes in tomato sauce. Tamarillo jelly can be made from tomatoes with a 50°C Brix sugar concentration, sucrose, and 2% pectin.

25. Conclusions

Tamarillo is a very unique fruit with not only a distinct flavor but also a fascinating history. This fruit is immensely popular in New Zealand, despite the fact that it is native to South America. Tamarillo grows best in a subtropical climate, although it can also be grown in areas where citrus crops are grown. Although it is currently grown in many countries, it is only commercially farmed in New Zealand and a few parts of South America. It rarely bears fruit in low-lying tropical places. It grows quickly and bears fruit in about 18 months. It reaches a height of around two meters and has a lifespan of about 7 years. Seeds or cuttings are commonly used for propagation, with plant trimming for efficient production varying according on the propagation method. Tamarillo plants are wind-sensitive and should be planted in naturally sheltered regions or protected by windbreaks. It is a really attractive fruit with a smooth and shiny skin on the outside. Vitamins, minerals, fiber, and antioxidants abound in this fruit. It has a very low-calorie count. It can be eaten after harvesting, much like most fruits. Juices, concentrates, jams, gelatins, and desserts are all made from it. It could be exported as fruit pulp or concentrate if processing facilities and suitable transportation are available. The tamarillo provides a chance to diversify fruit production in many subtropical fruit production locations as a high-value cash crop, with fine fruits fetching premium prices in niche markets in Europe, North America, and Japan.

Conflict of interest

The authors declare no conflict of interest.

Author details

Rafiq Ahmad Shah[1*], Parshant Bakshi[2], Hamidullah Itoo[1] and Gaganpreet Kour[2]

1 Ambri Apple Research Center, Sher-e-Kashmir University of Agricultural Sciences and Technology Kashmir, India

2 Division of Fruit Science, Sher-e-Kashmir University of Agricultural Sciences and Technology Jammu, India

*Address all correspondence to: rafiq_masoodi@rediffmail.com

References

[1] Correia SI, Lopes ML, Canhoto JM. Somatic embryogenesis in tamarillo (*Cyphomandra betacea*): Recent advances. Acta Horticulturae. 2009;**839**:157-164. DOI: 10.17660/ActaHortic.2009.839.18

[2] Bohs L. *Cyphomandra* (*Solanaceae*). Flora Neotropica Monogr 63. New York, USA: The New York Botanical Garden; 1994. p. 175

[3] Bohs L. Transfer of *Cyphomandra* (Solanaceae) and its species to *Solanum*. Taxon. 1995;**44**:583-587

[4] Morton JF. Tree tomato. In: Julia F, Morton, editors. Fruits of Warm Climates. Miami, FL: Miami. Julia F. Morton; 1987. pp. 437-440

[5] Ordonez RM, Cardoso ML, Zampini IC, Isla MI. Evaluation of antioxidant activity and genotoxicity of alcoholic and aqueous beverages and pomace derived from ripe fruits of *Cyphomandra betacea* Sendt. Journal of Agricultural and Food Chemistry. 2010;**58**:331-337

[6] Correia SI, Canhoto JM. Biotechnology of tamarillo (*Cyphomandra betacea*): From *in vitro* cloning to genetic transformation. Scientia Horticulturae. 2012;**148**:161-168

[7] Ministerio de Agricultura y Desarrollo Rural. Observatorio agrocadenas Colombia [internet]. 2006. Available from: http://www.agrocadenas.gov.co. [Accessed: May 18, 2018]

[8] Stangeland T, Remberg SV, Lye KA. Total antioxidant activity in 35 Ugandan fruits and vegetables. Food Chemistry. 2009;**113**:85-91

[9] Tamarillo Growers Association. Available from: http://www.tamarillo.com [Accessed: May 20, 2018]

[10] Correia SI. Somatic embryogenesis in *Cyphomandra betacea* (Cav.) Sendt (tamarillo) optimization and molecular analysis [thesis]. Portugal: University of Coimbra; 2011

[11] Acosta-Quezada PG, Martinez-Laborde JB, Prohens J. Variation among tree tomato (*Solanum betaceum* Cav.) accessions from different cultivar groups: Implications for conservation of genetic resources and breeding. Genetic Resources and Crop Evolution. 2011;**58**:943-960

[12] Prohens J, Nuez F. The tamarillo (*Cyphomandra betacea*): A review of a promising small fruit crop. Small Fruits Review. 2000;**1**:43-68

[13] Vasco C, Avila J, Ruales J, Svanberg U, Kamal-Eldin A. Physical and chemical characteristics of golden-yellow and purple-red varieties of tamarillo fruit (*Solanum betaceum* Cav.). International Journal of Food Sciences and Nutrition. 2009;**60**:278-288

[14] Schotsmans WC, East A, Woolf A. Tamarillo (*Solanum betaceum* (Cav.)). In: Postharvest Biology and Technology of Tropical and Subtropical Fruits. Oxford, UK: Woodhead Publishing Limited; 2011. pp. 427-441

[15] Mertz C, Gancel A, Gunata Z, Alter P, Dhuique-Mayer C, Vailant F, et al. Phenolic compounds, carotenoids and antioxidant activity of three tropical fruits. Journal of Food Composition and Analysis. 2009;**22**:381-387

[16] Hurtado NH, Morales AL, Gonnzalez Miret ML, Escudero-Gilete ML, Heredia FJ. Color, pH stability and antioxidant activity of anthocyanin rutinoside isolated from tamarillo

fruit (*Solanum betaceum* Cav.). Food Chemistry. 2009;**117**:88-93

[17] Mertz C, Brat P, Caris-Veyrat C, Gunata Z. Characterization and thermal lability of carotenoids and vitamin C of tamarillo fruit (*Solanum betaceum* Cav.). Food Chemistry. 2010;**119**:653-659

[18] Valente A, Albuquerque TG, Sanches-Silva A, Costa HS. Ascorbic acid content in exotic fruits: A contribution to produce quality data for food composition database. Food Research International. 2011;**44**:2237-2242

[19] Rodriguez-Amaya DB, Bobbio PA, Bobbio FO. Carotenoid composition and vitamin a value of the Brasilian fruit *Cyphomandra betacea*. Food Chemistry. 1983;**12**:61-65

[20] Kou M, Yen J, Hong J, Wang C. 2009. *Cyphomandra betacea* Sendt. Phenolics protect LDL from oxidation and PC12 cells from oxidative stress. LWT—Food Science Technology. 2009;**42**:458-463

[21] Kong JM, Chia L, Goh NK, Chia TF, Brouillard R. Analysis and biological activities of anthocyanins. Phytochemistry. 2013;**64**:923-933

[22] Osorio C, Hurtado N, Dawid C, Hofmann T, Heredia-Mira FJ, Morales AL. Chemical characterisation of anthocyanins in tamarillo (*Solanum betaceum* Cav.) and Andes berry (*Rubus glaucus* Benth.) fruits. Food Chemistry. 2012;**132**:1915-1921

[23] Vasco JR, Kamal-Eldin A. Total phenolic compounds and antioxidant capacities of major fruits from Ecuador Catalina. Food Chemistry. 2008;**111**:816-823

[24] Morton JF. Fruits of Warm Climates. Miami: Creative Resource Systems Inc.; 2013

[25] Ramakrishnan Y, Khoddami A, Gannasin SP, Muhammad K. Tamarillo (*cyphomandra betacea*) seed oils as a potential source of essential fatty acid for food, cosmetic and pharmacuetical industries. Acta Horticulturae. 2013;**1012**:1415-1421

[26] Lister CE, Morrison SC, Kerkhofs NS, Wright KM. The Nutritional Composition and Health Benefits of New Zealand Tamarillos (Report No. 1281). New Zealand: New Zealand Institute for Corp and Food Research Limited [internet]; 2005. Available from: http:// www.tamarillo.com/vdb/document/153 [Accessed: May 18, 2018]

[27] Guimaraes ML, Tome MC, Cruz GS. *Cyphomandra betacea* (Cav.) Sendt. (tamarillo). In: Bajaj YPS, editor. Biotechnology in Agriculture and Forestry. Berlin: Trees IV Springer Verlag; 1996. pp. 120-137

[28] Duke JA, duCellier JL. Handbook of Alternative Cash Crops. Boca Raton, F.L: CRC Press; 1993

[29] de Vincenzi M, Maialetti P, Dessi MR. Monographs on botanical flavouring substances used in foods. Part IV. Fitoterapia. 1995;**66**:203-210

[30] Bohs L. Crossing studies in *Cyphomandra* (Solanaceae) and their systematic and evolutionary significance. American Journal of Botany. 1991;**78**:1683-1693

[31] Brucher H. Tropische nutzplanzen. Berlin: Springer-Verlag; 1997

[32] Lester RN, Hawkes JG. *Solanaceae*. In: Hanelt P, editor. Mansfelds Encyclopedia of Agricultural and Horticultural Crops (Except Ornamentals). Berlin, Germany: Institute of Plant Genetics and Crop Research Springer; 2001. pp. 1790-1856

[33] Bohs L, Nelson A. *Solanum maternum* (Solanaceae), a new Bolivian relative of the tree tomato. Novon. 1997;7:341-345

[34] Fletcher WA. Growing tamarillos. In: Bulletin 306. Ministry of Agriculture and Fisheries, New Zealand; 1975

[35] Hewett EW. New horticultural crops in New Zealand. In: Janick J, Simon JE, editors. New Crops. New York: John Wiley and Sons; 1993. pp. 57-64

[36] Grau A. Aptitud climatica del noroeste argentino para el cultivo de frutales tropicales y subtropicales: Un analisis comparativo con otras areas subtropicales del mundo. Revista Industrial y Agricola de Tucuman. 1994;**71**:31-39

[37] Calabrese F, De Michele A, Barone F, Mirto I, Panno M. Comportamento agronomico-qualitativo di cultivar di tamarillo in Sicilia. Rivista di Frutticoltura. 1995;**57**:43-46

[38] Prohens J, Ruiz JJ, Nuez F. El tomate de arbol (*Cyphomandra betacea*): Un nuevo cultivo para el area mediterranea. Informacion Tecnica Economica Agraria. 1996;**17**:185-194

[39] Richardson A, Patterson K. Tamarillo growth and management. The Orchardist of New Zealand. 1993;**66**:33-35

[40] Sanchez-Vega I. Frutales and inos: Tomate de arbol *(Cyphomandra betacea)*. In: Hernandez-Bermejo JE, Leon J, editors. Cultivos Marginados: Otra Perspective. Rome, Italy: FAO; 1992, 1992. pp. 183-186

[41] Nuez F, Morales R, Prohens J, Fernandez de Cordova P, Soler SV, Solorzano E. Germplasm of Solanaceae horticultural crops in the south of Ecuador. Plant genetic resources news letter. 1999;**120**:44-47

[42] Bohs L. *Solanum allophyllum* (Miers) Standl. and the generic delimitation of *Cyphomandra* and *solanum* (Solanaceae). Annals of the Missouri Botanical Garden. 1989;**76**:1129-1140

[43] Pringle GJ, Murray BG. Interspecific hybridization involving the tamarillo, *Cyphomandra betacea* (Cav.) Sendt. (Solanaceae). New Zealand Journal of Crop and Horticultural Science. 1992;**19**:103-111

[44] Pringle GJ, Murray BG. Polyploidy and aneuploidy in the tamarillo, *Cyphomandra betacea* (Cav.) Sendt. (Solanaceae) II. Induction of tetraploidy, interploidy crosses and aneuploidy. Plant Breeding. 1992;**108**:139-148

[45] Kalloo G. Vegetable Breeding. Vol. 1. Boca Raton, Florida: CRC Press; 1988

[46] Barghchi M. In vitro regeneration, plant improvement and virus elimination of tamarillo (*Cyphomandra betacea* (Cav.) Sendt.). In: Davey MR, Alderson PG, Lowe KC, Power JB, editors. Tree Biotechnology towards the Millennium. Nottingham, UK: Nottingham University Press; 1998. pp. 173-185

[47] Lobo M, Medina CI, Cardona M. Resistencia de campo a la antracnosis de los frutos (*Colletotrichum gloeosporioides*) en tomate de arbol (*Cyphomandra betacea* (*Solanum betaceum*) Cav. Sendt.). Revista Facultad Nacional de Agronomía Medellín. 2000;**53**:1129-1141

[48] Pringle GJ, Murray BG. Polyploidy and aneuploidy in the tamarillo, *Cyphomandra betacea* (Cav.) Sendt. (Solanaceae) I. Spontaneous polyploidy and features of the euploids. Plant Breeding. 1992;**108**:132-138

[49] Correia SI, Lopes ML, Canhoto JM. Somatic embryogenesis induction

system for cloning an adult *Cyphomandra betacea* (Cav.) Sendt. (Tamarillo). Trees (Berl.). 2011;**25**:1009-1020. DOI: 10.1007/00468-011-0575-5

[50] Obando M, Jordan M. Regenerative responses of *Cyphomandra betacea* (Cav.) Sendt. (tamarillo) cultivated in vitro. In: IV International Symposium on In Vitro Culture and Horticultural Breeding. Vol. 560. 2000. pp. 429-432

[51] Lopes M, Ferreira M, Carloto J, Cruz G, Canhoto J. Somatic embryogenesis induction in tamarillo (*Cyphomandra betacea*). In: Jain S, Gupta P, Newton R, editors. Somatic Embryogenesis in Woody Plants. Vol. 6. Dordrecht: Springer; 2000. pp. 433-455

[52] Canhoto J, Lopes M, Cruz G. Protocol of somatic embryogenesis: Tamarillo (*Cyphomandra betacea* (Cav.) Sendtn.). In: Jain S, Gupta P, editors. Protocol for Somatic Embryogenesis in Woody Plants. Dordrecht: Springer; 2005. pp. 379-389

[53] Merkle SA, Parrot WA, Flinn BD. Morphogenic aspects of somatic embryogenesis. In: Thorpe TA, editor. In Vitro Embryogenesis in Plants. Dordrecht: Kluwer Academic Publishers; 1995. pp. 155-203

[54] Canhoto JM, Correia SI, Marques CI. Factors affecting somatic embryogenesis induction and development in *Feijoa sellowiana* Beig. Acta Horticulturae. 2009;**839**:147-156

[55] Atkinson RG, Gardner RC. Regeneration of transgenic tamarillo plants. Plant Cell Reporter. 1993;**12**:347-351

[56] Eagles RM, Gardner RC. Incidence and distribution of six viruses infecting tamarillo (*Cyphomandra batacea*) in New Zealand. New Zealand Journal of Crop and Horticultural Science. 1994;**22**:453-458

[57] Prohens J, Ruiz JJ, Nuez F. El tomate de arbol, un cultivo prometedor para regiones de clima mediterraneo. Agricola Vergel. 1997;**14**:209-214

[58] Romero-Rodriguez MA, Vazquez-Ochoa ML, Lopez-Hernandez J, Simal-Lozano J. Composition of babaco, feijoa, passion-fruit and tamarillo produced in Galicia (NW Spain). Food Chemistry. 1994;**49**:251-255

[59] Boyes S, Strubi P. Organic acid and sugar composition of three New Zealand grown tamarillo varieties (*Solanum betaceum* (Cav.)). Journal of Crop and Horticultural Science. 1997;**25**:79-83

[60] Morton JF. The tree tomato, or tamarillo a fast-growing, early-fruiting small tree for subtropical climates. Proceedings of the Florida State Horticultural Society. 1982;**95**:81-85

[61] Torrado A, Suarez M, Duque C, Krajewski D, Naugebauer W, Schreier P. Volatile constituents from tamarillo (*Cyphomandra betacea* Sendtn.) fruit. Flavour and Fragrance Journal. 1995;**10**:349-354

[62] Wong KC, Wong SN. Volatile constituents of Cyphomandra betacea Sendtn. Fruit. Journal of Essential Oil Research. 1997;**9**:357-359

[63] Hoyos RA, Afanador L. Sistemas biotecnologicos para la seleccion acelerada del tomate de arbol (*Solanum betaceum*) por su resistencia a la antracnosis. In: 2nd Seminario Frutales de Clima Templado, Manizales. Colombia: Colombian Agricultural Research Corporation; 12-14 August 1998. pp. 40-46

[64] Prohens J, Ruiz JJ, Nuez F. Variabilidad en el comportamiento en postcosecha del tomate de arbol. In:

Chamarro J, Pozueta J, editors. IV Simposio Nacional, I Iberico sobre maduracion y post-recoleccion de frutos y hortalizas. Valencia, Spain: Servicio de Publicaciones de la Universidad Politecnica de Valencia; 1996. pp. 151-154

[65] Pratt HK, Reid MS. The tamarillo: Fruit growth and maturation, ripening, respiration, and the role of ethylene. Journal of the Science of Food and Agriculture. 1976;**27**:399-404

[66] El-Zeftawi BM, Brohier L, Dooley L, Goubran FH, Holmes R, Scott B. Some maturity indices for tamarillo and Pepino fruits. Journal of Horticultural Sciences. 1988;**63**:163-169

[67] Prohens J, Ruiz JJ, Nuez F. Advancing the tamarillo harvest by induced postharvest ripening. Hortscience. 1996;**31**:109-111

[68] The Indian Agriculture Information Wing. Micropropagation of Tamarillo. Agricultural newsletter. Shillong, India: Agricultural Information Offset Press, Fruit Garden Meghalaya; 2009. p. 8

[69] Waweru B, Ishimwe R, Kajuga J, Kagiraneza B, Sallah PYK, Ahishakiye V, et al. In vitro plant regeneration of Cyphomandra betacea (tamarillo) through nodal culture. Rwanda Journal. 2011;**24**:58-66

[70] Espinosa O, John A, Gonzalez TO, Sanchez HRA, Kafrri A, Londono C. Potential of in vitro propagation for the tomato tree parthenocarpic *Cyphomandra betacea* Cav. (sendt). Revista Facultad Nacional de Agronomía Medellín. 2005;**58**:2685-2695

[71] Kahia WJ. In vitro propagation of the new Coffea arabica cultivar-Ruiru 11 [thesis]. UK: University of London; 1999

[72] Carnevali A. Tamarillo: Una nuova possibilita per la frutticoltura meridionale. Frutticoltura. 1974;**36**:31-37

[73] Lewis DH, Considine JA. Pollination and fruit set in the tamarillo *Cyphomandra betacea* (Cav.) Sendt. I. Floral biology. New Zealand Journal of Crop and Horticultural Science. 1999;**27**:101-112

[74] Heatherbell DA, Reid MS, Wrolstad RE. The tamarillo: Chemical composition during growth and maturation. New Zealand Journal of Crop and Horticultural Science. 1982;**25**:239-243

[75] Pongjaruvat W. Effect of Modified Atmosphere on Storage Life of Purple Passion Fruit and Red Tamarillo. Palmerston North: New Zealand, Massey University; 2008

[76] Paes AS, Quast LB, Quast E, Raupp DS. Development of Tamarillo Light Jelly with High Pulp Content. Revista Ciencias Exatas e Naturais. 2015;**17**:293-306

[77] Bohs L. Ethnobotany of the genus Cyphomandra (Solanaceae). Economic Botany. 1989;**43**:143-163

[78] Sale PR. Tamarillos: Orchard management. New Zealand Ministry of Agriculture and Fisheries Aglink HPP. 1983;**297**:4

[79] Bakshi P, Kour G, Ahmad R. Tamarillo (*Cyphomandra betacea*). In: Ghosh SN, editor. Underutilized Fruit Crops: Importance and Cultivation. 1st ed. Delhi: Jaya Publications; 2016. pp. 1271-1294

[80] Vizuete B, Insuasti ML, Ochoa J, Ellis M. Biological and Serological Characterization of Tree Tomato Virus Diseases in Ecuador. INIAP: Ohio State University; 1990

[81] Jaramillo M, Gutierrez PA, Lagos LE, Cotes JM, Mauricio M. Detection of a complex of viruses in tamarillo (*Solanum betaceum*) orchards in Andean region of Colombia. Tropical Plant Pathology. 2011;**36**:150-159

[82] Zapata MMJ, Alvarez JA, Montoya MM. Characteristics of the viruses related to virosis of tamarillo (*Solanum betaceum*) in Colombia. Revista Lasallista de Investigación. 2012;**9**:115-127

[83] Rheinlander PA, Jamieson LE, Fullerton RA, Manning MA, Meier X. Scarring in tamarillo fruit (*Solanum betaceum*). New Zealand Plant Protection. 2009;**62**:315-320

[84] Phillips PA, Faber BA, Morse JG, Hoddle MS. UCIPM Pest Management Guidelines: Avocado. University of California ANR Publication 3436; 2007 . Available from: http://www.ipm.ucdavis.edu/PMG/C008/m008fpfrtdmg.html [Accessed: May 18, 2018]

[85] Yearsley CW, Huang BY, McGrath HJW, Fry J, Stec MGH, Dale JR. Red tamarillos (*Cyphomandra betacea*): Comparison of postharvest strategies for control of fungal storage disorders. New Zealand Journal of Experimental Agriculture. 1988;**16**:359-366

[86] Ribas PG, Pessatto C, Zaika WR, Quast E, Quast LB, Ormenese RSC, et al. Development of tamarillo jams containing whole pulp. Brazilian Journal of Food. 2012;**15**:149-141

Chapter 8

The Production and Marketing Issues of Pineapple (*Ananas comosus*) under Humid Tropical Conditions in the State of Tabasco and Way-out

Maritza Alejo Jeronimo, Edward Manuel Arevalo de la Cruz, Hortensia Brito-Vega, Armando Gomez-Vazquez, Jose Manuel Salaya-Dominguez and Edmundo Gomez-Mendez

Abstract

Pineapple cultivation has had the greatest impact on the market and that has increased world production in the recent decades in all tropical and subtropical areas. It is one of the crops that best adapts to these environmental conditions. In Mexico, its production has decreased significantly. Tabasco pineapple producers have been facing various problems that have further worsened. The main causes of the crisis are increase in input costs, lack of provision of technical advice to small and medium producers, little support of field programs by the government, and the growth of imports of industrialized pineapple. At the same time, the problem is seen during the cultivation of pineapple in the field. The producer sows the plant with fertilization without having chemical tests of the soil and irrigation waterdue to the costs and no interpretation of the results of these soil analyses. This affects the harvest and quality of the pineapple in its sale price.

Keywords: fertility, phenological phases, price and soil

1. Introduction

Pineapple (*Anana comosus* (L.) Merr.) is a species of high commercial demand. MD-2 is among the most promising pineapple varieties, which has captured the consumer appeal in recent years [1]. Pineapple is very demanding in mineral elements such as nitrogen, phosphorus, potassium, boron, calcium, zinc, magnesium, magnesium, copper and humic-fulmic acids, which are applied in the region with chemical fertilization, with minimal use of compost or organic materials such as cattle or sheep manure existing in the region. The large number of inputs used includes the use of insecticides, nematicides, acaricides, and fungicides that contribute to contaminating the region's soils. The pineapple clones that are currently enjoying the greatest growth are MD2 and the smooth cayenne cultivars in the state of Tabasco [2].

The Mexican tropics present appropriate agroecological conditions for the development of tropical crops such as pineapple (*Ananas comosus*). Its fruit is highly demanded due to its pleasant flavor and aroma, as well as its content of vitamins A, B and C, is highly demanded in various markets of the world [3]. Maximum growth potential is expressed in subtropical, warm and humid climates. Therefore, its production is mainly distributed between latitudes 30° north latitude N and S [4]. The second place in the world production of tropical fruit trees, only surpassed by the mango [5], although it is native to Brazil and Paraguay, especially from the Paraná River basin [6]. Currently the main pineapple producing countries are Costa Rica, Brazil, the Philippines and Thailand [7].

Regarding the alternatives put in place by some farmers in the municipality of Huimanguillo, Tabasco, to improve their land, there is the use of compost that undergoes a biological process through which it is possible to convert organic waste into stable organic matter (mature compost), due to the action of various microorganisms. The most common applications of composting include the treatment of agricultural waste and garden waste, mainly and fertilizers from animal farms. As stated above, the objective of the present study is focused on the production of national and international commercial pineapple and some problems that arise during the production process in the field.

2. Pineapple

Pineapple (*A. comosus* L.) is a tropical fruit native to Brazil. The ancestors called it *Ananas*, which means "excellent fruit". Pineapple is a fruit of the Bromeliaceae family, it is non-climacteric that produces small amounts of ethylene in terms of ripening [8]. It is a perennial plant with a base formed by the compact union of several leaves forming a rosette. The concavities of the leaves can give rise to shoots with small basal rosettes (**Figure 1**), which facilitate the vegetative reproduction of the plant. They have a stem after 1–2 years that grows longitudinally and forms an inflorescence at the end. Its leaves are thorny that measure 30–100 cm long, its flowers are pink and have three petals that grow in the axils of pointed bracts and hypogynous ovary. Its flowers are grouped in spike inflorescences of about 30 cm in length with a thickened stem. The flowers bear fruit without need for fertilization and from the hypogynous ovary berry-shaped fruits develop, which together with the axis of the inflorescence and the bracts, give rise to a fleshy infructescence [8].

The main varieties are classified into six groups according to their growth habits, fruit shape, flesh characteristics and leaf morphology and have spread throughout the world based on their ability to adapt to local pedoclimatic conditions: 1) Cayenne (**Figure 2**), 2) Española, 3) Queen, 4) Pernambuco, 5) Perolera and 6) Gold "Extra Sweet" MD-2 [9].

Pineapple (*A. comosus* (L.) Merr.) has been for years one of the economic resources for export in many countries, is the United States (over 95% of exports), Argentina, Arabia Saudita, Chile, Egipto, Emiratos Árabes Unidos, the European Union and now Canada especially the cultivar Gold "Extra Sweet" MD-2, which due to its content of soluble solids, aroma and color has been preferred and has remained number one in world markets. This plant is fast growing and has a shorter production cycle; in addition, the production yields very sweet and pineapple juice extraction, although it is recognized that it is susceptible to mechanical damage and Phytophthora [10].

Figure 1.
Plant with a base formed by the compact union of several leaves forming a rosette.

Figure 2.
*The pineapple (*Ananas comosus* (L.) Merr.) that is produced in the soil of the sabana municipality of Huimanguillo, Tabasco.*

Replacing the low-yielding cultivars with the better ones is a difficult task, considering that pineapple is one of the fruit trees with a high planting density, around 60,000 propagules per hectare for 'MD-2' and, at the same time, is the one that produces fewer propagules naturally [11].

3. National and international pineapple production

In Mexico, pineapple was produced in 14 states during year 2019. Although 80% of the production was concentrated in the Bajo Papaloapan region, which include seven municipalities in the state of Veracruz: Isla, Juan Rodríguez Clara, José Azuela, Chacaltianguis, Medellín, Alvarado, Tlalixcoyan and two municipalities in the state of Oaxaca: Loma Bonita and Tuxtepec. Both the area planted and pineapple production have increased from 13.93 thousand ha to 376.15 thousand t in 1980 to 44.18 thousand ha and 1041.16 thousand t in 2019, which represents an average annual growth rate of 3% and 2.6%, respectively, Tuxtepec [12]. Therefore, pineapple is considered as the second most important tropical crop in the world after bananas, contributing more than 20% of the world volume of tropical fruits. Until 2002, Mexico occupied the seventh position in the world in pineapple production, contributing 4% of the total world volume Tuxtepec [13].

Worldwide, there are 14 countries that dominate pineapple cultivation, in which Mexico is the leader in yield with 42.8 t ha^{-1}, but its production is surpassed by Thailand, the Philippines, Brazil, and China, who report yields of 23.6 t ha^{-1}, 37.5 t ha^{-1}, 24.3 t ha^{-1} and 36.2 t ha^{-1} respectively Tuxtepec [14]. Although there are other countries that has greater production compared to Mexico, however, these are not considered important since their participation in the world market is less than 2%. The cultivated area of pineapple in Mexico in 2008 was 29.46 thousand hectares, which produced 718.29 thousand tons, with a value of $94.92 million dollar USD [14].

The state of Tabasco in 2008 ranked fourth in planted area with 1287 hectares, and third in production with 42,400 tons, which reached a value of $6.17 million dollar USD [14]. The largest importer of pineapple worldwide is the United States with 696.82 thousand tons. Although Canada, another trading partner of Mexico, in the same year imported 102.06 thousand tons [15], even though Mexico is one of the most important trading partners for the US and Canada, it does not export pineapple. Which is due, among other things, to the area dedicated to this crop [16].

4. Economic importance

Although pineapple is grown in the state of Tabasco, the government, through various institutions, carried out agroecological zoning studies to find out areas having the greatest production potential for pineapple cultivation. The world pineapple market is 2.49 million tons, which is imported by 123 countries. 74.31% of the world pineapple market is dominated by eight countries listed below in order of importance: United States (696,820 t), Belgium (292,499 t), Netherlands (200,026 t), Germany (167,416 t), Japan (165,794 t), United Kingdom (116,730 t), Spain (113,182 t) and Canada (102,064 t) [15].

The pineapple cultivation is profitable when more than 45.8 t ha^{-1} fruit is produced. 5.96 thousand ha area was identified with great potential to produce pineapple mainly located in Campeche, Chiapas, Oaxaca, Puebla, Quintana Roo, Tabasco, Veracruz and Yucatán. It is concluded that pineapple production is profitable in those regions that present ideal agroecological conditions for its production. The main marketing channels for fresh pineapple in Mexico are: 1.- Central de Abasto de la Ciudad de Mexico: In this center it is acquired between 30 and 35% of the total volume of national production that is channeled to the fresh pineapple market. However, all this volume is not consumed in Mexico City and the metropolitan area rather an important part is redistributed to other supply centers in the interior of the country. 2.- Central de Abasto de Monterrey: In this center, 20% of the produced crop volume is acquired. This plant supplies the market in the north of the country. 3.-Central de Abasto de Guadalajara: This center acquires an average of 10% of the national production destined for the fresh market. 4.- Second-order markets such as Puebla, Chihuahua, Tamaulipas, Yucatán and Michoacán, acquire 10% and 5.- Smaller places and self-service chains consume the remaining 30–35% [15, 16].

Figure 3.
Pineapple cultivation: a) Crop with problems of nutritional deficiency and excess water, b) crop in full growth with the proper management of fertilization doses and c) Development and growth of pineapple maguey without deficiencies and adequate drainage.

5. Limiting factors in production

The factors that limit pineapple production are potential of the soils including soil fertility, acidity, excess or deficit of water, clay content, soil erosion, nutritional deficiencies, presence of Na, alkalinity and low CIC that, alone or in groups, influence the detriment of soil fertility [17].

The acid soils of the Sabana de Huimanguillo are characterized by high phosphorus fixation, deficiencies of zinc, boron, calcium, magnesium and potassium, low rate of formation of ammonium and nitrates, in addition to a high percentage of aluminum saturation [18]. These restrictive soil fertility conditions are manifested in foliar deficiencies that affect the yield and quality of citrus and pineapple fruits. For this reason, acid soils are classified with the methodology of the Integrated System for There are programs with advice and recommendations for pineapple production such as cultural work, prevention and management of diseases and pests, fertilization doses to avoid macro and micronutrient deficiencies, depending on the pineapple variety, but these programs do not reach the small producers, they empirically produce even the fruit, and a third party buys at the foot of the field at a lower price than in the market [19].

Pineapple cultivation affects the natural environment (**Figure 3**). In many producing countries, there is a fruit sector of small producers whose plots are modest in size and have very little impact on the environment. However, industrial production, which produces most of the fruit destined both for fresh export and for processing, has important consequences for the environment.

6. Conclusions

The pineapple cultivation in Tabasco has been affected by the little support and technological interest which consequently increased the total costs of production. Furthermore, due to disproportionate increase in the application of chemical inputs, the cultivation surface and little or no technical advice, has made it difficult to place surplus production in the internal market. The situation is further aggravated by entry of the processed pineapple into the national territory through unfair trade practices. The small producers that produce the pineapple fruit without programs, recommendations to obtain better management and quality of the fruit, there is a national and international market with a good acceptability in flavor, consistency and special in the juice and syrup of the fruit of the same.

Acknowledgements

I thank the Universidad Juarez Autonoma de Tabasco for allowing us to participate in the Piña project.

Conflict of interest

The authors declare no conflict of interest.

Notes/thanks/other declarations

No declarations just to thank the pineapple producers of Chontalpa de Huimanguillo, Tabasco, Mexico. Place.

Author details

Maritza Alejo Jeronimo, Edward Manuel Arevalo de la Cruz, Hortensia Brito-Vega*, Armando Gomez-Vazquez, Jose Manuel Salaya-Dominguez and Edmundo Gomez-Mendez
Division Academica de Ciencias Agropecuarias, Universidad Juarez Autonoma de Tabasco, Teapa, Tabasco, México

*Address all correspondence to: hortensia.brito@ujat.mx

References

[1] Somogyi LP, Ramaswamy HS, Hui YH. Processing Fruits Science and Technology. Vol. 1 ed. CRC Press; 1996. p. 510. Available from: https://fruitandvegetable.ucdavis.edu/files/217079.pdf

[2] Hernández-Barbosa G. Caracterización social y técnica del cultivo de la piña criolla (*Ananas comosus*). 2018;**4**(1):1-9. Available from: https://revista-agroproductividad.org/index.php/agroproductividad/article/view/570

[3] Agencia de Servicios a la Comercialización y Desarrollo de Mercados Agropecuarios (ASERCA). La producción de piña en México, historia de un patrimonio regional. Claridades Agropecuarias. 2010:3-41. Available from: http://www.infoaserca.gob.mx/claridades/revistas/086/ca086.pdf

[4] Malézieux E, Côte F, Bartholomew DP. Crop environment, plant growth and physiology. In: Bartholomew DP, Paul R and Rohrbach KG. (Eds.). The Pineapple: Botany, Production and Uses. London, UK: CAB Publishing; 2003. pp. 69-107. DOI: 10.1079/9780851995038.0069

[5] Uriza ÁDE, Torres ÁA, Aguilar ÁJ, Santoyo CVH, Zetina LR, Rebolledo MA, La piña mexicana frente al reto de la innovación. Avances y retos en la gestión de la innovación; Colección Trópico Húmedo. Chapingo, Estado de México. Universidad Autónoma Chapingo (UACH). 2018. p. 484. Available from: https://ciestaam.edu.mx/libro/la-pinia-mexicana/

[6] Coppens-d'Eeckenbrugge G, Uriza ADE, Rebolledo MA, Martínez RL. The cascajal block: Another testimony of the antiquity of pineapple in Mexico? Pineapple News. 2011;**18**(1):47-48. Available from: https://www.academia.edu/35958609/The_Cascajal_Block_another_testimony_of_the_antiquity_of_pineapple_in_Mexico

[7] FAOSTAT. Datos de cultivos. Organización de las Naciones Unidas para la Alimentación y la Agricultura (FAO). 2020. Available from: http://www.fao.org/faostat/es/#data/QC

[8] Rodríguez R, Becquer R, Pino Y, López D, Rodríguez RC, Lorente G, et al. Producción de Frutos de Piña (Ananas comosus (L.) Merr.) MD-2 A Partir de Vitroplantas. Cultivos Tropicales. 2016;**37**:40-48. Available from: https://www. redalyc.org/pdf/1932/193246189006.pdf

[9] Delgado-Huertas H, Arango-Wiesner L. Morphoagronomic characterisation of pineapple (*Ananas* spp.) genotypes in high-terrace soil near Villavicencio. ORINOQUIA. 2015;**19**(2):154-165. Available from: http://www.scielo.org.co/pdf/rori/v19n2/v19n2a02.pdf

[10] Oculi BB, Ocwa A. Reactions of pineapple cultivars to pineapple heart rot disease in central Uganda. Crop Protection. 2020;**135**:105213. DOI: 10.1016/j.cropro.2020.105213

[11] Rebolledo-Martínez A, del Ángel PAL, Laureano R-M, Becerril-Román AE, Uriza-Ávila D. Fruit yield and quality of pineapple cultivars under planting densities. Revista Fitotecnia Mexicana. 2006;**29**(1):55-62

[12] Siap-Sagarpa. Servicio de información agroalimentaria y pescaSecretaría de Agricultura, Ganadería, Desarrollo Rural, Pesca y Alimentación. Anuario Estadístico de la Producción Agrícola. 2009. Available from: http://www.siap.sagarpa.gob.mx

[13] Organización de la Naciones Unidas para la Agricultura y la Alimentación (FAO). Statistical Database. Tropical Foods Commodity Notes. 2002

[14] SIAP. Sistema de Información Agropecuaria y Pesquera. Cierre de la producción agrícola por estado. SIAP-SAGARPA. 2020. Available from: http://www.siap.gob.mx/agrícultura-produccion-anu al

[15] FAO. (Organización de la Naciones Unidas para la Agricultura y la Alimentación). 2009. Available from: http://faostat.fao.org/site/567/DesktopDefault.aspx?PageID=567#ancor

[16] Vélez-Izquierdo A, Espinosa-García JA, Uresti-Gil J, Jolalpa-Barrer JL, Rangel-Quintos J, Uresti-Duran D. Estudio técnico-económico para identificar áreas con potencial para producir piña en el trópico húmedo de México. Revista mexicana de ciencias agrícolas. 2020;**11**(7):1619-1632. DOI: 10.29312/remexca.v11i7.2594

[17] Betancourt YP, Montilla I, Hernández C, Gallardo E. Nitrogen fertilization in pineapple crop (*Ananas comosus* L. Merr) at Paramo Negro community, Iribarren, Lara state. Revista de la Facultad de Agronomía, 2005;**22**(4):382-393. Available from: http://ve.scielo.org/scielo.php?script=sci_arttext&pid=S0378-78182005000400006&lng=es&tlng=es

[18] Murillo-Hernández FE, Córdova-Sánchez S, Salgado-García S, Bolio-López GI, De la Cruz-Burelo P, Sánchez-Gutierrez F. Determinación nutrimental de piña criolla (*Ananas comosus* L. Merril) en la sub-región de la Chontalpa Tabasco, México. Agroproductividad. 2019;**12**(7):1-4. Available from: https://revista-agroproductividad.org/index.php/agroproductividad/article/view/1472

[19] Alejandro-Martínez P, De la Cruz-Morales M, Palma-López DJ, Megía-Vera HJ, Palma-Cancino DJ. Effect of land use change on edaphic properties in la sabana, Huimanguillo, Tabasco, México. Agroproductividad. 2019;**12**(7):91-100. Available from: https://revista-agroproductividad.org/index.php/agroproductividad/article/view/1476

Chapter 9

Cassava Production Enterprise in the Tropics

Raufu Olusola Sanusi, Deola-Tayo Lordbanjou, Azeez Olalekan Ibrahim, Mohammad Babakatcha Abubakar and Oluwole Olalekan Oke

Abstract

Cassava, a tropical root crop, provides the staple food for millions of people around the world. It is one of the tuber crops that could be cultivated on a small scale in an environment with erratic rainfall, and without necessarily needing heavy equipment and machineries. Cassava could be successfully cultivated by resource-poor farm family. Farmers' productivity could be as much as 70 tonnes per hectares under favourable conditions. However, smallholder farmers do among other things improve productivity through proven cultural practices and a mix of organic and inorganic measures. Irrigation is very necessary for achieving bumper harvest in areas with shortage of rainfall and insufficient soil moisture content. The concept of sustainability in the practice of agriculture has been on the front burner world over in recent time. Therefore, the cultivation of cassava with the aim of increased productivity without jeopardising the factors of production meant for future time is encouraged. Practices that combine traditional knowledge with modern technologies that are adapted to the needs of small-scale farmers are on the increase around the world. Depending on the purpose, cassava could be harvested anytime from eight month. Cassava leaves could serve as vegetable and the stems use as fire wood.

Keywords: cassava, farmers, food, sustainability, smallholder

1. Introduction

Cassava (*Manihot esculenta*), a shrub that could survive up to 3 years or more is planted mainly in tropic and the sub-tropic regions of the world. It is a food crop cultivated for the consumption of its roots (**Figures 1** and **2**) and other various end products. Cassava is a very important staple crop in Africa, Asia and Latin America. Due to its low cold tolerant nature, it does not do well in temperate regions of the world. It is a crop cultivated by majority of resource-poor farm families. Cassava is a crop that can withstand vagaries of weather condition; it can survive in a poor environment where other crops could hardly survive. It can survive in an acidic soil where other crops could hardly survive. There is a mutual relationship between cassava root and soil fungi which enable it to take phosphorus and other micronutrients from the surroundings. The whitish liquid (hydrogen cyanide) from cassava is deadly

Figure 1.
A typical cassava farm/plant/root.

Figure 2.
Cassava root and a leaf.

to both human and livestock. And as such it must be fermented and properly drained before fed to livestock or consume by man. Cassava maximizes available soil moisture content. Also, it is hardy and resistant to common pests and diseases of crops. With limited inputs, farmers can achieve a lot in term of output. The root is about 65–70% water but when processed, the dry matter could be as much as 3350 kg per tonne depending on the cultivar. It is a common staple food that could be afforded by the poor almost everywhere around the globe because it is relatively cheap. It is better harvested any moment from 6 month when it is to be consumed as food. The longer it stays in the soil the higher the starch concentration. It implies that those who cultivate it solely for starch would get more when it stays longer in the soil before harvesting.

Cassava is very rich in carbohydrate, and the calorie is high. It is energy given food which seriously help to mitigate the incidence of famine among the rural poor in sub-Sahara Africa and other places where it is cultivated. Also, cassava is rich in vitamin C, thiamine, riboflavin and niacin [1]. It is normally peeled and cooked to remove the cyanide acid.

The cyanide gas is volatile and would escape in the course of processing, making it and its bye-products fit for consumption. Relatedly, cassava mash is processed (by

Figure 3.
Cassava leaves.

drying, roasting or boiling) into coarse flour and other food products. Furthermore, cassava could be cultivated for the sole purpose of harvesting its leaf (**Figure 3**).

According to [2] the leaf contain about 27% protein when dried. Both the leaves and the roots can be fed to livestock, and the stem could serve as firewood. Also, starch, which is one of the by-products of cassava serves as raw materials in food manufacturing, pharmaceuticals, textiles, plywood, paper and adhesives, and for the production of ethanol.

2. Cultural practices

Cultural practices are the activities involved in the cultivation of cassava from the decision to plant it, site selection and right up to the harvesting and post harvesting operations. There have been campaigns worldwide against practices that are inimical to the human environment. Farmers are being enjoined to embrace eco-friendly agriculture. It has been observed overtime that the traditional ways of farming is not sustainable. Traditional methods such as plowing, harrowing, ridging and other operations which disturb soil structure and disrupts soil micro-organic activities are being replaced by environmental-friendly agriculture (zero or minimum tillage). Also, the use of bio-nutrients such as organic fertilizers, mulching and integrated pest management (IPM) are to be chosen instead of mineral fertilizers and chemical pesticides. Mineral fertilizers are volatile and as such release harmful gases into the air. Also, the leaching of the mineral fertilizers into water below the soil and runoff by erosion cause pollution to the water bodies. Also, the residue of the mineral fertilizer is toxic to the soil and the environment.

Moreover, in order to mitigate the vagary of challenges associated with agriculture, farmers engage in the mix of different crops on the same plot of land.

This strategy helps to reinforce soil fertility, and lessen the perennial problem of market and or price instability peculiar to agriculture and its products. For instance, having a mixture of nutrient-demanding and nutrient-giving crops such as cassava and any leguminous crop helps to stabilize the soil nutrient. Also, crop specific pests and diseases will not have freedom of self-perpetuation. The intercrops among other things enriched soil organic matter and reduce if not eliminate soil erosion and leaching of nutrients beyond the reach of plants' roots. Having more than a crop on a plot of land is a form of diversification which enhances food security.

Cassava requires soil with a loose texture to allow for initial root penetration and strengthening. It's susceptible to weed competition and too much moisture in the soil. Because of these factors, it is typically planted on soil that has been loosened and weed-free. Conventional tillage makes it easy to fix stakes in degraded and unstructured soils and provides well-drained, aerated conditions for the root system [3]. Crop yields on the other hand, are determined by soil conditions rather than tillage. Cassava stakes can also be planted in non-tilled soil and give good yields, as long as the soil is healthy, well-structured, and free of compaction. Soils that are pliable and rich in organic matter are the best for its cultivation.

Farmers usually plant stem cuttings (planting materials) on manually created mounds or ridges where soils possess weak physical qualities. Conventional plowing, especially with tractor-mounted plows, harrows, and other heavy machines, bury the protective cover of the soil, kills soil microorganisms, promotes fast decomposition of organic matter, and damages soil structure by pulverizing soil aggregates. Season after season of plowing or hoeing the soil at the same depth results into a compacted soil layer commonly located below the topsoil, and that makes it difficult for water and roots to penetrate. For ongoing crop production in such soils, mechanical loosening will be required, but at the expense of increased soil degradation. Growing cassava without tillage in the same soil may result in poorer yields in the first few years. However, in the long run, by decreasing mineralization, erosion, and water loss, organic matter may build up while also ensuring soil aggregate stability and internal drainage. Zero tillage enhances root function to the greatest extent possible. Once soil health has been restored, untilled land can generate high yields at a cheaper cost to both the farmer and the farming system's natural resource base [4].

3. Cover crops and mulching

Another fundamental strategy for enjoying the full benefits of conservation tillage is maintaining a continuous ground cover. Because cassava's initial development is slow, the soil is exposed to direct rain at the first few months of its growth, and the wide spacing between planted stakes favors the appearance of weeds. Therefore, ground cover is very crucial in cassava cultivation. Farmers cover the soil surface with mulch, such as crop residues, or grow cover crops, to protect the soil surface, reduce runoff and erosion, and inhibit weed growth. With little or no effort, cassava stakes can be planted simply through the mulch cover. Even during lengthy droughts, mulch cover protects the soil, reducing daily temperature changes and water loss. It raises the organic matter content of the soil and creates a favorable environment for soil microorganisms and wildlife below ground. It favors higher yields by improving physical soil conditions: lower soil temperatures, higher levels of moisture, increased water infiltration capacity, and lower evaporation [5].

4. Mixed cropping

Cassava is widely cultivated as a single crop in Thailand and southern Brazil, but intercropping is done by small-scale farmers in many parts of the tropics. Small scale farmers do normally produce early crops such as common beans, mung beans, peanuts, corn, upland rice, and various types of grain legumes between the Cassava rows. This method has many advantages. It protects the soil from the direct effects of rain, reduces soil erosion due to runoff, and limits weed growth in the early stages of cassava development. Intercropping also produces crops that can be harvested at different times of the year, increasing total net income per unit area and reducing the risk of total crop failure. For example, in southwestern Nigeria, corn and cassava are often cultivated at the beginning of the twice-yearly rainy season. Corn is harvested during a short rain break, after which cassava continues alone. The two plants have different pest and disease and growth requirements, so if one fails, the other can survive.

5. Planting materials and species

Cassava does well on poor soils, and can withstand erratic rainfall. Its ability to produce good yields without fertilizer/agrochemicals and or other external resources makes it one of most widely grown staple. However, cassava's potential will not be realized until some important production constraints are addressed by high yield and well adapted cultivars. Cassava are more affected by biological restrictions than drought and high temperatures [3]. As the importance of cassava as a food, animal feed and industrial feedstock grows worldwide, there is a growing demand for varieties with specific characteristics and adaptation to different ecosystems. In Africa, new varieties are being developed as cultivation expands to dry savanna, semi-arid and subtropical regions and the transition to market-oriented production accelerates. Providing high-yielding, adapted cassava varieties to small-scale farmers via a specific system is very crucial. The system consists of three parts: conservation and distribution of genetic resources, variety development, production of high-quality and healthy planting materials and delivery to farmers.

6. Breeding improved varieties

The early introduction of cassava to Africa and Asia presented a limited gene choice that limits the diversity available to farmers to select new varieties. For instance, a single clone was cultivated by majority of the farmers in Thailand until the 1990s [5]. As researchers across different institutes and several domestic breeding programs take advantage of the vast national breeding programs, they have excellent combinations of many useful traits. The availability of varieties has improved significantly in recent decades the genetic diversity available in gene banks. Breeding of high-yielding varieties with resistance or tolerance to biological and non-biological stress contributes to a significant increase in cassava yield and overall production.

7. Varieties and planting material

Stakes cut from healthy stems free of pests and diseases have a higher rate of sprouting and produce higher root yields. As a result, many farmers do not save cassava stems for planting and frequently source cuttings from neighbors or in local markets; under such conditions, assuring the quality of planting material is practically impossible. Effective systems for routine multiplication and distribution of disease-free planting material of improved varieties is essential for sustainable intensification. Although several protocols have been developed for the rapid multiplication of cassava, and could be scaled up for the dedicated production of material that meets quality standards [6], very few countries have a formal seed system for cassava multiplication.

For the production of cassava, it is essential to maintain genetic purity and use high quality planting materials that are free of diseases and pathogens. Because cassava propagates vegetatively (**Figure 4**), diseases and pests can continue for several generations.

This is a negligible problem with plant seeds. In addition, cassava cuttings are perishable, bulky, and cumbersome to transport and require significant storage space. Subsistence farmers usually harvest in small portions over a year, so storing stakes until the next planting is logistically challenging. Stakes cut from healthy stems free of pests and diseases have high germination rates and high root yields. As a result, many farmers do not preserve cassava stalks for planting and often procure cuttings from their neighbors or local markets. In such situations, it is virtually impossible to guarantee the quality of the planting material. An effective system for the daily reproduction and distribution of disease-free planting materials of improved cultivars is essential for sustainable production. Several protocols have been developed for rapid breeding of cassava and can be extended for targeted production of materials that meet quality standards [6]. Few countries have a formal stem multiplication system for cassava breeding.

To increase the efficiency of cassava stem production, IITA and Nigeria's National Root Crops Research Institute have developed a rapid multiplication technology, which involves cutting cassava stems into stakes with 2 or 3 nodes, rather than the usual 5–7. With efficient field management, cassava stems can be harvested twice a year, at 6 and 12 months after planting, yielding around 50 times more stems than were used for planting [7]. In the absence of a national cassava seed system, cassava development programmes in a number of African countries have used a 3-tier

(a) (b)

Figure 4.
(a) Cassava stem (b) Planted cassava stem cuttings.

community-based system of rapid multiplication to supply farmers with improved, healthy planting material [8]. At the top level, material from breeders is multiplied under optimal agronomic conditions on research stations and government farms to produce disease-free foundation seed. The secondary level involves further multiplication on farms often run by farmer groups, community organizations and NGOs. Certified material is then distributed to tertiary multiplication sites, which are the main and most readily accessible source of stems [9].

High participation in grassroots growth was achieved through the Great Lakes Cassava Initiative, managed by the Catholic Relief Services Foundation and supported by the Bill & Melinda Gates Foundation. It established a network of 6500 small breeding plots with an average size of 0.3 ha, each serving about 350 local farmers and contributing to the breeding of a total of 33.6 million stems. This initiative also introduced a low-cost quality control protocol based on visual assessments to assess variety purity and pest and disease assessments. The use of poor-quality planting material will remain one of the major causes of low cassava yields, especially in Latin America and Africa for some time to come. In the absence of efficient systems of multiplication and distribution, farmers can help to improve the situation using some simple local practices:

a. Take stems from healthy plants that are 8–12 months old, free of pests and diseases, growing in fertile soil, and producing high root yields. The primary stems of late-branching types with long, straight primary stems are the best.

b. Store cut stems in the shade, erect, with the base of the stems resting on dirt that has been loosened with a hoe and is frequently watered. Stems that have only been preserved for 5 days before being cut into stakes will sprout faster.

c. Just before planting, cut stems into stakes 20 cm long with 5–7 nodes each. The stakes should have a diameter of at least 3 cm, and the pith should be less than half the diameter of the stem.

d. Soak the stakes in hot water for 5–10 minutes before planting to destroy any pests or disease-causing organisms that may be present. It's also easy to get the proper water temperature by mixing equal parts hot and cold water [10]. The stakes' mother plants should have been appropriately fertilized to achieve large yields. Cassava plants cultivated on low-nitrogen, low-phosphorus, and low-potassium soil produce stakes that are also poor in those nutrients. In addition, they are low in starch, reducing sugars, and total sugars. Plants produced from low-nutrient stakes have a reduced rate of sprouting, produce fewer stems, and have poorer root yields as a result [11]. Some plants develop faster and produce more roots than others, even in a uniformly treated field. Farmers can boost the quantity of their next cassava harvest by only using stems from plants with strong root yields as planting material.

Rainfall is the only source of water for almost 80% of the world's cropland. Rainfed cassava production accounts for up to 60% of worldwide agricultural output, and millions of the world's poorest farmers rely on it for their livelihoods and food security. Irrigated agriculture produces up to three times more from the same unit area of land due to higher cultivation intensities and average yields. Agriculture, both rainfed and irrigated, faces significant obstacles. Irrigation is under increasing pressure to produce more crops with fewer drops and to lessen its negative environmental implications, such as soil salinization and nitrate poisoning of drinking water, as competition for increasingly precious water resources intensifies. More precise water-saving methods, including drip and micro-irrigation, should be used. Rainfed agricultural production is in grave danger as a result of climate change. By 2050, most scenarios predict a 30% or more decrease in rainfall runoff across large areas of Sub-Saharan Africa, South Asia, and Latin America. Crop yields are expected to drop in many developing countries as water flows grow more erratic, and the frequency of droughts and floods rises [12]. Nonetheless, a comprehensive review of agricultural water management indicated that rainfed areas have the highest potential for productivity gains [13]. However, better and drought-tolerant cultivars should be cultivated. In addition, widespread adoption of conservation tillage, mulching, and other soil management measures, as well as land deterioration and irrigation reversal should be practiced. Cassava, unlike most other food crops, does not have a crucial period for blooming and seed formation during which adequate soil moisture is required. It also has various water-saving defense mechanisms, and its roots can reach enormous depths to access subterranean moisture stores [14]. Therefore, cassava can tolerate droughts for a long period of time [6].

8. Rainfed production

Cassava is nearly entirely a rainfed crop in most parts of the world. Rainfed cassava cultivation involves careful attention to planting dates, the use of planting methods and planting sites that make use of available soil moisture, and water-conserving soil management procedures. Cassava may be grown all year if rainfall is evenly distributed, but it cannot be planted during seasons of excessive rains or drought [15]. Farmers in locations where there is only one rainy season per year plant as soon as the rains begin, which is normally in April–May in the northern tropics and October–November in the southern tropics. As the topsoil begins to dry out with the coming of the dry season, new plants will grow deeper roots once established. Before the start of the 5-month rainy season in Andhra Pradesh, India, farmers plant cassava in well-watered nursery beds to induce sprouting and root development. The rooted stakes are relocated to the field as the rains begin. If the early rains fail to hold off and any of the transplanted stakes perish, they are replaced with newly sprouted stakes from the nursery beds. Farmers can make the most of the short wet season by using this method, which eliminates the need for irrigation. In lowland paddy fields, however, some farmers plant short-duration cassava in February, after the rice has been harvested and the soil is still wet.

In lowland paddy fields, however, some farmers plant short-duration cassava in February, after the rice has been harvested and the soil is still wet. During the dry months that follow, the crop benefits from the leftover soil moisture, and it is harvested after 8 months before the area is utilized again for rice. Because the plants receive adequate soil moisture throughout the most essential stage of their growth cycle, planting early in the rainy season will normally generate the largest yields. However, outputs depend on the cultivar of the crop planted. Also, the edaphic nature of the soil coupled with the maturity of the crop as well as the rainfall intensity reinforced to determine the harvest achieved by a farmer in a given year. Planting during the month of June for instance, resulted in yield of about

38 tonnes/hectare as against 26 tonnes/hectare at the beginning of dry season in October [15]. Later research at the same location in Thailand found that planting from August to November produced the highest average yield. A more recent experiment, this one conducted over 3 years, yielded a different outcome. Cassava root yields were highest when it was planted in December, early in the dry season, and harvested 11 months later, in November [16].

Under rainfed agriculture, planting practices must be adapted to the soil moisture levels. Plant stakes on the tops of ridges or mounds to keep the roots above the standing water when the soil is poorly drained and overly wet due to heavy rainfall. This will also help to prevent root rot. When cassava is planted on the flat land in Thailand during dry periods, the rates of stake sprouting and plant survival are much higher, owing to the somewhat increased soil moisture content in the top 30 cm of soil [17]. In heavy and wet soils, stakes should be planted at a shallow depth of 5–10 cm, but slightly deeper in light-textured and dry soils.

With minimal tillage, which enhances internal drainage, the risk of waterlogging is reduced. When tillage is employed, farmland is better worked during the time when the internal drainage of the soil is optimum. The advantage of this is that it gives room for the practice of zero tillage which further enhances the soil condition. Planting 2 months towards the end of rainy season is beneficial as it reduces weed menace.

9. Irrigated production

Cassava benefits from extra irrigation during rainless times when planted near the end of the rainy season or when the rainy season is relatively short. On level or almost flat land, flood or furrow irrigation can be used, but on sloping soil, overhead sprinklers or a spinning water cannon may be more practical. Irrigation at 100% of the crop's water needs increased the root production attained without irrigation. It also marginally enhanced the starch content of roots while significantly lowering the hydrogen cyanide concentration [18].

Drip irrigation, which saves water while keeping soil moisture at a level that is very beneficial to crop growth, is more successful in terms of water use efficiency. Drip irrigation saves water by giving modest and frequent water applications (it also allows the farmer to water the cassava plants but not the weeds). Drip irrigation of cassava generated roughly the same yields as flood irrigation in trials in the severely arid zone. When drip irrigation was employed with the same amount of water as flood irrigation, yields increased significantly, reaching 67.3 tonnes somewhere in India [19]. Experiments conducted in south-western Nigeria yielded similar results. Rainfed cassava yielded root yields of fewer than 5 tonnes per hectare during the growing season. In plots with supplemental drip irrigation, yields increased dramatically as the amount of water provided increased. Irrigation resulted in yields of about 30 tonnes at 100% rainfall.

10. Crop nutrition

Agriculture must literally return to its roots to attain the increased production required to fulfill present and future demand by recognizing the value of healthy soil, drawing on natural sources of crop nutrition, and properly applying mineral fertilizer. The overuse of mineral fertilizer in agricultural production has resulted in severe environmental consequences, such as soil acidification, water contamination, and air pollution. Fertilizer use that is more focused and sparing would save farmers money while also ensuring that nutrients reach crops and do not harm the air, soil, or rivers. The environmental impact of mineral fertilizer is a matter of management. In other words, the way with which fertilizers are used, particularly nitrogen (N) and phosphorus (P), affects whether this component of soil fertility management is beneficial to crops or harmful to the environment. Experience shows that crop nutrients from a mix of mineral fertilizer and organic sources, such as animal manure and trees and bushes, enrich the soil with nutrients, resulting in better and more sustainable yields of crops. Other biological relationships, such as those between plant roots and soil mycorrhizae, can improve crop nutrition. The foundation of a sustainable crop nutrition system that yields more is a mix of ecological processes and judicious application of mineral fertilizer [12]. On soils where many other crops would fail, cassava may flourish and generate reasonable yields. It has a great tolerance for low-phosphorus and can often thrive without the use of phosphorus fertilizer. This is because cassava has created a favorable relationship with a fungus group known as "vesicular-arbuscular mycorrhizae" [13]. Mycorrhizae, which may be found in almost all natural soils, penetrate the cassava root and feed on the sugars it produces. In exchange, the fungi's long filaments transfer phosphate and micronutrients to the plant from the surrounding soil. Cassava can absorb enough phosphorus for optimum growth because to this mutual relationship.

The plant tops contain the majority of the nutrients taken by cassava during its growth [6]. After the root harvest, returning stems and leaves to the soil as leaf litter or mulch nourishes the soil with new organic matter, and some of the nutrients are re-used by the following crop. When the plant tops are recycled, the root harvest eliminates less soil nutrients than most other crops [3]. A root yield of 15 tonnes per ha removes only about 30 kg of nitrogen, 20 kg of potassium (K), and just 3.5 kg of phosphorus [20]. Even after many years of continuous cassava production on the same land, there is little risk of phosphorus depletion. Cassava may be cultivated on very acidic and low-fertility soils due to its tolerance for low pH and the large levels of exchangeable aluminum that come with it. While maize and rice yields are typically negatively impacted when the soil pH is below 5 and aluminum saturation is above 50%, cassava yields are typically unaffected until the soil pH is below 4.2 and aluminum saturation is beyond 80%. As a result, cassava may not require a lot of lime on acidic soils where other crops would struggle to do well.

11. Mineral fertilizer

Cassava responds positively to mineral fertilizer application. Traditional methods of managing soil fertility, such as intercropping and mulching increase cassava requirement for fertilizer. The harvest removes considerable amounts of nitrogen and potassium when root yields are high and wastes are not returned to the soil. Cassava would require annual per hectare treatments of 50–100 kg nitrogen, 65–80 kg potassium, and 10–20 kg phosphorus to maintain both yields and soil fertility. The predominant nutrient constraint was lack of K in 12 trials, lack of N in five trials, and lack of P in just two trials, according to the results of 19 long-term fertility studies conducted over 4–36 years of continuous cassava planting on the same plots. When suitable amounts of mineral fertilizer (100 kg N + 22 kg P + 83 kg K) were supplied annually and plant foliage was returned to the soil before each new planting, high root yields of up to 40 tonnes per ha were maintained in Thailand. Due to nutrient depletion, notably of potassium, per hectare yields fell drastically when no fertilizer was provided and plant tops were removed from the field, from 30 tonnes in the first year to roughly 7 tonnes after 6 years. Similar effects have been observed in Colombia, India, Indonesia, Malaysia, Thailand, and Vietnam on a variety of soils [12].

Production of cassava on the same piece of land for several years would require adjustment in N-P-K balance to account for the removal of each nutrient during the root harvest. This can be accomplished by utilizing fertilizers with a 2:1:3 ratio of N, P2O5, and K2O, or any compound fertilizer high in K and N but low in P. Local fertilizer recommendations based on crop experiment outcomes and or simple fertilizer trials conducted in farmers' fields should be considered first. Compound fertilizers should be used either when the stakes are planted or, preferably, at or shortly after planting. N and K should be sprayed in two parts, one at or soon after planting and the other 2–3 months later, when cassava reaches its maximum growth rate. The majority of mineral fertilizers dissolve quickly in soil water. They should be planted in 20–30 cm long, 4–5 cm deep bands dug at a distance of around 6–10 cm from the cassava stake or plant. The fertilizers should be covered with soil after application to prevent N volatilization and nutrient losses due to runoff and erosion. The plant's roots will develop in the direction of the fertilizer solution to take up the nutrients.

12. Organic sources of nutrients

Mineral fertilizer can assist to reinforce yields. Nevertheless it cannot all alone sustain crop production for a long period of time on a depleted soil [21]. Farmers want to preserve and enhance soil best and fitness by the usage of different measures which include conservation tillage, alley cropping and manuring. Intercropping with grain legumes help fix atmospheric nitrogen to the soil. Although organic fixation cannot meet all of cassava's nitrogen needs, it is however very important. Combining *Leucaena* with fertilizer bring about yields of greater than 20 tonnes. However, the benefit of alley cropping is limited in tropical soils which are largely barren or less productive. The mix of shrubs in rows of cassava in such area might bring about bumper harvest [6].

13. Pests and diseases

By cultivating insect-resistant cultivar, maintaining and encouraging biological control agents as well as regulating crop nutrient levels to minimize insect reproduction, agricultural losses to insects are kept to an acceptable minimum. Diseases are controlled through the use of disease-free planting material, pathogen-suppressing crop rotations, and the removal of affected host plants. To reduce weed growth, timely hand weeding and the use of surface mulching are required for effective weed management. Low-risk selective pesticides can be employed for targeted control as necessary, in the right amount and at the right time. Because all pesticides have the potential to be dangerous to people and the environment, they must be locally registered and approved, with explicit instructions on how to handle and use them safely. Cassava, like all important crops is susceptible to pests and diseases that can result in significant yield losses. In Africa, their impact is very severe. Asia had few severe pest and disease concerns until recently, but that may be changing as the crop is produced more intensively over bigger regions and planted all year for industrial processing. When pest or disease management measures are required, a non-chemical control plan should be examined before deciding to use pesticides. Pesticides are frequently inefficient and rarely cost-effective because cassava is a long-season crop that is exposed to pests and diseases for a longer period of time. As a result, insecticides should only be used in short-term, localized applications in areas where the pest is first noticed, and only when the pest is still in its early stages (vulnerable stage) of development.

A variety of non-chemical methods can assist farmers in reducing pest and disease losses while also safeguarding the agro-ecosystem [19]. First, planting material should come from mother plants that are free of disease symptoms and insect attacks, as well as types that have tolerance or resistance to the most common cassava diseases and pests. Stem cuttings can be soaked in hot water as an extra precaution to eliminate any pests or disease-causing organisms that may be present. Also, cuttings may need to be soaked in a fungicide and pesticide solution in extreme circumstances. Farmers who do so, however, must have obtained pesticide training and should select herbicides based on the recommendations of local plant protection professionals. Mulching, planting hedges, and intercropping are examples of ecosystem-based techniques that can provide refuge for natural enemies of insect pests. Early in the cropping cycle, increasing soil organic matter enhances pest-regulating populations. Applying proper quantity of manure and or fertilizer help to improve crop resilience. Insecticides should be applied with caution as they possess the chemicals that are deadly to the natural enemies of pests and diseases. Insecticides kill those biological control agents and other predators that feed on cassava pests. When this is the case, pest population rises prompting farmers to use more pesticides, repeating and exacerbating the pest harm cycle. Whiteflies, mealybugs, and variegated grasshoppers can all be controlled with biopesticides like neem seed oil extract. Sticky traps and spraying plants with soapy water can also help to minimize the amount of whiteflies and mealybugs.

Although the majority of cassava diseases are found in Latin America and the Caribbean, where the plant originated, several are now prevalent in Sub-Saharan Africa and Asia as well. Some have evolved individually in Africa and Asia, and others have evolved together. Some have evolved in Africa and Asia separately and have yet to reach the Americas. One of the most common and dangerous cassava disease is bacterial blight. It is spread mostly by infected planting material or infected agricultural tools. Rain splash, as well as the movement of people, machines, or animals from infected to healthy fields, can transfer it from one plant to another. The bacterium affects the leaves initially, which become brown in big patches and eventually die, then the petioles and woody stems' vascular tissues. The impact of bacterial blight on yields varies according to region, variety, weather patterns, planting period, and planting material quality. Bacterial blight can jeopardize food security by lowering the yield of cassava leaves, a key source of vegetable protein in Central Africa. Despite its catastrophic potential, bacterial blight can be efficiently controlled by excellent agricultural techniques, viz.:

a. Use disease-free planting material or plants grown from meristem culture, rooted buds, or shoots

b. Soak stakes in hot water for about 50 minutes before planting. Stakes may be immersed in a cupper solution for 10 minutes in exceptional circumstances, and on the recommendation of a professionals.

c. Planting should be done towards the end of wet season

d. Infected tools should be sterilized

e. Plants should properly be fertilized, particularly in terms of potassium.

f. Burning infected plants and agricultural leftovers

g. Intercropping cassava with other crops to minimize diseases spread

h. Cassava should be rotated with other crops or left fallow in order to avoid disease transmission in the soil. The most common way for viral infections to spread is through the use of infected planting material.

In Sub-Saharan Africa, cassava mosaic disease (CMD) is endemic. Misshapen leaves, chlorosis, mottling, and mosaic are all common signs. Stunting and general decline occur in plants, and the more severe the symptoms are, the lower the root output. Corky necrosis in roots caused by cassava brown streak disease (CBSD) renders them unsafe for ingestion. Farmers may not realize their crops are infected until they harvest the roots because the signs of CBSD are not visible on the cassava leaves or stems. Because there are no visible indications above ground, disease-infected planting material is more likely to be used. Strict adherence to quarantine measures during international cassava germplasm exchange, as well as cultural methods, particularly the use of resistant or tolerant cultivars and virus-free planting material are two critical suggestions for controlling both CMD and CBSD. CMD and CBSD-free planting material has been developed and distributed with great success. In January 2012, the United Republic of Tanzania released four high-yielding cassava varieties that are resistant to CMD and tolerant to CBSD. Researchers at different institutes across the globe have been working to develop series of CMD-resistant lines [22]. Root rots are abundant in Africa, Asia, and Latin America, and they occur primarily in poorly drained soils during periods of heavy rain. They are caused by a variety of fungal and bacterial infections and result in leaf loss, stem and shoot death, and root degeneration as the crop matures or during post-harvest storage. Post-harvest farm implements and plant leftovers are frequently contaminated with disease-causing fungus and serve as sources of spores that infect new plants. Other cultural methods that control root rots include:

a. Immerse stakes in hot water for roughly 50 minutes if no disease-free planting material is available;

b. Plant on light-textured, fairly deep soils with good internal drainage.

c. Reduce tillage and use surface mulches to improve drainage.

d. Cassava should be rotated with cereals or grasses, and unhealthy plants should be uprooted and burned.

Immersion of the stakes in a suspension of Trichoderma viride is very efficient biological control for root rot [21]. Two groups of preserved cassava roots were injected with four pathogenic fungus in Nigerian tests. A culture filtrate of T. viride was also given to one of the groups. The incidence of rot in the group without T. viride ranged from 20 to 44% after 3 weeks; in the group inoculated with the biocontrol agent, there was a drastic reduction in the range and number of the target fungi after 3 weeks, with the incidence of rot ranging from 0 to 3%. T. viride inoculation eliminated the need for frequent synthetic fungicide application [23].

14. Weed management

Compared to several other crops, the initial growth of cassava is slow. As a result of this and the wide spacing between planted stakes, weed emergence and competition with the crop for available soil nutrients and sunlight is rife. In the first 4 months after planting, cassava can easily be overwhelmed by competition from weeds, and other leguminous plants. In East Africa, weeds are often a more serious production constraint than insect pests or diseases and may reduce yields by about 50% [24]. In Nigeria, farmers expend more resources controlling weed than other aspect of crop production. Once the cassava canopy has closed, it'll shade out most weeds and keep the sector almost completely weed-free (**Figure 5**). Six to eight months after planting, when cassava starts to shed many leaves (especially during the dry season), weeds may reappear, but this generally does not seriously affect yields. Excessive late weed growth may make harvesting harder, but also can protect the soil from erosion if post-harvest rains are heavy.

While cultural controls might not perfectly control weed, they are effective in reducing weed competition, and thus the necessity for mechanical or chemical weeding [25]. Cultural control begins with selection of high-quality planting material from varieties with vigorous early growth and tolerance or resistance to diseases and pests. High planting density and therefore the correct type and rate of fertilizer can stimulate early crop growth and rapid canopy closure. Planting within the season under drip irrigation also can encourage the expansion of cassava but not that of weeds. The soil should be covered with a thick layer of mulching material such as rice straw or maize residues to stop weed. Also, intercropping cassava with fast-growing plants, like melons, squash, pumpkins, common beans, groundnuts, soybeans, mungbeans and cowpeas proved to be effective in controlling weeds. Since those are short-duration crops, they will be harvested after about 3 to 4 months,

Figure 5.
Cassava canopy checks weeds.

when the cassava canopy closes and weeds are shaded out. While intercrops may reduce cassava root yields, they markedly reduce weed growth, and offer an eco-friendly and fewer expensive alternative to spraying with herbicides. A study in Nigeria of legume cover crops during a mixed cassava/maize system reported significant improvements in cassava root yields when velvet beans were grown to suppress weeds [26]. Common among the smallholder cassava farmers is mechanical control measures–by hoeing, starting after emergence. Research in Colombia found that with hand-weeding at 15, 30, 60 and 120 days after planting, cassava root yields were 18 tonnes per ha compared with only 8 tonnes/ha were obtained when weeds were controlled with herbicides. When weeds were not controlled in the least, yields fell to only 1.4 tonnes.

Weeds are often controlled with herbicides. Although many herbicides are highly toxic and, being water soluble and protracted within the environment, are often washed away to contaminate ground and surface water. Farmers got to exercise care within the choice of the herbicide to be used and follow the recommendation of local plant protection specialists. Pre-emergence herbicides do not kill existing weeds. Instead, they prevent weed seeds within the soil from emerging or, at least, reduce their rate of growth. Pre-emergence herbicides are either incorporated into the soil before planting or applied on the soil surface with a knapsack sprayer immediately after planting. Pre-emergence herbicides that are selective for cassava are often applied over the vertically planted stakes without affecting cassava sprouting or yield. The application of pre-emergence herbicides can maintain a cassava field almost weed-free for 6–8 weeks after planting. Farmers may apply a mix of two herbicides; one that controls the grassy weeds and the other on the broad-leaf weeds. A lower dosage is suggested on light-textured soils, while a higher dosage could be needed in heavy soils, like clay-loamy. Special care must be taken when cassava is grown in association with other crops, because the pre-emergence herbicides normally used for cassava may harm the intercrop. At about 2 months after planting, weeds may have to be controlled again to scale back competition with cassava. This is often usually done by hoeing or using an animal or tractor-mounted cultivator, counting on the peak of the growing cassava plants and therefore the extent of cover closure. When most of the weeds are grassy species, it's also possible to use a selective post-emergence herbicide, which kills grasses but does not affect the cassava plant. Post-emergence herbicides are often used about 4–5 months after planting, when some bottom leaves start to drop off. It is best done on a windless days and with a nozzle shield to stop spray from reaching the cassava stems or leaves.

15. Harvest

Cassava is due for harvesting any time from 6 month. The crop does not have a specific time or season for its harvest; it can be harvested all-year-round. The fact that it can stay long and be preserved in the soil gives it the utmost advantage of being harvested piecemeal over a long period. The root is cooked and consumed as a local delicacy. Also, it could be processed to give a varieties of products (**Figures 6–8**) as a result of value addition. Cassava leaves can be fed upon as vegetable, and it is used as such in many homes where they are planted in West African countries. Moreover, cassava leaves and root serve as a good source of nutrients for livestock. The leaves are rich in vitamins.

Figure 6.
Cassava coarse (grains) flour.

Figure 7.
Cassava (smooth) flour.

Figure 8.
Cassava end (food) products.

Cassava has a number of advantages, one of which is, it does not have a set harvesting season. They can be collected whenever needed during times of food scarcity, frequently one plant or even one root at a time. Harvesting for human consumption takes roughly 8–10 months; for industrial purposes, a longer growth time yields a higher root and starch output. Roots can be eaten directly by farm families, given to livestock, or sold for processing into a wide range of value-added products, from coarse flour ('Garri')to high-tech modified starch gels. The root of the plant is not the only portion that can be useful. The green section of the upper stem, which includes the leaves and petioles, is fed to cattle and buffaloes in several countries, while the leaf blades are fed to pigs and chickens. Fresh leaves are used to raise silkworms in China, Thailand, and Vietnam. Woody stems are crushed up and used as a substrate for growing mushrooms. Stumps are burned as fuelwood [12].

Cassava roots are typically collected by cutting the stems approximately 20 cm above ground and then dragging the entire root system out of the ground using the stump. If the soil is too hard or the roots are too deep, it may be necessary to remove the soil around the roots with a hoe, spade, or pick while avoiding injury to the roots. A harvesting blade mounted to a tractor is occasionally employed in heavy soils that can become quite hard in the dry season. The sword slashes through the soil material. The tractor's forward momentum pulls the root clusters to the surface as the blade slices through the dirt right below the roots. The roots are then removed from the stump and transported in baskets or bags.

Large cassava fields are frequently harvested by middlemen who employ teams of labourers and deliver the roots to marketplaces or processing plants via trucks. Plant tops are harvested after the root harvest. Plant tops are generally left to dry on the ground after root harvesting and then integrated into the soil to help preserve its fertility. However, by trimming the green tops every 3 months during the plant's growth cycle, farmers can considerably increase the total amount of cassava foliage available for feeding to animals. Within 2–3 months after each trimming, the remaining stems will sprout and produce a new crop of leaves. Cassava stakes should be planted at a closer spacing of about 60 × 60 cm for maximum foliage output. Young leaves harvested at regular intervals during the cassava growth cycle have a higher protein and lower fiber content than those gathered at the end of the cassava growth cycle, when plants are generally harvested between 11 and 12 months. Younger leaves are more pleasant and give better nutrition.

The ultimate root production decreased as the frequency of leaf cutting increased, from roughly 40 tonnes per ha when leaves were collected only once at the time of root harvest to less than 25 tonnes when leaves were removed 5 times [27]. This approach may or may not be cost-effective, depending on labour costs and the relative pricing of fresh roots and dry leaves. Harvesting the plant tops four or five times over a one-year growth cycle takes a substantial amount of nutrients, particularly nitrogen, from the field, and would be unsustainable unless large amounts of mineral fertilizer were applied to maintain soil fertility.

Author details

Raufu Olusola Sanusi[1*], Deola-Tayo Lordbanjou[2], Azeez Olalekan Ibrahim[3], Mohammad Babakatcha Abubakar[4] and Oluwole Olalekan Oke[5]

1 Federal University of Agriculture Abeokuta, Nigeria

2 Federal Ministry of Agriculture and Rural Development, Abuja, Nigeria

3 Federal College of Wildlife Management New Bussa, Nigeria

4 National Cereals Research Institute Badeggi, Nigeria

5 Federal College of Forestry Ibadan, Nigeria

*Address all correspondence to: unbeatable2013@gmail.com

References

[1] FAO. Feeding Pigs in the Tropics. FAO Animal Production and Health Paper 132. Rome: Ministry of sugar, Havana, Cuba; 1997

[2] Chavez AL, Bedoya JM, Sanchez T, Iglesias C, Ceballos H, Roca W. Iron, carotene, and ascorbic acid in cassava roots and leaves. Food and Nutrition Bulletin. 2000;**21**:4

[3] Hershey C, Álvarez E, Aye TM, Becerra LA, Bellotti A, Ceballos H, et al. Ecoefficient interventions to support cassava's multiple roles in improving the lives of smallholders. In: Ecoefficiency: From Vision to Reality. Cali, Colombia: CIAT; 2012

[4] FAO. FAOSTAT statistical database; 2013. Available from: http://faostat.fao.org

[5] Fukuba H, Igarashi O, Briones CM, Mendoza EMT. Determination and detoxification of cyanide in cassava and cassava products. Philippine Journal of Crop Science. 1982;**7**(3):170-175

[6] FAO. The Second Report on the State of the world's Plant Genetic Resources for Food and Agriculture. Rome; 2010

[7] CTA. In: Chikwado EK, editor. Cassava Stem Multiplication Technology: A Viable Option for Industry Development? Umudike, Nigeria: National Root Crops Research Institute; 2012

[8] ASARECA & TUUSI. Technology transfer mechanism: Threetier cassava multiplication approach complemented by and the seed voucher system. Lessons learnt and scaling up; 2007. Available from: http://www.asareca.org/tuusi/

[9] IFAD. Republic of Ghana root and tuber improvement programme interim evaluation. Report No. 1533- GH. Rome; 2004

[10] IITA. Impact Assessment Evaluation of Cassava Enterprise Development Project (CEDP). Ibadan, Nigeria; 2010

[11] Lopez J, El-Sharkawy MA. Increasing crop productivity in cassava by fertilizing production of planting material. Field Crops Research. 1995;**44**:151-157

[12] FAO. Save and Grow. A policymaker's Guide to the Sustainable Intensification of Smallholder Crop Production. Rome; 2012

[13] Jarvis A, Ramirez-Villegas J, Herrera Campo BV, NavarroRacines C. Is cassava the answer to African climate change adaptation? Tropical Plant Biology. 2012;**5**(1):9-29. DOI: 10.1007/s12042-012-9096-7

[14] Whyte J. Cassava: Lessons from investments since 2000. Document prepared for FAO. mimeo; 2012

[15] Kawano K. Thirty years of cassava breeding for productivity—Biological and social factors for success. Crop Science. 2003;**43**:1325-1335

[16] FAO. The Second Report on the State of the world's Plant Genetic Resources for Food and Agriculture. Rome; 2010

[17] Ceballos H, Sánchez T, Denyer K, Tofiño AP, Rosero EA, Dufour D, et al. Induction and identification of a small-granule, highamylose mutant in cassava (Manihot esculenta Crantz). Journal of Agricultural and Food Chemistry. 2008;**56**(16):7215-7222

[18] Sánchez T, Dufour D, Moreno IX, Ceballos H. Pasting and gel stability of

waxy and normal starches from cassava, potato, maize, and rice under thermal, chemical and mechanical stress. Journal of Agricultural and Food Chemistry. 2010;**58**:5093-5509

[19] Bellotti AC, Herrera CL, Hernandez MP, Arias B, Guerrero JM, Melo EL. Cassava pests in Latin America, Africa and Asia. In: Howeler RH, editor. The Cassava Handbook—A Reference Manual Based on the Asian Regional Cassava Training Course, Held in Thailand. Cali, Colombia: CIAT; 2012. pp. 199-257

[20] Sarakarn S, Limsila A, Watananonta W, Suparhan D, Suriyapan P. Cassava breeding and varietal dissemination in Thailand major achievements during the past 25 years. In: Howeler RH, Tan SL, editors. Cassava's Potential in Asia in the 21st Century: Present Situation and Future Research and Development Needs. Proceedings of the 6th Regional Cassava Workshop Held in Ho chi Minh City, Viet Nam, 21-25 February 2000. CIAT; 2000. pp. 161-166

[21] CIAT. Improved cassava for the developing world. Annual Report 2006. Project; 2006

[22] Araham K, Edison S, Sreekumari MT, Sheela MN, Unnikrishnan P, Pillai SV. Recent progress in cassava breeding in India. In: Howeler RH, editor. A New Future for Cassava in Asia: Its Use as Food, Feed and Fuel to Benefit the Poor. Proceedings of the 8th Regional Workshop, Held in Vientiane, Lao PDR. October 20-24, 2008

[23] Ubalua AO, Oti E. Antagonistic properties of Trichoderma viride on post-harvest cassava root rot pathogens. African Journal of Biotechnology. 2007;**6**(21):2447-2450

[24] Leihner D. Agronomy and cropping systems. In: Hillocks RJ, Thresh JM, Bellotti AC, editors. Cassava, Production and Utilization. Wallingford, UK: CAB International; 2002. pp. 91-113

[25] Leihner DE. Cultural control of weeds. In: Weber EJ, Toro JC, Graham M, editors. Cassava Cultural Practices. Proceeding of a Workshop, Held in Salvador, Bahia, Brazil. March 18-21, 1980. 1980. pp. 107-111

[26] Olorunmaiye PM. Weed control potential of five legume cover crops in maize/cassava intercrop in a southern Guinea savanna ecosystem of Nigeria. Australian Journal of Crop Science. 2010;**4**(5):324-329

[27] Lutaladio NB, Ezumah HC. Cassava leaf harvesting in Zaire. In: Terry E, Oduro K, Caveness F, editors. Tropical Root Crops: Research Strategies for the 1980s. Ibadan, Nigeria: IITA; 1981. pp. 134-136

Chapter 10

Identification of Native *Dendrobium* Based on Morphological and Anatomical Characters in Liwa Botanical Garden

Mahfut

Abstract

Orchid is the most popular ornamental plant. One of the many orchid genus collected and known to have high morphological variations in the Liwa Botanical Garden is *Dendrobium*. But until now many collections have not been identified. This study aims to determine variations and identification of *Dendrobium* based on characters of morphological and stomata anatomical at Liwa Botanical Garden. Total collections of five *Dendrobium* accessions were namely CAT140, CAT 144, CAT 271, CAT 274, and IR015. The result showed leaf organs had high variations based on observation of 11 morphological characters. The phenetic relationship showed five accessions of *Dendrobium* can be classified into 2 main groups formed with a similarity index value of 0.813 based on the Gower similarity value and the Unweighted Pair Group Method with Arithmetic Mean (UPGMA) method. Meanwhile, the results of the observation of 9 anatomical characters on the upper and lower surfaces of the leaves showed accessions IR015, CAT 140, and CAT 274 have stomata only on the lower surface, while CAT 144 has stomata on the upper and lower surfaces. The results of this study are expected to provide basic information in identifying natural orchids and conservation efforts in Liwa Botanical Garden.

Keywords: anatomical, *Dendrobium*, Liwa Botanical Garden, morphological, UPGMA

1. Introduction

Indonesia is known for its biodiversity of flora and fauna. One species of flora that has a high level of diversity is the orchid plant. The world's orchid species consist of 20,000 species spread over 900 genera. Orchids have a variety of variations that are located in the morphology, such as the shape of the flower, the number of florets, the size and color of the florets, the diversity of leaf and stem shapes (pseudobulb). One of the most orchid species, namely *Dendrobium*, with a total of 1500 species spread very widely throughout the world, from Japan, China, India, the Malacca Peninsula, Indonesia, the island of Papua, to Australia. This orchid has a charming flower and its types are also among the most [1–6].

Dendrobium comes from the words "dendro" (tree) and "bios" (life). *Dendrobium* means orchid that grows on a living tree. The advantages of *Dendrobium* because they have a variety of shapes, sizes and colors of flowers. Flowers that have bloomed can last more than 30 days (still in pots) and each stem has more than 20 flower buds arranged neatly and beautifully. When they are adults, *Dendrobium* can remove more than two flower stems at the same time throughout the year. Easy to adapt so easy to maintain. *Dendrobium* growth will be optimal at locations less than 400 meters above sea level. Even so, maintenance in areas over 400 m above sea level can still grow and flower, but not optimum. This orchid has a relatively cheap price. This makes many people tempted to hunt orchids because of high economic value [6].

Lampung is one of the places on the island of Sumatra which has a flora conservation area, which is located in the Liwa Botanical Gardens, West Lampung Regency. The Liwa Botanical Garden has many species of orchids that have not yet been identified, given the high increase in exploitation due to economic reasons, this can threaten the existence of natural orchid plants that cause loss of their natural habitat and natural damage resulting in the extinction of existing flora species, especially on plants orchid. Moreover, orchid plants have high economic value because of the beauty of the various forms of flowers. This makes the reason people can just hunt for existing natural orchids. For this reason, the existence of the Liwa Botanical Garden is expected to guarantee the preservation of natural orchid species that can be utilized sustainably. However, until now there are several types of natural orchids that are not known with certainty what natural orchid species exist in the Liwa Botanical Garden [7–11].

Considering the importance of preserving and preserving orchids in the region, there is a need for further action. One way to do that is by identifying the types of natural orchids, especially *Dendrobium* species which have a high diversity compared to other orchid species. The identification results will be addressed based on leaf morphology and leaf stomata anatomical structure, given the morphological character is one of the approaches that play an important role in the plant's taxonomic and systemic basis. This study aims to determine variations in morphological characters and phenetic relationships and identification of *Dendrobium* based on morphological characters in the Liwa Botanical Garden. The results of this study are expected to be basic information in the identification of natural orchids and conservation efforts in the Liwa Botanical Garden.

2. Identification of stomata morphological characters

Sample collection was carried out on *Dendrobium* leaves in the Liwa Botanical Garden. Overall, the sample accession is a native orchid of flora from Lampung. All samples were tabulated and documented with photos. The sample collection stage was conducted in December 2019–February 2020 at the orchid green house in the Liwa Botanical Garden. *Dendrobium* samples were chosen based on orchid data that could not yet be identified. Collection results obtained 5 accessions of *Dendrobium* samples with sample codes CAT140, CAT 144, CAT 271, CAT 274, and IR015 (**Table 1**). Overall, the sample accession is a native natural orchid of flora from Lampung.

The morphological identification research phase was carried out by direct observation when sampling in the field. Leaf morphology characters identified included leaf shape, length (P) and width (L) of leaf, leaf tip shape, leaf cross section, leaf arrangement, leaf edge shape, leaf surface texture, leaf symmetry, and leaf sitting [12, 13].

No. Acc.	Species	Origin location
CAT140	*Dendrobium* sp.	Bukit Barisan Selatan National Park
CAT144	*Dendrobium* sp.	Seminung Forest
CAT271	*Dendrobium* sp.	Bukit Barisan Selatan National Park
CAT274	*Dendrobium* sp.	Bukit Barisan Selatan National Park
IR015	*Dendrobium* sp.	Bukit Barisan Selatan National Park

Table 1.
List of accessions of Dendrobium *samples in the Liwa Botanical Garden.*

Based on observations of morphological characters in the field, orchid plants have a high variation. These variations were found in habitus, pseudobulb, leaves, and flowers [13]. In this research the character of the flower is not done because the variation in habitus is seen in plant height, which ranges from 50–125 cm. Plant height can be categorized into 2, namely ≤ 100 cm (short) and> 100 cm (height) (**Figure 1**).

Figure 1.
Habitus accession of Dendrobium *samples in the Liwa Botanical Garden: (A) CAT 274, (B) CAT 144, (C) CAT 140, (D) IR 015, and (E) CAT 271.*

No. Acc.	Leaf pictures
CAT140	
CAT144	
CAT271	
CAT274	
IR015	

Bar = 1 cm.

Table 2.
Leaf type of accession of Dendrobium *samples in the Liwa Botanical Garden [14].*

Leaves of *Dendrobium* have most varied organs based on field observation, (**Table 2**). Variation of leaf shape i.e. shape, length (P) and width (L), tip shape, cross section, arrangement, edge shape, surface texture, symmetry, and sitting of leaf [14].

Five accessions of *Dendrobium* from Liwa Botanical Garden showed a different morphological characterization of leaves, i.e. cross-sectional characteristics and sitting of leaf. In the sample with accession number CAT 274, CAT 140, IR 015 and CAT 271 the leaf cross section is zigomorphic (symmetry), whereas on accession number CAT 144 the leaf cross section is tight. In addition, intermittent leaf sitting was found in CAT 274 accession number, IR 215 and CAT 271, in contrast to samples with CAT accession number 144 characters of intermittent leaf sitting and meeting almost filled the plant stem and CAT 140 characters sat leaf intermittent intermittent interspersed and tightly located near the end of the stem [14].

Although the morphological characterization of the leaf shape was the same in all samples, namely the lanceolate/javelin shape, the length and width of the leaves differed from one sample to another. In CAT 274 samples (P: 4.5 cm and L: 1 cm), CAT 144 samples (P: ± 1.7 cm and L: 0.5 cm), CAT 140 samples (P: 9.5 cm and L : 1.5 cm), IR 015 samples (P: 8 cm and L: 2 cm), and CAT 271 samples (P: 8.5 cm and L: 1.5 cm).

Analysis of phenetic was done using cluster analysis and principal component analysis (PCA) methods. The first step is the scoring morphological character using cluster analysis, then calculated the value of Gower similarity (Gower's General Similarity) which results in a matrix of similarity between accessions. Data of matrix similarity using the UPGMA method was done by agglomerative hierarchial clustering, then the result was displayed on dendrogram.

Samples that have a longer leaf length morphological character CAT 140 (P: 9.5 cm and L: 1.5 cm), IR015 (P: 8 cm and L: 2 cm) and CAT 271 (P: 8.5 cm and L: 1.5 cm) will have a higher plant height habitus compared to samples that have shorter leaf morphological characters in CAT 274 samples (P: 4.5 cm and L: 1 cm) and CAT 144 (P : ± 1.7 cm and L: 0.5 cm). Furthermore, there is also a striking difference in the morphological character of the lowest plant leaves, namely in CAT 144 (P: ± 1.7 cm and L: 0.5 cm). Other character differences also have the lowest plant height habitus and have a thicker leaf thickness than the others. The result of morphology character identification of the *Dendrobium* leaves from Liwa Botanical Garden is presented in **Table 3**.

Based on **Table 3**, it is known that most of the accessions of Dendrobium samples in the Liwa Botanical Garden show the same morphological variation in the leaves:

1. The shape of lanceolate/javelin-shaped leaves.

2. The shape of the leaf tip is sharp/pointed/sharp to the tip.

3. Dual leaf arrangement.

4. The edge of the leaf is frayed (even).

5. Leaf surface texture is bald (smooth).

6. Symmetry of leaves in the form of symmetry.

In other characters, namely the form of pseudobulb and the place of growth, it is known that the entire accession of *Dendrobium* samples did not form pseudobulb and

Morphology character	CAT 274	CAT 144	CAT 140	IR 015	CAT 271
Shape	Shape of lanceolate/ eye javelin	Shape of lanceolate/eye javelin	Shape of lanceolate/eye javelin	Shape of lanceolate/ eye javelin	Shape of lanceolate/ eye javelin
Length (P) and width (L)	P: 4.5 cm and L: 1 cm	P: ±1.7 cm and L: 0.5 cm	P: 9.5 cm and L: 1.5 cm	P: 8 cm and L: 2 cm	P: 8.5 cm and L: 1.5 cm
Tip shape	Taper/pointed/sharp to the tip	Taper/pointed/sharp to the tip	Taper/pointed/sharp to the tip	Taper/pointed/sharp to the tip	Taper/pointed/sharp to the tip
Cross section	Zigomorph (symmetry)	Double	Zigomorph (symmetry)	Zigomorph (symmetry)	Zigomorph (symmetry)
Arrangement	Double	Double	Double	Double	Double
Edge shape	Frayed (flat)	Frayed (flat)	Frayed (flat)	Frayed (flat)	Frayed (flat)
Surface texture	Hairless (smooth)	Hairless (smooth)	Hairless (smooth)	Hairless (smooth)	Hairless (smooth)
Symmetry	Symmetry	Symmetry	Symmetry	Symmetry	Symmetry
Sitting	Intermittent	Intermittent and close to almost meet the stems of plants	Intermittently intermittent and close to the end of the stem	Intermittent	Intermittent

Table 3.
Leaf variation of morphology character identification result of accession of Dendrobium *samples from Liwa Botanical Garden.*

epiphytic antibiotic types. Character types of habitats in general all *Dendrobium* have the same type, epiphytes, according to natural conditions where the sample collection of the Liwa Botanical Garden is a natural orchid taken from its natural habitat, such as the Bukit Barisan Selatan National Park and Seminung Forest which has low humidity (dry) at an altitude of 800–900 m above sea level. Epiphytic orchids grow in their natural habitat requires the intensity of indirect sunlight. In accordance with the type of growth attached to the host [13].

2.1 Phenetic analysis

Phenetic analysis on five accession of *Dendrobium* is done through cluster analysis and principal component analysis (PCA) methods. The first steps, scoring of the morphological character using cluster analysis, then calculated the Gower similarity value (Gower's General Similarity) between five accessions using matrix of similarities. Then UPGMA methods to calculate the similarity matrix data using clustering of agglomerative hierarchial. The dendrogram results of analysis cluster of five *Dendrobium* accessions based on the morphological characters are presented in **Figure 2**.

Grouping the sample based on the level of similarity between accessions calculated using the gower coefficient formula and UPGMA was chosen for the clustering technique to produce a dendogram showing 2 main groups formed with a similarity index value of 0.813 marked as group A and group B. Group A consists of CAT 144 which has a distinguishing character that distinguishes from group B, namely the cross section of the double leaf character (Figure CAT 144). Group B consists of CAT 140, CAT 271, IR 015, and CAT 274 which have symmetrical cross-section characters (Figure CAT 140, CAT 271, IR 015, and CAT 274). Group B is divided into 2 sub-groups with a similarity index value of 0.861 marked with B1 and B2 on the dendogram. Characters that show the difference between the two, namely the ratio of the length and width of the leaf and leaf sitting. CAT 140 consists to subgroup B1, meanwhile CAT 271, IR 015, and CAT 274 consist to subgroup B2. Subgroups B2 are divided into 2 based on differences in leaf length and width ratios namely B2a and B2b. IR 015 and CAT 274

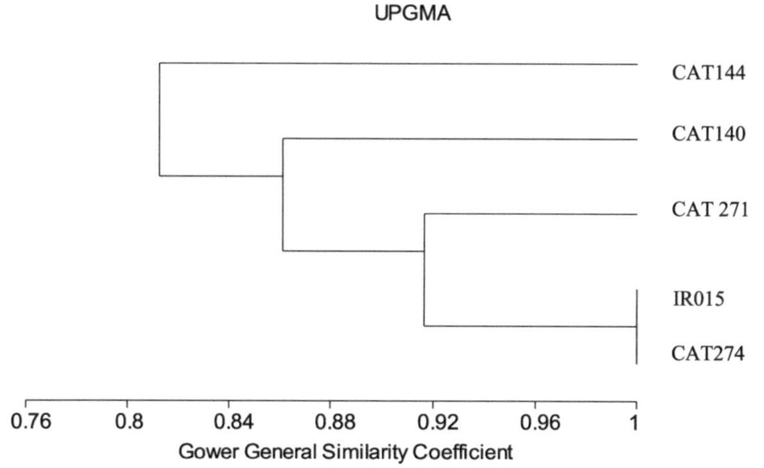

Figure 2.
Dendrogram five Dendrobium *accession of samples from Liwa Botanical Garden using UPGMA.*

in one group B2b with a similarity level of 100% were indicates the same type. Based on the PCA values, it can be seen that characters have a large influence on grouping are the ratio of length and width (PLD), cross section (PMD), and sitting (DKD). Variability of orchid leaf characters can be possible due to the hybridization and outcrossing processes.

3. Identification of stomata anatomical characters

The research stage of leaf stomata identification was carried out by direct observation using a miscropcope shortly after sampling in the field. The anatomical characters of leaf stomata identified include the average number of stomata, average length of stomata, average stomata width, average number of epidermis, average length of epidermis, average epidermal width, stomata index, stomata type, shape epidermis on the upper and lower surfaces of the leaves [13].

The epidermis is a system of cells, varying in structure and function that protects the primary plant body. The leaf epidermis is the outermost layer of cells, generally only one layer. In the leaf epidermis, there are usually epidermal derivatives in the form of stomas (plural: stomata). A stoma is a gap in the epidermis that is bounded by

No.	Parameter	IR 015	CAT 140	CAT 144	CAT 274
1.			Leaf upper surface		
	Average number of stomata	—	—	2.80	—
	Average length of stomata (μm)	—	—	2.45	—
	Average stomata width (μm)	—	—	2.17	—
	Average number of epidermis	29	41	45	42.8
	Average epidermal length (μm)	5.22	5.49	4.31	4.64
	Average epidermal width (μm)	4.22	2.79	2.65	2.88
	Stomata index (%)	—	—	5.86	—
	Stomata type	—	—	Tetracytic (1 stomata surrounded by 4 neighboring cells)	—
	Epidermis shape	Irregular, pentagonal, hexagonal	Irregular, pentagonal	Irregular, quadrilateral, pentagonal	Irregular, quadrilateral, pentagonal

No.	Parameter	IR 015	CAT 140	CAT 144	CAT 274
2.			Leaf Lower Surface		
	Average number of stomata	5.2	2.50	2.50	3.4
	Average length of stomata (μm)	2.75	2.78	2.33	2.57
	Average stomata width (μm)	1.96	2.45	2.07	1.82
	Average number of epidermis	32.2	30.75	43.50	42.2
	Average epidermal length (μm)	3.54	5.13	4.42	3.74
	Average epidermal width (μm)	2.72	3.03	3.06	2.45
	Stomata index (%)	13.90	7.52	5.43	7.46
	Stomata type	Tetracytic (1 stomata surrounded by 4 neighboring cells)	Tetracytic (1 stomata surrounded by 4 neighboring cells)	Tetracytic (1 stomata surrounded by 4 neighboring cells)	Tetracytic (1 stomata surrounded by 4 neighboring cells)
	Epidermis shape	Irregular, pentagonal, hexagonal	Irregular, pentagonal	Irregular, quadrilateral, pentagonal	Irregular, quadrilateral, pentagonal

Table 4.
Anatomical characters on the upper and lower surfaces of leaves.

two special epidermal cells, namely the covering cell that functions to widen or narrow the gap. The stoma is surrounded by cells that can be the same or different from other epidermal cells called neighboring cells [13].

In this study, observations were made on 4 types of *Dendrobium*, namely *Dendrobium* IR015, *Dendrobium* CAT 140, *Dendrobium* CAT 144, and *Dendrobium* 274. Observations were made on the upper and lower surfaces of leaves (**Table 4**).

Based on **Table 4**, it is known that the types of IR015, CAT 140, and CAT 274 have stomata only on the lower surface, while CAT 144 has stomata on the upper and lower surfaces. This difference is due to the position of the leaves attached to the stem at CAT 144 forming an angle of 45°C, while the others open horizontally. This causes the top and bottom sides to be the same. The anatomical characters on CAT 144 on the upper and lower surfaces of the leaves, such as the number of stomata, length of stomata, width of stomata, number of epidermis, length of epidermis, width of epidermis, index of stomata are not much different. Stomata are kidney-shaped, and belong to the tetracytic type, namely in the form of stomata surrounded by 4 neighboring cells. In IR015, CAT 140, and CAT 274 the upper surface of the leaves is composed only of the epidermis which is mostly irregular and pentagonal in shape. The highest number was in CAT 274 because the epidermis was smaller than IR015 and CAT 144.

On the lower leaf surface, it is known that the number of stomata at IR015 is the highest and the number of epidermis is the least, so that the stomata index is the largest. The stomata index looks the largest because the stomata index shows the number of stomata divided by the number of stomata plus the number of epidermis and multiplied by 100%. The types of stomata in all types are the same, namely tetracytic in the form of stomata surrounded by 4 neighboring cells. The stomata are kidney-shaped and the epidermis is irregular, pentagonal, and hexagonal in shape.

4. Conclusions

Five accessions of *Dendrobium* from Liwa Botanical Garden were identification based on morphological characters and phenetic relationships. The result of analysist 11 morphological characters showed leaf of *Dendrobium* have high variations. The phenetic relationship showed five accessions of *Dendrobium* from Liwa Botanical Garden can be classified into 2 main groups formed with a similarity index value of 0.813. The morphological characters that have a large influence on grouping are the ratio of length and width, cross section, and sitting of leaf based on PCA values. Variability of *Dendrobium* leaf characters can be possible due to the hybridization and outcrossing processes. The resulting phenetic dendrogram topology is supported by the morphological character classification. The results of the observation of 9 anatomical characters on the upper and lower surfaces of the leaves indicate that the leaf organs have high variations. Accessions IR015, CAT 140, and CAT 274 have stomata only on the lower surface, while CAT 144 has stomata on the upper and lower surfaces. This difference is due to the position of the leaves attached to the stem at CAT 144 forming an angle of 45C, while the others open horizontally. IR015, CAT 140, and CAT 274 the upper surface of the leaves is only composed of epidermis which is mostly irregular and pentagonal in shape. The highest number was in CAT 274 because the epidermis was smaller than IR015 and CAT 144. The types of stomata in all types were the same, namely tetracytic in the form of stomata surrounded by 4 neighboring cells. The stomata are kidney-shaped and the epidermis is irregular, pentagonal, and hexagonal in shape.

Acknowledgements

The author thanks to Lembaga Penelitian dan Pengabdian Masyarakat (LPPM), Lampung University was funded this research through the DIPA BLU 2020 Grant with contract number 1492/UN26.21/PN/2020. Thank you to the Liwa Botanical Garden who facilitated the research based on collaboration program with Department of Biology, Faculty of Mathematics and Natural Science, Universitas Lampung.

Author details

Mahfut
Department of Biology, Faculty of Mathematics and Natural Science, University of Lampung, Lampung, Indonesia

*Address all correspondence to: mahfut.mipa@fmipa.unila.ac.id

References

[1] Mahfut M, Daryono BS, Somowiyarjo S. Identifikasi Molekuler DNA Kloroplas Pada Anggrek Terinfeksi Odontoglossum ringspot virus (ORSV) di Magelang, Jawa Tengah. In: Proceedings of Seminar Nasional Pengendalian Penyakit Pada Tanaman Pertanian Ramah Lingkungan II Perhimpunan Fitopatologi Indonesia Komisariat Daerah Yogyakarta, Solo, dan Semarang; 27 August 2016. Indonesia: Yogyakarta; 2016. pp. 354-360

[2] Mahfut M, Daryono BS, Indrianto A, Somowiyarjo S. Plant-virus interaction on orchids infected odontoglossum ringspot virus (ORSV) in Bogor Botanical Garden, Indonesia. In: Proceedings of the EUDL The 1st International Conference on Science and Technology (ICoST). Makassar, Indonesia: EUDL; 2019. pp. 1-8

[3] Mahfut M. Indonesia Darurat Konservasi: Sudah Amankah Kebun Raya Kita. In: Proceeding of the Seminar Nasional Biodiversitas Indonesia. Indonesia: Makassar; 2019. pp. 1-6

[4] Mahfut M, Indrianto A, Somowiyarjo S, Daryono BS. Molecular phylogeny of orchids mycorrhiza isolated from native tropical orchids in Indonesia. Malaysian Journal of Microbiology. 2020;**16**(1):68-72. DOI: 10.21161/mjm.190425

[5] Munawaroh E, Yuzammi, Solihah SM, Suhendar. Koleksi Kebun Raya Liwa, Lampung Tumbuhan Berpotensi sebagai Tanaman Hias. Jakarta: LIPI Press. p. 133

[6] Solihah SM. Koleksi, Status, dan Potensi Anggrek di Kebun Raya Liwa. Warta Kebun Raya (Semi-Popular Magazine). 2009;**13**(1):15-23

[7] Kumalawati AD, Abdullah S, Daryono BS, Mahfut M. Study on genetic diversity and conservation of orchids in Wonosadi forest, Gunung Kidul based on molecular analysis. In: Proceedings of the International Conference on Biological Science. Indonesia: Yogyakarta; 2011. pp. 72-79

[8] Mahfut K, Daryono BS. Deteksi Odontoglossum ringspot virus (ORSV) Terhadap Anggrek Alam di Hutan Wonosadi, Gunung Kidul. Biogenesis: Jurnal Ilmiah Biologi. 2014;**2**(2):101-108

[9] Mahfut M, Joko T, Daryono BS. Molecular characterization molecular of odontoglossum ringspot virus (ORSV) in Jawa and Bali, Indonesia. Asian Journal of Plant Pathology. 2016;**10**(1-2):9-14. DOI: 10.3923/ajppaj.2016.9.14

[10] Daryono BS, Joko T, Somowiyarjo S. Survei odontoglossum ringspot virus (ORSV) yang Menginfeksi Anggrek Alam Tropis di Indonesia. Jurnal Perlindungan Tanaman Indonesia. 2016;**20**(1):1-6. DOI: 10.22146/jpti.16617

[11] Mahfut M, Daryono BS, Somowiyarjo S. Deteksi Odontoglossum ringspot virus (ORSV) yang Menginfeksi Anggrek Asli Koleksi Kebun Raya di Indonesia. Jurnal Fitopatologi Indonesia. 2017;**13**(1):1. DOI: 10.14692/jfi.13.1.1

[12] Adi MC, Yulika R, Nugraha I, Virditha UA. Buku Saku Kebun Raya Liwa. Badan Penelitian dan Pengembangan Kabupaten Lampung Barat. Indonesia: UPTD Pengelola Kebun Raya Liwa; 2019. p. 41

[13] Arif A, Ratnawati R. Hubungan Kekerabatan Anggrek Dendrobium Berdasarkan Karakteristik Morfologis dan Anatomis Daun. Jurnal Prodi Biologi. 2018;**7**(3):213-222

[14] Mahfut M, Handayani TT, Wahyuningsih S, Sukimin S. Identification of *Dendrobium* (Orchidaceae) in Liwa Botanical Garden based on leaf morphological characters. Journal of Tropical Biodiversity and Biotechnology. 2021;**6**(1):1-6. DOI: 10.22146/jtbb.59423

Chapter 11

Holistic Pest Management Strategies in Tropical Plant Species

John Samuel Kennedy and Jeeva Kasiviswanathan Lekshmi

Abstract

The tropical climate shift is causing herbivores to emerge almost ceaselessly throughout the year in certain regions exhibiting homodynamic cycles and unbalanced biodiversity. Crop management and pest management are being viewed as separate activities, with recent focus on sustainability. Even though there is a great deal of information on crop loss assessment, systems analysis, systems modelling, individual pest sciences, and pest management, the Integrated Pest Management (IPM) concept is not frequently deployed. The IPM system is a multi-tactic approach to pest management in agricultural production that takes into account economic, environmental, ecological, and human health implications. This paper provides an overview of key achievements in the development of management strategies, including the transition from a specific level of pest control that focuses on the suppression of target pests to an eco-friendlier and/or systems approach to pest management that employs a variety of non-chemical options as well as the judicious use of pesticides. The agroecological protection techniques and their integration to sustainably minimise pest risks are also reviewed here and describe technological advances in tropical pest management using host resistance, semiochemicals, natural enemies, selective pesticides, ecological engineering and habitat management which promotes sustainable pest management.

Keywords: tropical crops, pest, synthetic insecticides, integrated pest management

1. Introduction

Changes in pest populations and natural enemies in the tropics are more difficult to understand and manage due to various factors such as climate change and non-judicial use of control techniques [1]. The crop health and production are thus, irreparably harmed across the world in addition to the time-worn exploitation of synthetic organic pesticides that rooted an increased chemical pressure in agroecosystems ensuring technical, environmental and health consequences [2]. As a result, there is a rising awareness of the harmful impacts of chemical pesticides, leading to the notion of integrated pest management, which is esteemed as an ecosystem-based pest control philosophy aimed at achieving protracted pest or damage prevention through a holistic approach [3]. The concept of IPM was established as an alternative to pesticides and the strategy entails a combination of techniques in a coordinated manner to keep insect populations below the threshold at which they cause significant loss [4]. The

IntechOpen

key components of this technique include compatible cultural, mechanical, biological and chemical methods of controlling insect pests and rodents [5]. Managing the diversity of natural enemies to promote crop productivity could help minimise the usage of synthetic pesticides in agriculture [6]. Landscape management, on the other hand, has been shown to increase the prevalence of natural enemies, hence improving biological control and perhaps reducing pesticide use [7]. They are well-known for causing significant crop yield losses by reducing plant survival, growth, and reproduction. Global yield losses are estimated to be between 7.9% and 15.1% [8]. Along with the notion of Integrated Pest Management, pest control practices are divided into two categories: human input-based practices and biodiversity-based practices [9]. The IPM strategy entails using a variety of techniques in a coordinated manner to keep insect populations below the threshold at which they cause significant loss and reduces the use of chemical pesticides, minimising their negative consequences [10]. The farmers' perception is that pesticides can give a quick and certain control within a short period. They lack knowledge of other management strategies and considered them as not practical because of lack of training. In this chapter, we have briefly described the important tropical crops and their integrated management which will be useful for training personnels and researchers.

2. Integrated pest management in cereals

2.1 Rice

2.1.1 Important pests of rice

Yellow stem borer (*Scirpophaga incertulas*), Brown plant hopper (*Nilaparvata lugens*) and White backed plant hopper (*Sogatella furcifera*), Leaf folder (*Cnaphalocrocis medinalis*), Gundhi bug (*Leptocorisa acuta*), Gall midge (*Orseolia oryzae*), Termite (*Odontotermes obesus*), Rice Hispa (*Dicladispa armigera*), Climbing cutworm/Rice Ear Cutting Caterpillar/ Armyworm (*Mythimna separata*), Caseworm (*Nymphula depunctalis*), Mealy bug (*Brevennia rehi*), Black bug (*Scotinophara coaractata*).

2.1.2 Integrated pest management approach

Clipping the tops of bundled seedlings is a typical method for eliminating eggs and larvae of stem borer, hispa, thrips and leaf folder [11]. Plant hoppers, leaf hoppers, leaf folders, gall midges and cutworms are all kept at bay by keeping enough space between plants. Aquatic insects such as whorl maggot, root weevils, yellow stem borer, and case worm are suppressed when fields are drained for 1–2 days. Crop rotation is used to combat gall midge, stem borer, and termites. Infestations of white-backed plant hoppers can be controlled by planting rice crops at the proper time and synchronising transplants. Deep summer ploughing of fields reduces insect pest populations by exposing them to bird predation and parasitization. Flooding the field shortly after harvest has been mostly utilised to combat stem borer [12]. Mechanically trapping or crushing insects with a hand, tool, or machine is a traditional method to protect the rice crop from pests [11]. Pheromones were particularly effective in the control of the yellow stem borer, where they were employed for both species monitoring and management via mating disruption or mass catching of males. The bulk of paddy pests is combated by a range of natural enemies like *Tetrastichus* spp., *Telenomos* spp., and *Trichogramma* spp. *Xanthopimpla flavolineata* was found as a prominent rice leaf folder pupal parasitoid. In various hoppers, more than 100 parasitoid species and 72 predatory species have been. Spiders are the most prevalent predators in the rice environment. The wolf spider, *Lycosa pseudoannulata*, and *Oxyopes* sp. destroyed up to 90% of 130 borer larvae in a single day. This spider was also dangerous to stem borer moths. Neem seed kernel extract (NSKE) 5% and neem oil 3% were effective against brown plant hopper, as were botanical powder formulations using NSKE, *Vitex negundo*, *Prosopis juliflora*, and *Ipomoea carnea* leaf extract 5% for earhead insect and black bug [13]. The plant-derived natural compound chrysoeriol7 can potentially thus be used to develop environmentally-friendly pesticides [14]. Chemical insecticides, such as carbofuran 3% CG or carbosulfan 6% G or carbosulfan 25% EC for gall midge, are applied based on need. Cartap hydrochloride 4% granules or cartap hydrochloride 50% SP, carbofuran 3% CG or monocrotophos 36% SL for stem borer. Carbofuran 3% CG or monocrotophos 36% SL, apply cartap hydrochloride 4% granules, cartap hydrochloride 50% SP @, monocrotophos 36% SL, or chlorpyrifos 1.5% DP for leaf folder Spray imidacloprid 70% WG or imidacloprid 30.5% m/m SC @ or ethofenoprox 10% EC or acephate 75% SP or buprofezin 25% SC @ or ethofenoprox 10% EC or acephate 75% SP or buprofezin 25% SC for brown plant hopper, WBPH and other sucking pests [15].

2.2 Maize

2.2.1 Important pests

Maize stem borer (*Chilo partellus*), Pink stem borer (*Sesamia inferens*), Shoot fly (*Atherigona spp.*), White grub (*Holotrichia consanguinea*), Cut Worm (*Agrotis ipsilon*), Hairy caterpillar (*Amsacta albistriga*), Aphid (*Rhopalosiphum maidis*), Army worm (*Mythimna separata*), Pyrilla (*Pyrilla perpusilla*), Thrips (*Anaphothrips sudanensis*), Termites (*Microtermes obesi*), Chafer beetle (*Chiloloba acuta*), Fall armyworm (*Spodoptera frugiperda*).

2.2.2 IPM approaches

To minimise pest populations, remove and destroy crop leftovers, any substitute host plants after harvest, and cut stems harbouring diapausing larvae. Growing recommended hybrids and composites and seeding with the first rain to minimise borer attack. Crop rotation with non-hosts proved highly efficient. Insects become more common as plant density increases. Insect incidence rises as plant density rises; so, the suitable plant population should be kept in the fields to prevent insect incidence. *Trichogramma* spp., a parasitic egg parasite can be used to keep stem borers away. Many spider species have been discovered in plant whorls feeding on stem borer eggs and early larval stages. Adult *Menochilus sexmaculata* Fab. and *Coccinella septumpunctata* L. coccinellid beetles feed on newly emerging stem borer larvae [16]. To control one of the annoyance pests, the fall armyworm (FAW), single cross maize hybrids were selected, as well as heavy ploughing before each crop season to open up the soil and expose FAW pupae to sunshine and predators. The FAW trap crop was sown in the form of Napier grass in the border rows. According to the researchers, seeds treated with

Cyantraniliprole 19.8% + Thiomethoxam 19.8% @ 4 ml per kg seed offered protection for up to 2–3 weeks after germination [17]. If staggered seeding is required, spray the crop with 5% NSKE or azadirachtin 1500 ppm @ 5 ml/l at weekly intervals, as in peri-urban baby corn and sweet corn cultivation. Release 50,000 *Trichogramma pretiosum* or *Telenomus remus* seeds per acre at weekly intervals beginning a week after germination and continuing until harvest. Instal 5 FAW pheromone traps per acre before or during crop germination [18]. To keep population growth under control, mass capture male moths using traps. Need-based pesticides should be used to efficiently remove the pest, viz. carbofuran 3% CG for stem borer, shoot fly and thrips, dimethoate (30% EC for Stem borer and Shoot fly), imidacloprid (48%) FS, monocrotophos (36% SL), oxydemeton (25%) EC for shoot fly, thiamethoxam 30% FS for stem fly, thiamethoxam 70% WS for shoot fly & aphids, thiamethoxam 12.6% + lambda cyhalothrin 9.5% ZC for aphid, shoot Fly, stem borer [19].

2.3 Wheat

2.3.1 Important pests

Termite (*Odontotermis obesus, Microtermes obesi*), Wheat aphid (*Sitobian avenae*), Army worm/cut worm (*Mythimna separata*), American pod borer (*Helicoverpa armigera)*, Brown mite (*Petrobia latens*), Pink stem borer: (*Sesamia inferens*), Shootfly (*Atherigona naqvii*).

2.3.2 IPM approaches

Deep summer ploughing should be done in May and June to expose nematodes, rodents, and pupating larvae to radiation and predation termite management will be facilitated by the use of well-rotten farmyard manure, while it will be lowered by planting late and treating seeds with chlorpyriphos at a rate of 4 ml per kg of seeds prior to sowing. *Trichogramma* sp. was released at a rate of 50000 per acre for 2 to 3 times to suppress lepidopteran pests [13]. Need based insecticides viz., bromadiolone 00.005% RB for Indian field mouse, field rat, carbofuran 3% CG for ear cockle nematode, cereal cyst nematode, cypermethrin 10% EC for shoot fly, fipronil 0.3% GR for termites, imidachloprid 48% FS for aphids and termites, quinalphos 25% EC for aphid, ear head caterpillar and mite, thiamethoxam 70% WS for termites & aphids, thiamethoxam 25% WG for aphids were effectively utilised [20].

3. Integrated pest management in millets

3.1 Sorghum

3.1.1 Important pests

Shoot fly (*Atherigona soccata*), Stem borer (*Chilo partellus*), Midge (*Stenodiplosis sorghicola*), White grub (*Holotrichia consanguinea*), Armyworm (*Mythimna separata*), Cutworm (*A. ipsilon*), Grasshopper (*Hieroglyphus sp*), Pyrilla (*Pyrilla perpusilla*), Shoot bug: (*Peregrinus maidis*), Earhead caterpillars (*Helicoverpa armigera*), Earhead bug (*Calocoris angustatus*), Sugar cane aphid (*Rhophalosiphum maidis*), Spider mite (*Oligonychus indicus*).

3.1.2 IPM approaches

Sowing dates must be altered based on the population dynamics of the main pests to keep the pest population under control. Sowing the same cultivar early and consistently across large areas reduced shoot fly, midge, and head bug damage. Cultural practices such as high seed rate, balanced fertiliser treatment, field cleanliness, weeding, and intercropping with legumes reduced damage from shoot fly, stem borer, armyworm, and sorghum midge. Pest resistant cultivars should form the foundation of any sorghum pest control approach. P 311, SPV 1015, M-35-1, Swati, and CSV 14R are resistant to shoot fly and are suitable for cultivation throughout the rainy and post-rainy seasons [21]. In locations where head bugs and caterpillars are common, cultivars with loose panicles, such as ICSV 88032, can be grown. Insecticides derived from plants, such as those obtained from *Azadirachta indica* (neem), *Annona squamosa*, *Acrorus calamus*, *Catharanthus roseus*, and others, were exploited in IPM [22]. Sorghum shoot fly eggs have been parasitized by *Trichogrammatoidea bactrae* and *Trichogramma simmondsi*; *Trichogramma evanescens*, *Trichogramma kalkae*, T*richogramma chilonis* and *Trichogramma. australicum Aprostocetus spp*. and *Neotrichoporoides* (*Tetrastichus*) *nyemitawus* were the most important shoot fly larval parasitoids. Need-based insecticides viz., malathion 50% EC for earhead midge, Oxydemeton – methyl 25% EC for shoot fly, phenthoate 2% DP for red spider mite, pink mite, purple mite, scarlet mite, phosalone 4% DP for ear head midge, quinalphos 5% G for stem borer, quinalphos 25% EC for Mite and Shoot fly, quinalphos 1.5% DP fo earhead bug, earhead midge, thiamethoxam 30% FS for shoot fly.

3.2 Pearl millet

3.2.1 Important pests

Cutworm (*A. ipsilon*), white grub (*Holotrichia* consanguinea), shoot fly (*Atherigona soccata*), stem borer (*Chilo partellus*), grasshopper (*Hieroglyphus* spp.), white ant (*Chrotogonus* sp), grey weevil (*Myllocerus* sp), earhead bug (*Calocoris angustatus*), hairy caterpillar (*Spilosoma obliqua*), earhead worm (*Cryptoblabes gnidiella*), blister beetle (*Mylabris pustulata*), chaffer beetle (*Rhizotrogus majalis*).

3.2.2 IPM approaches

The pest management strategies recommended for sorghum might likewise be used for pearl millet. Long-term pearl millet pest control requires the development of high-yielding insect-resistant cultivars and hybrids. Among the possible sources of resistance discovered for numerous pests were MP-16, MP-19, MP-53, MP-67, MH-49, MH-52, MH-9, MH-82, MH-99 and MH-105 [16]. Shoot fly may wreak havoc on late-planted kharif crops. As a result, sowing should begin soon after the monsoon begins, or no later than 10–15 days after the first monsoon rain. Adult beetles of White grubs were gathered and exterminated immediately after the first shower in the endemic zones after mating on trees such as neem or Acacia.

4. Integrated pest management in pulses

4.1 Important pests

Pod borer (*Helicoverpa armigera*), spotted pod borer (*Maruca vitrata*), spiny pod borer (*Etiella zinckenella*), blue butterfly (*Lampides boeticus*), grass blue butterfly (*Euchrysops cnejus*), bihar hairy caterpillar (*Spilosoma obliqua*), stem fly (*Ophiomyia phaseoli*), pod weevil (*Apionam plum*), bean Aphid (*Aphis craccivora*), leaf hopper (*Empoasca kerri*), podfly (*Melanagromyza obtuse),* lab bug(*Coptosoma cribraria*), whitefly (*Bemisia tabaci*), thrips (*Megalurothrips usitatus*), blister beetle (*Mylabris spp*), stem fly (*Melanagromyza sojae*), tobacco caterpillar (*Spodoptera litura*), green semi-loopers (*Chrysodeixis acuta, Gesonia gemma and Diachrysia orichalcea*), girdle beetle (*Obereopsis brevis*), pod borer (*Helicoverpa armigera*), white fly (*B. tabaci*).

4.2 IPM approaches

Plant spacing, sowing time, intercropping, and soil activities can all be adjusted to reduce *H. armigera* harm Chickpea germplasm with low to high insect pest resistance has been established. Deep ploughing of fields in the summer and leaving the land for solarization are widespread cultural practices in black gram and green gram pre-sowing [23]. Avoiding waterlogging, judicious fertiliser usage, and other common cultural methods for stem fly, pod weevil, pod fly, blister beetle, white grub and grass butterfly management. Installing light traps in and around fields to minimise crop stress [24], enhancing parasitic activity by avoiding chemical spray collecting and destroying eggs and early-stage larvae, handpicking older, gregarious caterpillars and cocoons during early stages and using yellow/blue pan water/sticky traps @ 4–5 traps/acre [25] light traps @ 1/acre, and pheromone traps @ 4–5/acre for monitoring adult moths' activity. Cleaning of infected stubbles followed by deep summer ploughing, optimal fertiliser application, timely sowing, proper seedbed conditions and depth of sowing, optimum seeding rate and plant population, regular scouting, rogueing, and destruction of infected crop/plant parts, elimination of collateral/alternate and reservoir hosts, crop rotation and intercropping, cultivation of soybean only during the rainy season, and agronomic practices to avoid pests are all common cultural practices in soybean [26]. Collecting and eliminating girdle beetle-infested plant parts, egg masses, and gregariously feeding hairy caterpillar and tobacco caterpillar larvae should be prioritised. Ten to twelve bird perches will be installed on each acre. Pheromone traps should be used to track the spread of *S. litura* and *Helicoverpa armigera*, as well as Castor as a tobacco caterpillar trap crop and Dhaincha as a girdle beetle trap crop. It is recommended to intercrop soybean with asafoetida (early maturing variety), maize, or sorghum in a 4-row soybean with 2-row intercrop sequence. Increased biodiversity will aid natural biocontrol fauna such as coccinellid beetles, *Chrysoperla*, and others. In girdle beetle and semilooper endemic areas, intercropping with maize or sorghum should be avoided. *Campoletis chlorideae*, an ichneumonid, is the most prominent chickpea larval parasitoid of *Helicoverpa armigera* Six parasitoid species have been identified in Helicoverpa pupae collected in the field and potential biocontrol agents for *B. pisorum* have been reported *Chrysopa spp*., *Chrysoperla spp*., *Nabis spp*., *Geocoris spp*., *Orius spp*., *Polistes spp*., and species belonging to the Pentatomidae, Reduviidae, Coccinellidae, Carabidae, Formicidae, and Araneidae, respectively, are the most prevalent predators of insect pests The entomopathogenic fungus *Nomuraea* *rileyi*caused 90–100% larval mortality, while *Beauveria bassiana* Balsamo caused only 6% chickpea damage compared to 16.33% damage in untreated control plots Spraying *Bacillus thuringiensis* (Bt) (Berliner) formulations later in the day yields greater control than spraying earlier in the day. Vegetable oils, neem oil, and karanj oil are excellent in protecting pulses from bruchid damage. For the management of *Helicoverpa armigera*, strategies for deploying Bt genes in transgenic chickpea have been devised. In order to deal with the Bihar hairy caterpillar, preventing preharvest infestation by irrigating once to forestall a prolonged mid-season drought. Dig 1-inch-deep holes between the fields and dust them to kill the larvae. To control the larvae, spray Quinalphos 25% EC 600 ml diluted in (black gram) or Phenthoate 50% EC diluted in (blackgram & greengram). Rates of sowing and seeding the ideal seed rate should be used depending on seed size. After every 15 rows, a one-row break should be provided to enable for spraying in a standing crop. Conservation of spiders, coccinellid beetles, tachinid fly, praying mantids, dragon fly, damsel fly, *Chrysoperla*, and meadow grasshoppers by limiting the use of wide-spectrum pesticides and releasing *Telenomus remus* at a rate of 50000/ha against *S. litura*. Spraying *B. thuringiensis var. kurstaki*, Serotype H-39, 3b, Strain Z-52 at a rate of for semilooper complex management (*Chrysodeixis acuta*, *Gessonia gemma*, *Diachrysia orichalcea* and *defoliators*). *Trichogramma chilonis*, *Tetrastichus*, and *Telenomus* are egg parasitoids, while *Ichneumon promissorius*, *Carcelia sp*, and *Diglyphus isaea* are larva parasitoids of *Spodoptera* and *Helicoverpa*; *Xanthopimpla flavolineata* is both a larval and a pupa parasitoid of an adult wasp; *Encarsia formosa*, *Eretmocerus sp*. Removal and destruction of damaged plant parts, as well as two applications of monocrotophos 36 WSC at one and three weeks of crop age. Need-based insecticides viz., Monocrotophos 36% SL, Chlorantraniliprole 18.5% SC, Lufenuron 5.4% EC, Thiodicarb 75% WP (*Helicoverpa* spp.) & (*Maruca* spp.), Novaluron 05.25% + Indoxacarb 04.50% SC, Azadirachtin 0.03% (300 PPM) for pod Borer, Flubendiamide 39.35% w/w SC for fruit borer, Flubendamide 20% WGfor *S. litura*, *Maruca spp*, Pod borer, Quinalphos 25% EC for Bihar hairy Caterpillar, *Beauveria bassiana* 1.0% WP for gram pod borer (*Helicoverpa armigera*), *Metarhizium anisopliae* 1.15% WP for *Heliothis armigera*, NPV OF *Helicoverpa armigera* 2.0% AS Strain for pod borer (*Helicoverpa armigera*), Bromadiolone 0.005% RB, Indian house rat, field Rat, Chlorantraniliprole 18.5% SC for green semi looper, stem fly, girdle beetle [27, 28]. Bioefficacy of flubendiamide 24% w/v + thiacloprid 24% SC w/v against shoot and fruit borer and its sucking pests and its safety to non-target organisms in brinjal was also proved.

5. Integrated pest management in oilseeds.

5.1 Groundnut

5.1.1 Important pests

Aphid (*A. craccivora*), bruchids (*Caryedon serratus*), jassid (*Empoasca kerri*), leaf miner (*Aproarema modicella*), termite (*Odontotermes spp*.), thrips (*Scirtothrips dorsalis, Thrips palmi*), tobacco caterpillar (*S. litura*), white grub (*Lachnosterna (Holotrichia) serrata* and *Lachnosterna (Holotrichia) consanguinea*), Bihar hairy caterpillar, (*Spilosoma/Diacrisia obliqua*), gram pod borer, (*Helicoverpa armigera*), jewel beetle (*Chrysochroa fulgidissima*).

5.1.2 IPM approaches

Semi loopers, capsule borers, and hairy caterpillars in groundnut can be controlled by planting one row of pigeonpea following groundnut. The following methods are used to manage *Helicoverpa armigera*: Setting up pheromone traps at a rate of 10 per hectare to monitor *Helicoverpaa armigera / S. liturai* [29], When a considerable number of eggs and early instar larvae are seen, HaNPV @ 250 LE/ha or SlNPV @ 250 LE/ha or Bt. 1 kg/ha or 5% NSKE will be sprayed if not reduced [30]. Pre-monsoon planting reduces damage from white grub and bud necrosis (if protective irrigation is available). The plants' cowpea or soybean function as trap crops for leaf miner. The major predators are a wide variety of spiders, *Odynerus punctum*, *Coccinella septempunctata*, *Chrysoperla carnea*, *Rhynocoris marginatus*, reduviid bug, praying mantis, fire ants, big eyed bugs (*Geocoris* sp.), pentatomid bug (*Eocanthecona furcellata*), earwigs, ground beetles, rove beetles etc. *Ovomermis albicans*, a nematode etc. and parasitoids are *Chelonus blackburni*, *Bracon spp*, *Brchymeria spp.*, *Apanteles spp.*, *Goniozus spp.*, *Elasmus spp.*, *Stenomesius*, *Sympiesis* and *Tetrastichus*, *Trichogramma chilonis*,*Tetrastichus spp.*, *Telenomus spp.*, *Chelonus blackburni Carcelia spp.*, *Campoletis chlorideae*, *Bracon spp*. etc. [31]. If the pest is severe, use the prescribed dosages of a safe pesticide viz., carbofuran 3% CG for pod borer and white grub, chlorpyrifos 20% EC for aphid root grub, deltamethrin 2.8% EC for leaf miner, flubendiamide 20% WG for *spodoptera litura*, imidacloprid 17.8% SL for aphid and jassid, lambda-cyhalothrin 5% EC for thrips, leaf hopper, leaf miner, methomyl 40% SP for *spodoptera litura*, methoxyfenozide 21.8% w/w SC for leaf eating caterpillar groundnut leaf minor and pod borer, oxydemeton – methyl 25% EC for aphid/ leaf minor, phenthoate 50% EC for leaf webber, quinalphos 1.5% DP for thrips, jassids and red hairy caterpillar, quinalphos 20% AF for spodoptera, quinalphos 25% EC for leaf hopper, leaf miner, thrips, thiamethoxam 75% W/W SG for termites etc. [32].

5.2 Castor

5.2.1 Important pests

Tobacco caterpillar (*S. litura*), castor semilooper (*Achaea janata*), shoot and capsule borer (*Conogethes (Dichocrocis) punctiferalis*), red headed hairy caterpillar (*Amsacta albistriga, A. mooreii*), Bihar hairy caterpillar (Spilosoma obliqua), whitefly (*Trialeurodes ricini*), thrips (*Retithrips siriacus, S. dorsalis*), hairy caterpillars (*Euproctis fraternal*), castor spiny caterpillar (*Ergolis merione*), castor slug (*Parasa lepida*).

5.2.2 IPM approaches

Summer ploughing to expose the hibernating pupae to predatory birds or hot sun [33] and selection of triple or double bloom castor cultivars viz., DCH-519, GCH-4, GCH-5, GCH-7, YRCH-1 which are tolerant to leafhopper. Springer, Cham. Castor varieties/hybrids with non-spiny capsules (Jwala) or semi-compact spike (GCH-4, GCH-7) are less damaged by capsule borer. In areas where red hairy caterpillar (RHC) is a problem, using a light trap (200-watt mercury lamp covers 10 ha area) on a community basis with the first monsoon rains to attract and kill the adult moths. In situations, where operating electric light trap is not feasible, a petromax light of 200 candle power is also effective in attracting moths, covering 4–6 ha area [34]. Sowing cucumber along field borders preferably before sowing of castor attracts the migrating caterpillars of RHC [35]. Using vegetative twig traps (*Jatropha* or *Ipomoea* or *Calotropis*) for collection and killing of migrating larvae of red hairy caterpillars in endemic areas [36]. Sex pheromone trapinstalled for *S. litura* @ 10 traps/ha for monitoring and implementing timely control measures. Hand-picking and destruction of gregarious stages of *S. litura* and hairy caterpillars along with damaged leaves are effective for the management of defoliators in castor, which keep the defoliation level usually less [35]. Manipulation of parasitoid activity by avoiding spraying of insecticides, when 1–2 cocoons of larval parasitoid (*Microplitis maculipennis*) observed per plant [37]. If the damage by the insect pests exceeds ETL any of the following insecticides could be sprayed. Dimethoate 30% EC - jassids, mites, semi looper, malathion 50% EC. jassids, mites, semi looper, *B. thuringiensis var. Kurstaki*, Serotype H-39, 3B, Strain Z-52 for hairy caterpillar, *Achea Janata* [33].

5.3 Sunflower

5.3.1 Important pests

Tobacco caterpillar (*S. litura*), head borer (Helicoverpa armigera), jassids (Amrasca biguttula), thrips (*S. dorsalis*), green semilooper: Thysanoplusia orichalcea, cabbage semilooper (*Trichoplusia ni*), Bihar hairy caterpillar (Spilosoma obliqua), cutworm (*A. ipsilon*), termite (Odontotermes obesus).

5.3.2 IPM approaches

Close spacing, particularly if the rainfall is heavy, mixed cropping of sunflower with cotton, Studies on groundnut and sunflower intercropping system. Removing nearby weeds that may serve as a host for aphids before planting sunflowers can slow or prevent a serious infestation [38]. To manage whiteflies, installing yellow sticky traps, which are coated with grease/sticky oily materials may be effective. Flooding of orchard with water in the month of October to kill the eggs, ploughing of orchard in November, raking of soil around tree trunk to expose the eggs to natural enemies and sun and removal of weeds. Fastening of alkathene sheet (400 gauge)/grease band of 25 cm wide afterwards mud plastering of trunk at 30 cm above the ground in the middle of December and in July –August destruction of fallen leaves infested with scales. Bihar hairy caterpillar could be managed by pre-monsoon deep ploughing (two to three times) which expose the hibernating pupae to sunlight and predatory birds and timely sowing and clean cultivation [39]. Use of well rotten manure, intercropping with pigeon pea at a row ratio of 2:1 is effective in reducing the insect attack. Tobacco caterpillar will be controlled by intercropping sunflower with pigeon pea and spraying 5% neem seed kernel extract preferably in the evening or spraying SlNPV @ 100LE/acre/spraying *Clerodendrum inerme* dust (25%) and plant extracts (10%) [35]. For head borer management, intercrop with pigeon pea, groundnut, finger millet and soybean along with sowing trap crops like marigold at 50 plants/acre. The use of pheromone traps (4 traps/acre) to trap the male moths and setting of light traps (1 light trap/5 acre) to know the range of pest incidence as well as to kill moths' population is also effective method [38]. Spraying dichlorvos 76% EC, thiamethoxam 30% FS for jassids and thrips, thiamethoxam 70% WS for jassids & thrips, cypermethrin 10% EC for Bihar hairy caterpillar, imidacloprid 48% FS for jassid, whitefly, imidacloprid 70% WS for jassid and whitefly, imidacloprid 17.8% SL for aphid and jassid, malathion 50% EC for white fly.

6. Integrated pest management in commercial crops

6.1 Sugarcane

6.1.1 Important pests

Early shoot borer (*Chilo infuscatellus*), pink borer (*Sesamia inferens*), top shoot borer (*Scirpophaga excerptalis*), root borer: (*Emmalocera depressella*), internode borer (*Chilo sacchariphagus indicus*), stalk borer (*Chilo auricilius*), white woolly aphid (*Ceratovacuna lanigera*), black bug (*Cavelerius sweeti*), whitefly (*Aleurolobus barodensis*), pyrilla (*Pyrilla perpusilla*), mealybug (*Saccharicoccus sacchari*), (*Oligonychus sacchari*), termites (*Odontotermes spp*).

6.1.2 IPM approaches

Expose the grub stages by deep ploughings for predation. Destroying the termitarium present on the bunds and nearer to the field. Sugarcane woolly aphid (SWA) reduced in Paired or wider row planting. Selection of infestation free stalks and the discarding of seed stalks, and leaves left after seed preparation reduce scale insect, mealy bugs, white flies, borers, sugarcane woolly aphids [40]. Collection and destruction of beetles from neem trees during nighttime immediately after first heavy showers for white grub control. For top shoot borer, the egg masses should be destroyed and the affected canes along with pest stages will be removed [41]. Avoid excess use of N fertilisers before earthing up [42]. For *Pyrilla* releasing 1000 viable cocoons of *Epiricania* parasites per ha. is effective [43]. Syrphid fly @ 1000 larvae or cocoons per ha [44]. Internode borer will be managed by releasing *Trichogramma chilonis* parasitized eggs in suitable instalments and the use of pheromone traps. Placing of pheromone sleeve traps @ 25 per hectare for *Chilo infuscatellus* control by destroying adult males. Soil application of 10 G phorate or 2% methyl parathion dust. For Pyrilla Releasing 1000 viable cocoons of *Epiricania* parasites per ha. White grub (L) will be managed by collection and destruction of adults from sugarcane and application of 10 G phorate. At planting (January) to manage termite, shoot borer and root borer drenching of 20 EC chlorpyriphos. Drenching the sets immediately after planting in-furrow with Chloranthriniliprole 35%WG and Chloranthriniliprole 35%WG, were proved superior by recording the lowest average per cent dead hearts by early shoot borer [45]. Application of *Metarhizium anisopliae* (Metschnikoff) Sorokin (Ma-1) against sugarcane white grub, *Holotrichia serrata* (Hope) at $1x10^{13}$ conidia ha^{-1} was found next best to chlorpyriphos. Spinetoram 12 SC were significantly effective in minimising, number of termites per colony [46]. For Rodents control bromadiolone cake 0.005% will be kept in rodent burrows or bait stations continuously for two days. Field evaluation of anticoagulant rodenticides, bromadiolone and difethialone in sugarcane fields of Cauvery delta. For Internode borer, Spot spraying of biopesticide like *Verticillium* Grasshopper Dusting of 2% methyl parathion dust in sugarcane and on bunds. 210–240 days after planting (August to September). Removal of 2–3 leaves containing egg and pupal stages and spraying 0.08% DDVP or monocrotophos with addition of 2.5% N in spray solution or spraying neemark.

6.2 Tobacco

6.2.1 Important pests

Leaf eating caterpillar (*S. litura*), whitefly (*B. tabaci*), stem Borer (*Scrobipalpa heliopa*), gram pod borer/bud worm/ capsule borer (*Helicoverpa armigera*), grass hopper (*Acrida exultata, Cyrtacanthacris tartarica, Atractomorpha crenulate*).

6.2.2 IPM approaches

Deep summer ploughing, growing of castor as trap crop for oviposition, collection and destruction of egg masses and early instar larvae, removal of weeds are the common cultural practices of leaf eating caterpillar. Stem borer will be managed by removal of infested plants, use light trap. Whitefly and grass hoppers will be managed by field sanitation and rogueing of alternate hosts, planting tall border crops to reduce white fly infestations, Using yellow sticky traps or cards, Conserving the available natural enemies such as *Encarsia formosa*, *Eretmocerus spp*., *Dicyphushe sperus*, *Chrysocharis pentheus*, spiders, coccinellids, lacewings etc. *B. thuringiensis var. kurstaki*, Serotype H-3a, 3b, Strain Z-52, NPV of *S. litura* 0.5% AS (1x109 POB/ml), Release parasitoids viz., *Trichogramma chilonis*, *Tetrastichus spp*., *Telenomus spp*, Spraying NSKE 5% against eggs and first instar larva, *Ichneumon promissorius*, *Bracon sp*, *Carcelia spp*, *Chaetopthalmus*, *Campoletis chloridae*, Lissopimpla excels, *Ichneumon promissorius*, Neem extract containing 5% azadirachtin W/W [47, 48].

6.3 Cotton

6.3.1 Important pests

Leaf hopper (*Amrasca devastans*), whitefly (*B. tabaci*), thrips (*Thrips tabaci*), aphids (*Aphis gossypii*), mealybugs (*Phenacoccus solenopsis*), tobacco caterpillar (*S. litura*), pink bollworm (*Pectinophora gossypiella*), spotted and spiny bollworm (*Earias vittella*) & (*Earias insulana*), Helicoverpa bollworm (*Helicoverpa armigera*), leaf roller (*Sylepta derogata*), red cotton bug (*Dysdercus cingulatus*), dusky cotton bug (*Oxycarenus hyalipennis*), semi-looper (*Anomis flava*), stem weevil (*Pempherulus affinis*), shoot weevil (*Alcidodes affaber*).

6.3.2 IPM approaches

Summer deep ploughing is used to reveal the soil's insect population's inhabiting/resting phases. Crop rotation can help to limit the occurrence of many pests of cotton The crop should be kept weed-free for at least 8–9 weeks following sowing, or until the canopy begins to close in due to timely inter-culture. Intercropping cotton with pigeon pea, groundnut and pulse crops is encouraged, as is the use of trap/border crops such as okra (for shoot weevil), cannabis, castor, marigold, early pigeon pea, jowar, and maize crops [49]. To suppress main perennial weeds, a hoeing in between crop rows should be performed following the appearance of cotton seedlings.

Allowing animals to graze after the last picking is advised for reducing the carryover population of bollworms. Growing of *Setaria* as intercrop to serve as live bird perches and installing 8–10 bird perches per ha after 90 days of crop growth for the benefit of predatory birds Hand-picking and destruction of various insect stages viz., egg masses and gregarious larvae of *S. litura*, grown-up larva of *Helicoverpa armigera*, affected plant parts, rosetted flowers due to pink boll worm and rotted bolls. Growing maize interspersed with cowpea on border to attract predators and parasitoids [50]. Only sucking pest tolerant Bt cultivars should be used for endemic areas. For bollworm and *Spodoptera*, *Bacillus thuriengiensis var kurstaki* is recommended. [Only suitable for non-Bt cotton]. Chemical control strategies under IPM need need-based, rational, and safe pesticide use like 50WP Diafenthiuron and Diflubenzuron 25 WP for whiteflies, aphids, thrips, and jassids, diflubenzuron 25 WP for tobacco caterpillar, dinotefuran 20 SG for bollworms white flies, jassids, aphids, and Thrips, 5 SG emamectin benzoate for boll worms, pink american boll worm, spotted and spiny, fenvalerate 0.4 DP and Fipronil 5 for Aphid, Jassid, Thrips, White fly, spotted bollworm and pink bollworm, flonicamid 50 WG for boll worms aphids, jassids, thrips, and whiteflies; flubendiamide 20 WG and flubendiamide 39.35 SC for american bollworm, Fluvalinate 25 EC for bollworms (American and Spotted bollworm), imidacloprid 70 WG for aphids, jassids, red cotton bug, bollworm jassids, aphids and thrips. Avoid combining two or more pesticides in the same tank [51, 52]. Using pesticides like pyrethroids, which cause sucking bugs to resurface should be avoided.

7. Integrated pest management in vegetables

7.1 Okra

7.1.1 Important pests

Shoot and fruit borer (*Earias vitelli, E. insulana*), gram pod borer: (*Helicoverpa armigera*), jassids (*Amrasca biguttula biguttula*), aphids: (*A. gossypii*), whitefly (*B. tabaci*), red spider mite (*Tetranychus spp),* red cotton bug (*Dysdercus cingulatus*), ash/grey weevils (*Myllocerus subfaciatus*), stem fly (*Melanagromyza hibisci*).

7.1.2 IPM approaches

Growing maize/sorghum on borders as a barrier/trap crop for the entry of shoot & fruit borer adults and set up yellow sticky and delta traps for white fly etc. Erection of bird perches @ 10/acre in the field for facilitating bird predation [53]. Removal and destruction of borer affected shoots and fruits. Sprinkler irrigation to reduce the whitefly population Application of botanical insecticides. Inundative release of natural enemies such as *Trichogramma brasiliensis* against *Earias vittella* and *H. armigera* and *Chrysoperla zastrowi sillemi* for sap feeders [54]. Two to three sprays of NSKE @ 5% alternating with sprays of pesticides, if needed, for leaf hopper, white fly, mites and aphids etc. [55]. Leaf hopper, if crosses ETL (5 hoppers/plant), spray imidacloprid 17.8 [56]. This will be effective in controlling other sucking pests as well. Installation of pheromone traps @ 2/ acre for monitoring of *Earias vittella* moth emergence. Replace the lures after every 15–20-day interval. Releasing egg parasitoid *Trichogramma chilonis* @ 1–1.5 lakh/ ha starting from 30 to 35 days after sowing, 4–5 times at weekly interval for shoot & fruit borer. Need based application of chemical pesticides viz. imidacloprid 17.8 SL @ 150 ml/ha, cypermethrin 25 EC @ 200 g a.i/ha (0.005%), quinalphos 25 EC @ 0.05% or Propargite etc. 57 EC @ 0.1% for control of leaf hoppers, aphids, white flies, borers and mites.

7.2 Brinjal

7.2.1 Important pests

Fruit and shoot borer (*Leucinodes orbonalis*), jassids (*Amrasca biguttula biguttula*), hadda beetle (*Epilachna vigintioctopunctata*), whitefly (*B. tabaci*), aphids (*A. gossypii*), spider mites (*Tetranychus spp.*) grey weevils (*Myllocerus subfasciatus*), tobacco cut worm (*S. litura*), stem borer (*Euzophera perticella*), thrips (*T. palmi*), brinjal lacewing bug (*Urentius hystericellus*).

7.2.2 IPM approaches

Soil solarisation during June will help in reducing the soil-borne insects. However, care should be taken that sufficient moisture is present in the soil for its solarisation. Clipping of borer damaged shoots and collection and destruction of damaged fruits i.e., clean cultivation helps in management of borer and phomosis disease effectively. Seed of popular hybrids is sown in beds in the first week of July. Weeding should be done from time to time and infected seedlings should be rogued out from the nursery [57]. Bird perches @ 10/acre should be erected for facilitating field visits of predatory birds [58]. Blue/yellow sticky trap should be installed for hoppers, aphids, white fly etc. [59]. Give 2 to 3 sprays of 5% NSKE against sucking pests and borer. Neem oil (2%) application reduces borer infestation, though marginally. Pheromone traps @ 5/ acre should be installed for monitoring and mass trapping of shoot & fruit borer *Leucinodes orbonalis*. Replace the lures with fresh lures after every 15–20-day interval. Release egg parasitoid *T. brasiliensis* for shoot & fruit borer, 4–5 times at weekly intervals. Apply neem cake @ 250 kg/ ha (in two splits) in soil along the plant rows at 25 and 60 DAT for reducing nematodes and borer damage [60]. If the borer incidence crosses ETL (5% infestation), apply cypermethrin 25 EC (0.005%) or carbaryl 50 WP, if the incidence of leaf hopper and other sucking insect pests is still above ETL, then apply imidacloprid 17.8 SL.

7.3 Cabbage and cauliflower

7.3.1 Important pests

Diamondback moth (*Plutella xylostella*), head borer (*Hellula undalis*), leaf webber (*Crocidolomia binotalis*), cabbage aphid (*Brevicoryne brassicae*), cabbage butterfly (*Pieris brassicae*), tobacco caterpillar (*S. litura*).

Removal and destruction of plant remnants, stubbles, debris after harvest and ploughing the field. Sowing 2 rows of bold seeded mustard as a trap crop for every 25 rows of cabbage to attract moths to mustard [61]. Grow intercrops such as tomato, garlic, coriander and carrot in alternate rows with cabbage. Installing pheromone traps @ 4–5/acre for monitoring [62]. Release egg parasitoid, *T. chilonis/pretiosum* @ 20,000/acre 4–6 times at weekly interval and larval parasitoids [63], *Diadegma semiclausm* @ 1,00,000/acre (Hills – below 25–27°C) or *Cotesia plutellae* (plains) @ 20,000/acre from 20 days after planting. Fungal pathogens, for example, *Paecilomyces spp*. and Zoophthoraradican are effective. Cabbage borers collect and destroy caterpillars mechanically in the early stages of attack. Remove and destroy the webbed leaves with caterpillars and set up light traps @ 1/acre. Conserve parasitoids such as *Cotesia crocidolomiae* etc. for managing Cabbage leaf webber. For cabbage butterfly, fine-mesh netting in nursery will stop butterflies from reaching the crop and laying eggs. Collect and destroy eggs or caterpillars mechanically by hand- usually on the underside of the leaves. Release *Trichogramma spp*. and erect bird perches. Conserve parasitoids such as *Cotesia glomeratus* (larval), *Pteromalus puparum* (larval), *Aphidius colemani* (adult and nymph), *Diaeretiella spp*. (adult and nymph), *Aphelinus spp*. (adult and nymph) and predators such wasps, green lacewings, earwigs, ground beetles, rove beetles, spiders etc. [64]. Foliar spray with dimethoate 30% EC @ 264 ml in 200–400 l of water/acre or fenvalerate 20% EC @ 120–150 ml in 240–300 l of water/acre or phosalone 35% EC @ or acetamiprid 20% SP. Foliar spray with 5% NSKE or azadirachtin 0.03% (300 ppm) neem oil-based WSP. Spraying flubendiamide 20% WG or lufenuron 5.4% EC or spinosad 2.5% SC or indoxacarb 15.8% EC @ or emamectin benzoate 5% SG or fipronil 5% SC of water/acre (Last spray should be 15 days before harvesting) [65].

7.4 Tomato

7.4.1 Important pests

Gram pod borer (*Helicoverpa armigera*), tobacco caterpillar (*S. litura*), whitefly (*B. tabaci*), serpentine leaf miner (*Liriomyza trifolii*), thrips (*T. tabaci*, *Frankliniella schultzei*), red spider mite: (*Tetranychus spp*), cutworm (*A. ipsilon*), aphids (*Myzus persicae, A. gossypii, A. craccivora*), mealybug (*Phenacoccus solenopsis*).

7.4.2 IPM approaches

For managing Serpentine leaf miner, use yellow sticky traps or cards. Erecting of bird perches @ 20/acre for encouraging predatory birds such as king crow, mynah etc. [66]. Ecological engineering of tomato with growing of ovipositional trap crops such as castor [67]. For gram pod borer field sanitation, ecological engineering of tomato with growing intercrops such as cowpea, onion, maize, coriander, uradbean etc. [68] and growing sorghum or maize in 4 rows all around tomato crop as guard crop. Rotate the tomato crop with a non-host cereal crop, cucurbit, or cruciferous vegetable. Instal pheromone traps @ 4–5/acre for monitoring adult moths' activity [69], setting up of light trap @ 1/acre. Conserve parasitoids such as *Trichogramma chilonis* (egg), *Tetrastichus* spp. (egg), *Telenomus* spp. (egg), *Chelonus blackburni* (egg-larval), *Carcelia* spp. (larval-pupal), *Campoletis chlorideae* (larval), *Eriborusargentio pilosus* (larval), *Microplitis* sp. etc. Conserve predators such as lacewings, lady beetles, spiders and fire ants conserve predators such as *C. carnea*, coccinellids, King crow, common mynah, wasp, dragonfly, spider, robber fly, reduviid bug, praying mantis, fire ants, big-eyed bugs (*Geocoris* sp), pentatomid bug (*Eocanthecona furcellata*), earwigs, ground 34 beetles, rove beetles etc. Spraying NSKE 5% against eggs and first instar larva or azadirachtin 5% W/W neem extract concentrate @ [70]. spray dimethoate 30% EC. Leafhoppers, soil application of neem cake 100 kg/acre, Conserving predators such as ladybird beetles and green lacewings and parasitoids such as *Anagrus flaveolus* and *Stethynium triclavatum*. For serpentine leaf miner, spraying azadirachtin 1% (10000 ppm) neem-based EC or azadirachtin 5% W/W neem extract concentrate [71], apply entomopathogenic nematodes (EPNs) @ 20–120 crore infective juveniles of *Steinernemafeltiae*/acre, Spray with indoxacarb 14.5% SC or flubendiamide 20% WG or flubendiamide 39.35% M/M SC or novaluron 10% EC or carbaryl 50% WP or chlorantranilioprole 18.5% SC or lambda-cyhalothrin 4.9% CS or lambda-cyhalothrin 5% EC or phosalone 35% EC or quinalphos 20% AF or quinalphos 25% EC.

8. Integrated pest management in fruits

8.1 Mango

8.1.1 Important pests

Mango hopper (*Idioscopus clypealis*, *Idioscopus niveosparsus*), mango mealy bug (*Drosicha mangiferae*), fruit fly (*Bactrocera dorsalis*), stem borer (*Batocera rufomaculata*), bark eating caterpillar (*Indarbela quadrinotata*), stone Weevil (*Sternochetus mangiferae*), shoot borer (*Chlumetia transversa*), leaf webber (*Orthaga exvinascea*), shoot gall psylla (*Apsyllaci stellate*), red ant (*Oecophylla smaragdina*), Eriophyid mite (*Aceria mangiferae*),stone weevil (*S. mangiferae*), termites (*Odontotermes obesus, Microtermes obesi*).

8.1.2 IPM approaches

Collection and destruction of crop debris, insect-damaged plant parts, removing weed plants [72], timely irrigation, organic manure, fertiliser at the recommended dose, drainage, weeding, mulching, interculture and so on are examples of cultural practices. Handpicking gregarious caterpillars and cocoons discovered on stems and destroying them in kerosene mixed water are examples of mechanical techniques. Use yellow sticky traps at a rate of 4–5 traps per acre and a light trap at 1/acre and Instal pheromone traps @ 4–5/acre for monitoring adult moth activity (replace the lures with fresh lures every 2–3 weeks) [73]. Erecting bird perches @ 25/ha for encouraging predatory birds such as King crow, common mynah, etc. [74]. Common practices such as dense orchard pruning in December, orchard and field sanitation, rogueing and the application of bio-agents such as *Metarhizium anisopliae* @ 1x 108 cfu/ml or *Beauveria bassiana* @ 108 cfu/ml on tree trunk once during toffseason and twice at 7 days intervals during the flowering season. Chemicals such as buprofezin 25% SC/ deltamethrin 2.8% EC)/ Dimethoate30% EC, imidacloprid17.8% SL, malathion 50% EC, monocrotophos 36% SL, oxydemeton–methyl 25% EC, oxydemeton–methyl 25% EC used judiciously [75].

8.2 Banana

8.2.1 Important pests

Banana rhizome weevil (*Cosmopolitus sordidus*), banana stem weevil *Odoiporus longicollis)*, banana leaf-eating caterpillar (*S. litura*), Banana aphid (*Pentalonia nigronervosa*), flower thrips (*Thrips florum*), Banana lacewing bug (*Stephanitis typicus*), hard scale (*Aspidiotus destructor*), fruit fly (*B. dorsalis*), banana scab moth (*Nacoleia octasema).*

8.2.2 IPM approaches

Maintain sanitation in the orchard for banana aphids by following clean cultural practices [76]. Deep ploughing of the field is a helpful strategy for managing white grubs because it exposes the grubs to desiccation and insectivorous bird predation. During the night hours, collect and destroy beetles in kerosene mixed water and with light traps/pheromone traps. Addition of cover crop (inclusion of fallow in rotation sequences mass trapping and use of biological control agents in insect pest suppression may prove to be beneficial as alternative IPM strategies for the Banana rhizome weevil. Pheromone traps are used in mass trapping to keep insect populations under control Individual-based models (IBMs) were examined to depict the spatial dynamics of the banana weevil in relation to the cropping system. Crop fragmentation and mass trapping were considered tools for reducing insect numbers. The results showed that altering agricultural residues in the area around each pheromone trap increased trap efficiency significantly. Traps were most efficient at catching weevils escaping the fallow in an intensive banana plantation fallow when placed at the transition zone between the banana region and the fallow. Before planting the suckers, wash them and immerse them in a Chlorpyriphos 20 EC [77]. Cleanliness in the orchard is critical. On the white grub-infested host plants, spray carbaryl at a rate of 2 ml per lit. Plants are protected from a termite infestation by using chlorpyriphos 20EC @ 400 ml/gunta with irrigation water or intermittent irrigation. Spray the pseudostem and soak the base of the tree with chlorpyriphos 20 EC. Spray Malathion 50 EC after one week. The fungus *Beauveria bassiana*, in combination with entomopathogenic nematodes (*Steinernema spp*. and *Heterorhabditis spp.*), appears to be effective in nematode control. The use of biological control agents in pseudostem traps in combination with pheromone attractants could be future development in control procedures.

8.3 Citrus

8.3.1 Important pests

Aphid (*Toxoptera aurantii*), citrus psylla (*Diaphorina citri*), fruit sucking moth (*Eudocima fullonica*, *E. maternal*), citrus/lemon butterfly (*Papilio demoleus, P. polytes*), citrus blackfly *(Aleurocanthus woglumi)*, soft scale: *Coccus hesperidium*).

8.3.2 Integrated pest management practices

Intercropping, excessive irrigation, and nitrogen application should all be avoided to control citrus psylla since they increase humidity in the orchards, which is conducive to pest growth. Pruning the impacted and dry shoots as well as modifying the canopy structure to aid in optimum light interception is recommended. Curry leaf should not be grown near citrus orchards because it can serve as a breeding ground for psylla. Each flushing season, two releases of *Mallada desjardinsi* (Navas) (=*M. boninensis* (Okamoto) @ 30 larvae/tree lower citrus psylla infestation levels Pruning of all afflicted sections throughout the winter is recommended for managing citrus leaf miner and keeping the pest population under control in extreme cases. Pruning should be avoided during active growth phases because it causes more fresh flushes, allowing the pest to have more generations. The citrus blackfly can be controlled by not planting infested seedlings. Plant the orchards at the appropriate spacing. Close spacing, excessive nitrogen application, and waterlogging should all be avoided.

Alternative hosts of the pest include guava, sapota, and pomegranate, which should not be grown near citrus groves. Clipping and destroying the afflicted shoots is recommended [78]. Avoid twig cutting before the blackfly's predicted egg-laying season. Spraying neem oil @ 100 ml or Karanja oil @ 100 ml + Teepol 10 ml during egg-laying reduces blackfly egg-laying. Fruit Sucking Moths can be controlled by getting rid of fallen fruits, which attract the moths. To avoid pest development, orchard horticulture must be kept clean. Fruit bagging on a modest scale is effective. Smoke is produced in orchards in the late evening hours, which repels the bug The adults would be attracted to light traps and poison baiting with malathion 50 EC @ 10 ml + 100 g jaggery +100 ml mandarin juice +900 ml water (two bottles of poison bait/25–30 trees). Fruit flies in citrus can be controlled by submerging wooden blocks in a 6: 4: 1 solution of ethanol, methyl eugenol, and malathion for 72 hours. In the 24th week of August, instal PAU fruit fly traps at a rate of 16 traps per square foot. If necessary, refill the traps. The collection and destruction of fallen fruits at regular intervals would limit the growth of puparia, reducing the fly population for the next year. Use male attractive fly traps baited with 0.1% methyl eugenol and 0.05% malathion 20 EC @ 25 traps/ha beginning 60 days before fruit harvest and fresh solution every 7 days to control fruit fly. Citrus aphids can be controlled with horticultural mineral oil at a concentration of 1.25% (12.5 ml I″ water) [Quinalphos 25 EC @ 1 ml or petroleum spray oil @5.9 ml or novaluron 10 EC @ 0.55 ml or imidacloprid 17.8 SL @ 0.4 ml or thiamethoxam 25 WG @ 0.32 g or acetamiprid 20 SP or neem oil @ 1–5 ml or imidacloprid 17.8 SL @ 0.5 ml can be exploited for the control of various pests in citrus [79].

9. Integrated pest management in plantation crops

9.1 Tea

9.1.1 Important pests

Tea mosquito bug (*Helopeltis theivora*), thrips (*S. dorsalis*), jassid (*Empoasca flavescens*), aphids: (*Toxoptera aurantia*), leaf eating caterpillar (*S. litura*), Red spider mite (*Oligonychus coffeae*), tea looper complex (*Buzura suppressaria, Hyposidra talaca, H. infixaria*), shot hole borer (*Euwallacea fornicates*), wood eating termite (*Microcerotermes sp.),* Scavenging termites (*Odontermes sp*).

9.1.2 IPM approaches

Routine cultural activities like as plucking rounds, adjusting pruning cycles, modifying shade trees, and timely weed treatment can all be used as effective pest control measures in tea culture. Many foliar pests, such as tea mosquito bugs, aphids, jassids, scales, and leaf folding caterpillars like flush worms and leaf rollers, are removed or reduced by this technique. On the broken ends (stalks) of plucked shoots, tea mosquito bugs lay their eggs. The more eggs, larvae, and juvenile stages of pests are removed from the bushes, the shorter the plucking rounds must be. Intensive stalk removal during plucking will help to limit the prevalence of this insect. The intensity of plucking, on the other hand, is critical; the higher the intensity, the greater the pest population reduction. During a light pruning operation, most foliar pests such as the tea mosquito bug, flushworm, aphid, jassid, thrips, red Spider Mite (RSM), scarlet mite and purple mite are eradicated. The Light Skiff assists in the removal of unproductive shoots and

Helopeltis and thrips eggs *Helopeltis* is more likely to invade densely shaded regions Because the tea mosquito bug is a negatively phototropic pest, overgrown plantations should be thinned to provide for adequate sunshine and aeration. The insect is unable to endure sunshine, resulting in a reduction in the infestation. Several caterpillar pests have alternate hosts in shade trees such as *Indigofera* and *Albizzia*. As a result, the prescription for shade control will aid in the prevention of thrips, mites, and *Helopeltis* infestations. Sanitation in the field: Field cleanliness is important in the control of a variety of pests. Weeds provide ideal hiding places for *Helopeltis* and RSM, and they also act as alternate hosts for *Helopeltis* and RSM. RSM is controlled by weed-free agriculture and prohibits cattle, goats, and other animals from straying on RSM-infested fields. A trap crop also changes the habitat of an agro-ecosystem, which can be classified as an ecological engineering strategy. Marigold, on the other hand, is an attractive plant that may be used as a red spider mite trap crop in tea Removal by hand: Lepidopteran caterpillar collection and annihilation is cost-effective and useful for both small and large plantations [80]. Manual removal of larvae and pupae can greatly reduce the population of foliage-feeding caterpillars such as the looper caterpillar, faggot worms, flush worms and leaf roller. Solarization of the soil and heat treatment: The medium in which tea plants are grown in soil. Many insects, such as eelworms, cockchafer grubs, termites, and root mealy bugs, dwell or hibernate under or near the soil surface in ideal temperature and humidity conditions. In tea plantations, a light trap is a cost-effective and environmentally friendly monitoring method for lepidopteran pests The mechanical control method for destroying termitaria appears to be a viable termite management solution [80]. In Bangladesh, the elimination of isolated termitaria is a common practice in tea plantations. In tea, *Oligota flaviceps* have been recognised as a predator of the red spider mite. The two most prevalent predators of *Acaphylla theae and Calacarus carinatus are Amblyseius herbicolus and Euseius ovalis. Anthocorids of the genera Anthocoris and Orius,* as well as predatory thrips such as *Aelothrips intermedius and Mymarothrips garuda,* are natural thrips adversaries. *C. carnea* has recently been recognised as a thrips *and Helopeltis predator*. *Caloptilia theivora*, a leaf roller, is highly parasitized by the eulophid *Sympiesis dolichogaster*. The looper caterpillar, *Buzura suppressaria*, is parasitized by *Apanteles fabiae* and *Apantelesta probanae*. *Erythmelu shelopeltidis*, an egg parasitoid, was found to be effective against the tea mosquito bug, *Helopeltis theivora* (The percentage parasitism in the field ranged from 52 to 83%, and this is the first time this species has been found attacking *H. theivora*. *B. thuringiensis* bacterial pesticides have been successfully employed to combat looper caterpillars, cutworms, flushworms, and other lepidopterous pests [81]. *Verticillium lecani*, *Paecilomyces fumosoroseus*, and *Hirsutella thompsonii*, three entomopathogenic fungi, were tested and proved to be efficient against pink, purple, and red spider mites. The possible entomopathogenic fungi for the management of *Helopeltis* in tea were discovered to be *Cladosporium sp.*, *Aspergillus niger* and *Aspergillus flavus*. The most common entomopathogenic fungus, *Metarhizium anisopliae*, reduced the population of red spider mites, thrips, and live wood termites in tea *Azadirachtin*, an oxygenated triterpenoid derived from the seed kernel of the neem tree *A. indica*, is presently being tested against a variety of tea pests, including *Helopeltis*, Red spider mites, flushworm, and others. Plants suffering from root-knot nematodes, *Meloidogyne brevicauda*, were found to benefit from the application of neemcake at a rate of 2 kg/bush Furthermore, extracts from Mahogany, *Karanja*, *Datura*, Tobacco, *Bishkatali*, *Katamehedi*, *Lantana*, *Xanthium* and *Clerodendrum* may be useful against significant tea pests such as tea mosquito bugs and red spider mites. Neem and *Mahogani* cake drastically reduced the nematode population in the soil [82].

DOI: http://dx.doi.org/10.5772/intechopen.105104

9.2 Coffee

9.2.1 Important pests

White coffee stem borer (*Xylotrechus quadripes*), coffee berry borer (*Hypothenemus hampei),* coffee root mealybug (*Planococcus citri*) Shot hole borer (*Xylosandrus compactus)*, brown scale (*Saissetia coffeae*), green scale (*Coccus viridis)*, cock chafers or white grubs (*Holotrichia spp)*, hairy caterpillars (*Eupterote spp)*, coffee bean beetle (*Araecerus fasciculatus*), red coffee borer (*Zeuzera coffeae*).

9.2.2 IPM approaches

Destroying ant nests from shade trees and promoting favorable environmental conditions for the growth of the white halo fungus are two ways to manage coffee scales (*Verticillium lecanii*). Maintain optimal shade on the estates to control the White coffee stem borer (two-tiered shade tree system). Collar prune the infested plants, uproot if the borer has entered into the root, burn the affected plants immediately and remove the loose scaly bark of the main stem and thick primaries using a coir glove or coconut husk to eliminate the cracks and crevices which are used by the female beetle to place eggs on the stem. After removing susceptible plants by tracing, use a scrubbing or 10% lime coating or stem wrapping with empty fertiliser bags in hot spot areas, such as open patches and estate borders with poorly managed estates [83]. Pheromone traps can be set up in the field at a height of 1.8 to 2 metres above ground level. The traps should be laid out in a grid of 25 traps per hectare, with a 20-meter interval between them. During the summer, the shot hole borer will be controlled by cutting the affected twigs 2.5 to 7 cm below the shot hole and burning, removing, and killing all undesired / infested suckers, as well as keeping thin shade and providing good drainage in the estate. During the early part of the flight period, between April and October each year, spray Chloropyrifos 20 EC at a dosage of 600 ml in 200 l of water, coupled with 200 ml of any wetting agent on the main stem and thick primaries. Coffeeberry borer will be controlled through the use of cultural practises and phytosanitary measures, such as fumigation with aluminium phosphide under the supervision of a pest control agency or a technical expert, timely harvest, spreading gunny bags or polythene sheets at the time of harvest to reduce gleaning, removing gleanings and leftovers, dipping infested berries in boiling water for 2–3 minutes kills all stages inside, During the drying process, traps can be placed around the drying yard [84]. Dusting quinalphos 1.5% or methyl parathion 2% on afflicted patches and spraying with 4 litres of kerosene in 22 litres of water combined with 200 ml of any agricultural wetting agent reduced coffee mealybug. Drenching with roger 30EC at 3.3 ml per litre of water in the case of young plants (2–4 years).

9.3 Coconut

9.3.1 Important pests

Rhinoceros beetle (Oryctes rhinoceros), red palm weevil (Rhynchophorus ferrugineus), black headed caterpillar (Opisina arenosella), eriophyid mite: *Aceria guerreronis* (Acari: Eriophyidae), Termite: *Odontotermes sp.*

9.3.2 IPM approaches

Rhinoceros beetle and cock chafer beetle could be controlled by collecting and eliminating the various stages of the beetle's life cycle from manure pits (the pest's breeding site) whenever manure is removed. GI hooks can be used to extract the adult beetle from the palm crown during the peak phase of population growth [85]. Pheromone traps installed and gathered away from the main plantation are effective. Set up a pheromone trap for rhinoceros beetles at a rate of one trap per 100 hectares by attaching it to the plant at a height of 0.6 to 1 m to trap and kill the beetles [86]. Avoid cutting green leaves for red palm weevil, and if necessary, cut them about 120 cm away from the stem to prevent successful inward passage of the grubs via the cut end. Set up a pheromone trap and a trap with coconut logs: Set up attractant traps (dirt pots) containing sugarcane molasses 212 kg or toddy 212 l (or pineapple or sugarcane activated with yeast or molasses) + acetic acid 5 ml + yeast 5 g + longitudinally split tender coconut stem/logs of green petiole of leaves into 30 numbers in one acre to trap adult red palm weevils in large numbers. The discharge of baculovirus oryctes injected adult rhinoceros beetles at a rate of 6 insects per acre provides biological control by reducing the beetle's leaf and crown damage. To attract and kill the adults, soak one kilogramme of castor cake in five litres of water and place it in little mud pots in the coconut gardens. In the base of the three innermost leaves in the crown, apply a mixture of neem seed powder + sand (1: 2) @ 150 g/palm or neem seed kernel powder + sand (1: 2) @ 150 g/palm. Growing intercrop (sun hemp, four crops per year) and a shelterbelt of *Casuarina* around the coconut garden to prevent additional infiltration could help manage the coconut eriophyid mite. Fenpyroximate 5% EC was used as a chemical control (spray fluid volume as required). To control the Leaf Eating Caterpillar / Black Headed Caterpillar as a preventative precaution, clip and burn the first afflicted leaves at the start of the summer season. Cut the root at an angle and place it in a 7 × 10 cm polythene bag with an insecticidal solution containing monocrotophos 36% WSC + water 10 ml [87].

10. Conclusions

The pest management in tropical food crops with special emphasis on integration of cultural, mechanical, chemical and natural enemies is discussed in this chapter which will be useful for students, farmers, researchers and also entrepreneurs for updating their knowledge for future endeavours. The judicial and selective use of management strategies described could be helpful for sustainable pest management and production of these crops

Conflict of interest

I confirm there are no conflicts of interest.

Acronyms and abbreviations

Integrated Pest Management (IPM)
Neem seed kernel extract (NSKE)

Encapsulated granule (CG)
Granule (G)
Water-soluble powder (SP)
Soluble concentrate (SL)
Dustable powder (DP)
Fall armyworm (FAW)
Emulsifiable concentrate (EC)
Parts per million (ppm)
A mixed formulation of CS en SC (ZC)
Flowable concentrate for seed treatment (FS)
Bait [ready for use] (RB)
Water dispersible powder for slurry treatment (WS)
Water-soluble granule (SG)
Red hairy caterpillar (RHC)

Author details

John Samuel Kennedy* and Jeeva Kasiviswanathan Lekshmi
Department of Agricultural Entomology, Tamil Nadu Agricultural University, Coimbatore, India

*Address all correspondence to: jskennedy@tnau.ac.in

References

[1] Sharma HC. Climate change effects on insects: Implications for crop protection and food security. Journal of Crop Improvement. 2014;**28**(2):229-259

[2] Horrigan L, Lawrence RS, Walker P. How sustainable agriculture can address the environmental and human health harms of industrial agriculture. Environmental Health Perspectives. 2002;**110**(5):445-456

[3] Deguine JP, Aubertot JN, Flor RJ, Lescourret F, Wyckhuys KA, Ratnadass A. Integrated pest management: Good intentions, hard realities: A review. Agronomy for Sustainable Development. 2021;**41**(3):1-35

[4] Brewer MJ, Goodell PB. Approaches and incentives to implement integrated pest management that addresses regional and environmental issues. Annual Review of Entomology. 2012;**57**:41-59

[5] Abrol DP, Shankar U. History, overview and principles of ecologically based pest management. In: Integrated Pest Management: Principles and practice. Wallingford, England: CABI; 2012. pp. 1-26

[6] Hatt S, Lopes T, Boeraeve F, Chen J, Francis F. Pest regulation and support of natural enemies in agriculture: Experimental evidence of within field wildflower strips. Ecological Engineering. 2017;**98**:240-245

[7] Tscharntke T, Karp DS, Chaplin-Kramer R, Batáry P, DeClerck F, Gratton C, et al. When natural habitat fails to enhance biological pest control–Five hypotheses. Biological Conservation. 2016;**204**:449-458

[8] Oerke EC. Crop losses to pests. Journal of Agricultural Science. 2006;**144**:31-43

[9] Merle I, Hipólito J, Requier F. Towards integrated pest and pollinator management in tropical crops. Current Opinion in Insect Science. 2022;**50**:100866

[10] Samada LH, Tambunan USF. Biopesticides as promising alternatives to chemical pesticides: A review of their current and future status. Online Journal of Biological Science. 2020;**20**:66-76

[11] Litsinger JA. Biology and Management of Rice Insects. Philippines: International Rice Research Institute; 1994. pp. 549-584

[12] Singh A, Vasisht AK, Kumar R, Das DK. Adoption of integrated pest management practices in paddy and cotton: A case study in Haryana and Punjab. Agricultural Economics Researh Review. 2008;**21**(2):221-226

[13] Bragg DE, Silvia IR, John G, Uma S, Abrol DP. Integrated Pest Management in Tropical Cereal Crops. New Delhi: New India Publishing Agency; 2016. pp. 249-273

[14] Kim EG, Yun S, Park JR, Jang YH, Farooq M, Yun BJ, et al. Bio-Efficacy of Chrysoeriol7, a Natural Chemical and Repellent, against Brown Planthopper in Rice. International Journal of Molecular Sciences. 2022;**23**(3):1540

[15] Prakash A, Bentur JS, Prasad MS, Tanwar RK, Sharma OP, Bhagat S, et al. Integrated Pest Management for Rice. New Delhi: NCIPM; 2014. p. 43

[16] Kishore P, Sharma RK. Pest Management in Cereals: Entomology e Novel Approaches. New Delhi: New India Publishing Agency; 2007. pp. 245-266

[17] Suganthi A, Krishnamoorthy SV, Sathiah N, Rabindra RJ, Muthukrishnan N, Jeyarani S, et al. Bioefficacy, persistent toxicity, and persistence of translocated residues of seed treatment insecticides in maize against fall armyworm, *Spodoptera frugiperda* (JE Smith, 1797). Crop Protection. 2022;**154**:105892

[18] Sekhar JC, Soujanya PL, Suby SB, Rakshit S. Fall armyworm-Status and initiatives taken for its sustainable management in maize. Integrated Management Strategies for Major Crop Pests and Diseases in Developing Countries. 2019;**39**(44):3

[19] Sangle SV, Jayewar NE, Kadam DR. Efficacy of insecticides on larval population of fall armyworm, *Spodoptera frugiperda* on maize. Journal of Entomology and Zoology Studies. 2020;**8**(6):1831-1834

[20] Patil SD, Padhye AP, Katare S. Efficacy of insecticidal seed treatment against pests of wheat. International Journal of Plant Protection. 2016;**9**(2):372-380

[21] Kishore P. Strategies for pest management in sorghum in India. Journal of Entomological Research. 1990;**14**(1):16-20

[22] Mishra R, Bala S. Use of plant products as potential eco-friendly management alternatives for the control of rice weevil, Sitophilus oryzae in stored sorghum. Journal of Entomological Research. 2019;**43**(3):253-255

[23] Pooniya V, Choudhary AK, Dass A, Bana RS, Rana KS, Rana DS, et al. Improved crop management practices for sustainable pulse production: An Indian perspective. Indian Journal of Agricultural Sciences. 2015;**85**(6):747-758

[24] Abbas M, Ramzan M, Hussain N, Ghaffar A, Hussain K, Abbas S, et al. Role of light traps in attracting, killing and biodiversity studies of insect pests in Thal. Pakistan Journal of Agricultural Research. 2019;**32**(4):690

[25] Patidar RK, Dutta P, Patil J, Pathak M, Jeevan H, Euphema N, et al. Management of important insect pests of horticultural crops using traps. International Journal of Agriculture, Environment and Sustainability. 2021;**3**(1):34-38

[26] Sharma AN, Gupta GK, Verma RK, Sharma OP, Bhagat S, Amaresan N, et al. Integrated Pest Management for Soybean. New Delhi: National Centre for Integrated Pest Management; 2014. p. 41

[27] Sangamithra S, Vinothkumar B, Muthukrishnan N, Manoharan T. Evaluation of bioefficacy of flubendiamide 24% w/v+ thiacloprid 24% SC w/v against shoot and fruit borer and its sucking pests and its safety to non-target organisms in brinjal. Journal of Entomology and Zoology Studies. 2018;**6**(1):245-249

[28] Umashankar PKS, Kumar S, Alam MA. Estimation of the flower loss due to thrips (*Megalurothrips usitatus*) and grain loss due to tur pod bug (*Clavigralla gibbosa*) in pigeonpea crop. 2021

[29] Madhu TN, Shah VK, Prabhulinga T, Chakravarthy AK, Ashok Kumar CT. Optimization of pheromone trap densities and impact of insecticides on pheromone catches for mass trapping *Helicoverpa armigera* (Hubner) (Lepidoptera: Noctuidae) in chickpea. Journal of Entomology Zoological Studies. 2019;7:78-84

[30] Basappa H, Patil RK. Current status and potential of biopesticides in pest management in oilseeds. In: Biopesticides

in Sustainable Agriculture Progress and Potential. Jodhpur, Rajasthan: Scientific Publishers; 2014. p. 93

[31] War AR, Hugar SV, Rahman SJ, Sharma HC. Orientation behaviour of Helicoverpa armigera egg parasitoid, Trichogramma chilonis and the larval parasitoid, Campoletis chlorideae towards different groundnut genotypes. Biocontrol Science and Technology. 2016;**26**(10):1420-1431

[32] Nataraja MV, Jadon KS, Holajjer P, Thirumalaisamy PP, Jasrotia P, Dutta R. Integrated Pest and Disease Management in Groundnut. Junagadh: Indian Council of Agricultural Research; 2014

[33] Basappa H and Duraimurugan P. Management of pests of oilseed crops. In: Pests and Their Management. Sujatha M, Devi PV, Reddy TP. *Insect pests of castor (Ricinus communis L) and their management strategies*. BS Publications, CRC Press, 2011. 2018. pp. 223-240

[34] Narayanamma VL, Reddy KD, Reddy AV. Integrated pest and disease management practices in oilseeds–a critical review. Indian Journal of Entomology. 2013;**75**(1):34-56

[35] Rabeena I, Bose ASC, Sathyan T. Integrated pest management of red hairy caterpillar in groundnut. Biotica Research Today. 2021;**3**(3):200-201

[36] Geetha B, Venkatachalam S, Duraimurugan P. Evaluation of Field Efficacy of Microbial and Chemical Insecticides for the Management of Defoliator Pests in Castor. Hyderabad, India: The Indian Society of Oilseeds Research; p. 323

[37] Basappa H. 2004. Integrated pest management in sunflower: An Indian scenario. In Proc. 16th International Sunflower Conference, Fargo, North Dakota, USA (pp. 853-759)

[38] Ranasingh N, Mahalik JK. Insect pest management in sunflower. Orissa Review. 2008;**2008**:51-52

[39] Satyagopal K, Sushil SN, Jeyakumar P, Shankar G, Sharma OP, Boina DR, Sain SK, Reddy M, Rao NS, Sunanda BS, Asre R. AESA based IPM package for sugarcane. 2014

[40] Geetha MV, Kalyanasundaram M, Jayaraj J, Shanthi M, Vijayashanthi VA, Hemalatha D, et al. Pests of sugarcane. In: Pests and Their Management. Singapore: Springer; 2018. pp. 241-310

[41] Goebel FR, Nikpay A. Integrated pest management in sugarcane cropping systems. In: Rapisarda C, Massimino-Cocuzza GE, editors. Integrated Pest Management in Tropical Regions. Wallingford, England: CABI; 2017. pp. 113-133

[42] Virk JS, Sharma S, Joshi N. Management Approaches for Sugarcane Insect Pests. Integrated Pest Management. Jodhpur, Rajasthan: Scientific Publishers; 2013. p. 403

[43] Srikanth J, Singaravelu B, Kurup NK, Mukunthan N, Santhalakshmi G, Nirmala R. Predators as natural and applied biocontrol agents of sugarcane woolly aphid *Ceratovacuna lanigera* in india: An appraisal. Journal of Sugarcane Research. 2015;**5**(2):53-72

[44] Sunilkumar NM, Hosamani AK, Shobharani M, Jadhav RL. Bio efficacy of New Insecticide Molecules Chlorantraniliprole 35% WG against Sugarcane Early Shoot Borer *Chilo infuscatellus* (Snellen) and Internode Borer *Chilo sacchariphagus* indicus (Kapur). International Journal of Current Microbiology and Applied Sciences. 2018;**7**(12):3680-3368

[45] Muthukrishnan N, Vishnupriya M. Antifeedant activity and field evaluation of spinetoram 12 SC against termite, *Odontotermes obesus* on sugarcane. International Journal of Plant Protection. 2019;**12**(2):87-93

[46] Yadav AS. Farmers participatory on farm evaluation of IPM technology against semilooper and tobacco caterpiller in Soybean. Indian Journal of Entomology. 2014;**76**(3):250-251

[47] Mafi SA, Ohbayashi N. Toxicity of insecticides to the citrus leafminer, *Phyllocnistis citrella*, and its parasitoids, *Chrysocharis pentheus* and *Sympiesis striatipes* (Hymenoptera: Eulophidae). Applied Entomology and Zoology. 2006;**41**(1):33-39

[48] Mohan S, Monga D, Kumar R, Nagrare V, Gokte-Narkhedkar N, Vennila S, Tanwar RK, Sharma OP, Bhagat S, Agarwal M, Chattopadhyay C. Integrated pest management package for cotton. 2014

[49] Pokharkar DS, Nakat RV, Dhane AS, Tamboli ND. Enhancement of Natural Enemies Population by Habitat Manipulation in Rainfed Cotton. Journal of Agriculture Research and Technology. 2013;**38**(2):271

[50] Das SK. Scope and relevance of using pesticide mixtures in crop protection: A critical review. International Journal of Environmental Science and Toxicology. 2014;**2**(5):119-123

[51] Butter NS, Battu GS, Kular JS, Singh TH, Brar JS. Integrated use of *Bacillus thuringiensis* Berliner with some insecticides for the management of bollworms on cotton. Journal of Entomological Research. 1995;**19**(3):255-263

[52] Sahu PS, Khan HH, Ghongade DS, Naz H. Chapter-13 Integarted Pest Control Srategies against Okra Fruit and Shoot Borrer, *Earias vittella* (Fabricius). In: Latest Trends in Zoology and Entomology Sciences. New Delhi: AkiNik Publications; p. 233

[53] Mohankumar S, Karthikeyan G, Durairaj C, Ramakrishnan S, Preetha B, Sambathkumar S. Integrated Pest Management of Okra in India. In: Integrated Pest Management of Tropical Vegetable Crops. Dordrecht: Springer; 2016. pp. 167-177

[54] Sardana HR, Bambawale OM, Kadu LN, Singh DK. Development and validation of adaptable IPM in okra through farmers participatory approach. Annals of Plant Protection Sciences. 2005;**13**(1):54-59

[55] Parthiban SL, Ravikumar P, Chinniah C. Newer insecticidal molecules and their implication on pest management–eco-friendly insecticides for IPM. Advances in Life Sciences. 2016;**5**(23):10854-10859

[56] Javed H, Mukhtar T, Javed K. Management of Eggplant Shoot and Fruit Borer (*Leucinodes Orbonalis* Guenee) by integrating different non-chemical approaches. Pakistan Journal of Agricultural Sciences. 2017;**54**(1):65-70

[57] Nayak US, Rath LK, Baral K, Senapathy S. Field Efficacy of Different IPM Modules Against Brinjal Shoot and Fruit Borer (BSFB) and The Sucking Insect Pests of Brinjal. International Journal of Bio-resource and Stress Management. 2016;7(6):1315-1320

[58] Sehgal M, Malik M, Chakraborty A. Key Pests of Okra and Their Eco-Friendly Management. Biotica Research Today. 2021;**3**(5):348-350

[59] Suresh K, Rajendran R, Rani BU. Eco-friendly appraoch for managing

major sucking pests of brinjal. Journal of Entomological Research. 2007;**31**(1):19-22

[60] Moorthy PK, Kumar NK. Integrated Pest Management in Vegetable Crops. Indian Agriculture; 2004. p. 95

[61] Kunjwal N, Srivastava RM. Insect pests of vegetables. In: Pests and Their Management. Singapore: Springer; 2018. pp. 163-221

[62] Patra S, Samanta A. Toxicity of some new and conventional insecticides on *Trichogramma chilonis* and *Bracon brevicornis* under laboratory condition. Journal of Environmental Biology. 2017;**38**(5):753

[63] Shah R, Ahmed S, Poswal A. Population dynamics of insect pests, parasitoids and predators in cabbage and cauliflower agro-ecosystems. Journal of Entomological Research. 2013;**37**(2):129-137

[64] Singh G, Chahil GS, Jyot G, Battu RS, Singh B. Degradation dynamics of emamectin benzoate on cabbage under subtropical conditions of Punjab, India. Bulletin of Environmental Contamination and Toxicology. 2013;**91**(1):129-133

[65] Seni A. Insect pests of tomato and their management. In: Management of Insect Pests in Vegetable Crops: Concepts and Approaches. CRC Press, United States; 2020. p. 225

[66] Kumara A, Shakywar RC, Chand G, Hiremath R, Sehgal M, Malik M. Ecological engineering as a pest management tool in horticultural ecosystems. International Journal of Agriculture, Environment and Sustainability. 2021;**3**(1):8-16

[67] Saha T, Chandran N, Anu BC. Major insect pests of vegetable crops in Bihar and their management. In: Sustainable Agriculture: Advances in Technological Interventions. Palm Bay, Florida, USA: Apple Academic press; 2020. p. 6

[68] Kapoor B, Shankar U. Trap catches of Helicoverpa armigera Hubner and Maruca vitrata Fabricius and their natural enemies on black gram [*Vigna mungo* (L.) Hepper]. Journal of Entomological Zoological Studies. 2019;7:246-251

[69] Chavan RD, Yeotikar SG, Gaikwad BB, Dongarjal RP. Management of major pests of tomato with biopesticides. Journal of Entomological Research. 2015;**39**(3):213-217

[70] Chavan SM, Kumar S, Arve SS. Population dynamics and development of suitable pest management module against major insect pests of tomato (*Solanum lycopersicum*). Journal of Applied Horticulture. 2013;**15**(2):1673-1676

[71] Venkata Rami Reddy P, Gundappa B, Chakravarthy AK. Pests of mango. In: Pests and Their Management. Singapore: Springer; 2018. pp. 415-440

[72] Susanto A, Sudarjat S, Yulia E, Permana AD, Gunawan A, Yudistira DH. Effectiveness of modified traps for protection against fruit flies on mango. Jurnal Biodjati. 2020;**5**(1):99-106

[73] Reddy PP. Biointensive integrated pest management. In: Recent Advances in Crop Protection, Springer, New Delhi; 2012. pp. 223-244

[74] Chaudhari AU, Sridharan S, Sundar Singh SD. Management of mango hopper with newer molecules and biopesticides under ultra-high density planting system. Journal of Entomology and Zoology Studies. 2017;**5**(6):454-458

[75] Sharma DR, Batra RC. Recent advances in management of insect pests

of fruit crops. In: Entomology: Novel Approaches. New Delhi: New India Publishing Agency; 2007. p. 333

[76] Mathew M, Mani M. Root mealybugs. In: Mealybugs and their Management in Agricultural and Horticultural Crops. New Delhi: Springer; 2016. pp. 629-641

[77] Ahuja DB, Chattopadhyay C. Pests of Fruit Trees (Citrus, Banana, Mango, Pomegranate and Sapota). In: E-Pest Surveillance and Pest Management Advisory, ICAR-National Research Centre for Integrated Pest Management. Pune, Maharashtra: New Delhi and State Department of Horticulture, Commissionerate of Agriculture; 2015

[78] Shinde SS, Neharkar PS, Dhurve NG, Sawai HR, Lavhe NV, Masolkar DS. Evaluation of different insecticides against citrus leaf miner on Nagpur mandarin. Journal of Entomology and Zoology Studies. 2017;**5**(6):1889-1892

[79] Mamun MSA, Ahmed M. Integrated pest management in tea: Prospects and future strategies in Bangladesh. The Journal of Plant Protection Sciences. 2011;**3**(2):1-13

[80] Guastella D, Cocuzza GEM, Rapisarda C. 15 Integrated Pest Management in Tea, Cocoa and Coffee. Integrated Pest Management in Tropical Regions, CABI, Wallingford, England; 2017. p. 285

[81] Bordoloi M, Madhab M, Dutta P, Borah T, Nair SC, Pukhan I, Debnath S, Barthakur BK. Potential of entomopathogenic fungi management of *Helopeltis theivora* (Waterhouse). Abstract of World Tea Science Congress, Tocklai Experimental Station, tea cultivation, Tea Research Association, Assam, India. 2011. p. 62

[82] Mamun MSA, Ahmed M, Paul SK, Chowdhury RS. Evaluation of some indigenous plant extracts against tea mosquito bug, *Helopeltis theivora* Waterhouse (Hemiptera: Miridae) infesting tea. Tea Journal of Bangladesh. 2013;**42**:10-20

[83] Uma MS, Seetharama HG, Reddy GM, Kumar TM, Ashwitha TV, Raghuramalu Y. A novel approach to arrest white stem borer in coffee. Journal of Experimental Zoology, India. 2017;**20**(1):377-382

[84] Samy MA. The Potential of Neem, Pongamia Oil and Garlic Extracts Formulation for an Effective Management of Berry Borers in Organic Coffee Plants. International Journal of Food Bioscience. 2018;**1**:49-55

[85] Josephrajkumar A, Mohan C, Prathibha PS, Nalinakumari T, Nair CPR. Pest dynamics and suppression strategies. In: The Coconut Palm (*Cocos nucifera* L.)- Research and Development Perspectives. Singapore: Springer; 2018. pp. 557-634

[86] Chakravarthy AK, Chandrashekharaiah M, Kandakoor SB, Nagaraj DN. Efficacy of aggregation pheromone in trapping red palm weevil (*Rhynchophorus ferrugineus* Olivier) and rhinoceros beetle (*Oryctes rhinoceros* Linn.) from infested coconut palms. Journal of Environmental Biology. 2014;**35**(3):479

[87] Balaji K, Hariprasad Y. Evaluation of biopesticides against coconut perianth mite *Aceria guerreronis* Keifer under laboratory conditions. International Journal of Plant Protection. 2016;**9**(1):329-332

Chapter 12

Phytochemical Contents of Essential Oils from *Cymbopogon* Species: A Tropical Medicinal Plant

Margaret Ikhiwili Oniha, Eze Frank Ahuekwe and Sharon Oluwatobi Akinpelu

Abstract

Natural resources especially medicinal plants possess the potentials to sustain all existence on earth. *Cymbopogon*, a globally cultivated herb, possesses high contents of diverse essential oils for medicinal and economic purposes including treatment of malaria and candidiasis. Notable species include *Cymbopogon citratus* and *C. flexosus* having citral as the main chemical compound. Numerous compounds of these species include limonene, citronella, geranyl acetic derivatives, elemol, among others. Phytochemical analysis of these essential oils is usually done by the gas chromatography-mass spectrometry (GC-MS) method sequel to obtaining them through solvent extraction, hydrodistillation, supercritical CO2 extraction, chromatography among others. Although the supercritical CO2 extraction method gives greater quality yields void of toxic wastes with preserved thermal stability compared with other methods, its high-working pressure generates issues of safety risks and costs. Quantitative determination is done using spectrophotometric, chromatographic, and Folin-Ciocalteu methods. In comparison with other chromatographic techniques employed, gas chromatography exhibits greater efficiency by quantifying and determining the presence of various components at low concentrations. This prominently economical plant with potent ethnobotanical benefits hinged on the essential oils phytochemicals is faced with diverse extraction challenges; thus, improvement in the extraction and quantification techniques is key to the harvest of pure yields of lemon grass essential oils.

Keywords: *Cymbopogon*, essential oils, phytochemicals, plants, extraction, chromatography

1. Introduction

Medicinal plants play an important role in a healthy society. Restoration of practices and knowledge related to medicinal plant resources is part of an important strategy related to biodiversity conservation, knowledge of new drugs and improving the living standards of rural populations [1]. The Gramineae family includes the genus *Cymbopogon*, which encapsulates herbs that are globally recognized for possessing high essential oil content. Its species are broadly distributed across the globe where they are utilized for diverse purposes. Both the commercial and medicinal uses of its differential species have been well authenticated [2].

Additionally, the ethnopharmacology corroboration reveals the presence of an expansive array of properties possessed by these species, which establishes their utilization for pest control for cosmetics and anti-inflammatory media. Species of *Cymbopogon* may also envelope potentials as potent antitumor and chemopreventive drugs [3]. *Cymbopogon flexuosus* and *Cymbopogon citratus* are the two main species vastly farmed for their essential oils in various parts of the world [4]. It is cultivated in the subtropical and tropical regions of the world and widely used in the agriculture, cosmetics, flavor, food, pharmaceutical industries [1]. It is a member of the aromatic grasses containing essential oils with lemon flavor. Its species are tufted perennial C4 grasses with several hard stems emerging from a short, rhizomatous base [5] with a citrus flavor, dried to powder or freshly used. The name *Cymbopogon* is derived from the Greek words "kymbe" (boat) and "pogon" (beard), referring to the flower spike arrangement [5]. The species *C. citratus* is identified by many international common names, such as West Indian lemon grass or lemon grass (English), citronelle or verveine des indes (French), hierba limon or zacate de limón (Spanish), xiang mao (Chinese), capimcidrao, or capim-santo (Portuguese), and locally, there are more than 28 indigenous names identified from different countries of the world [4]. Other common names of *C. citratus* include lemongrass, barbed wire grass, citronella grass, fever grass, and tanglad [6]. *C. citratus* thrives best in sunny, warm, humid conditions of the tropics and grown in a wide range of soil types, from rich loam to poor laterite. Although calcareous and water-logged soils adversely affect growth [7], those cultivated on sandy soils have higher leaf oil yields and higher citral content [8]. *C. citratus* is believed to have originated from Malaysia, and it is now widely grown in Central and South America, regions in Africa, Southeast Asia, and the Indian Ocean Islands, both on subsistence and commercial scales particularly in South-east Asia. It is an aromatic, evergreen, perennial grass that produces multiple stiff stems emerging from a short rhizome-like rootstock and grows to approximately 1.5 m tall. Although it rarely produces florets, the leaflets are blue-green, erect, and linear and exude a characteristic lemon flavor when crushed [9]. The *C. citratus* is positioned as one of the most globally distributed genera that are usually utilized in all parts of the globe [3]. The plant, which can be dried and powdered, or used fresh, has been employed in diverse activities that include food flavoring, in teas, soups, with poultry, fish, beef, seafood, and curries. Reports have validated its global diverse health benefits, including the fact that lemongrass leaves and other parts can be infused to treat nausea, stomach aches, constipation, and a variety of stomach infections as well as to prevent ulcers [4]. *C. flexuosus* (Poaceae) is described as a native, tall perennial aromatic grass (sweet smelling sedge) with growth confined to specific patches of subtropical parts of Asia, Africa, and America. *Cymbopogon flexuous,* also known as the Cochin or Malabar grass, is native to Sri Lanka, India, Thailand, and Burma. It is naturalized in numerous parts of the tropical and subtropical Southeast Asia and Africa [10, 11]. Consequently, it has received significant global demand due its varied range of applications in differential industries. Reports reveal that *Cymbopogon* flexosus include more than 140 with 52 of them growing in Africa, 45 in India, 6 in Australia, 6 in South America, 4 in Europe (only in Montenegro), 2 in North America, and the others in South Asia. It is utilized as a medicinal tea, in preparation of soups, curries, and starting agent for vitamin A synthesis and has been known to be both perfumery and flavorful with the therapeutic characteristics [10]. *C. flexosus* is a C_4 grass endowed with industrial importance and abundant medicinal properties, and utilized for its essential oil (EO) production [11]. India is a significant exporter and the major producer of lemongrass oil. The essential oil comprises citral (i.e., a mixture of neral and geranial), geraniol, limonene, and geranyl acetate among others and is well-recognized for their antimicrobial, anticancer, and allelopathic activities [11, 12]. These essential oils are employed in the production of eco-friendly pesticides [3, 11]. In addition, lemongrass is an important source of several vitamins (A, B1, B2, B3, B5, B6, folate, and vitamin C) and essential minerals (calcium, copper, iron, magnesium, manganese, potassium, phosphorous, zinc) [11]. The above-listed properties cause lemongrass to be an industrially preferred crop due to its enormous potential in the fields of medicine, cosmetics, food, and biotechnology [13]. In furtherance, a couple of studies published that lemongrass essential oil can be utilized as biofuel; thus, it is regarded as an energy grass [14].

2. Phytochemicals of essential oils of *Cymbopogon* species

The essential oils of *Cymbopogon* are identified by monoterpene constituents including citral, limonene, geraniol, citronellol, elemol, b-carophyllene, citronellal, 1,8 cineole, linalool, methylheptenone, geranylformate, and geranyl acetic acid derivation. Essential oils are typically chemically characterized by GC-MS [1, 15–17]. The plant *C. citratus* is abundant in bioactive substances. Flavonoids, alkaloids, saponins, tannins, and phenolic compounds, such as quercetin, luteolin, apiginin, isoorientin 2'-O-rhamnoside, and kaempferol, have been isolated and identified from the plant's leaves [18, 19]. These phytochemicals have been reported to be beneficial, especially in the pharmaceutical, food, health, and agricultural industries [20, 21]. Alcohols, aldehyde, ketones, esters, and terpenes are predominantly the other compounds found in *C. citratus* [20]. It also consists of 1–2 percent essential oil on a dry basis with the chemical composition varying greatly depending on the habitat, genetic diversity, and agricultural treatment of the crops. Longifolene (V4) (56.67%) and selina-6-en-4-ol (20.03%) are the constituents of volatile oil from the roots [22]. Although the primary chemical constituent of lemongrass essential oil is citral, borneol, geranial, geraniol, β-myrcene, limonene, neral, geranyl acetate, alpha-terpeniol, estragole, methyleugenol, citronellal, careen-2, farnesol, (+)-cymbodiacetal, proximadiol, methyl heptenone, terpinolene, pinene, linalool, linalyl acetate, and β-caryophyllene have also been reported [5, 22]. Citral (3, 7-dimethyl-2, 6-octadien-2-al) refers to the natural mixture of two isomeric acyclic monoterpene aldehydes, that is, geranial (citral A or trans citral) and neral (citral B or ciscitral) [20], which have same molecular formula ($C_{10}H_{16}O$) but different structures [23–26]. The various components of *C. flexosus* are significantly recognized due to the high concentration of aromatic essential oil, which contain many secondary metabolites, particularly monoterpenes (citral) and sesquiterpenes (caryophyllene) [11, 27]. Lemongrass is used in a variety of traditional Asian dishes and beverages, and also in high-end perfumes, pharmaceuticals, and biomedical applications [28]. The antibacterial, insecticide, larvicide, antitumoral, and cytotoxic characteristics of C. flexuosus' essential oil make it popular in alternative medicine [20]. The main constituents of the essential oil of C. flexuosus are Z-citral (-citral), geraniol, and -geranial (-citral), with citral contributing significantly to the oil's antibacterial properties.

3. Economic importance of essential oils of *Cymbopogon*

C. citratus Stapf. (lemongrass) is a spice commonly used in tropical regions, particularly in Southeast Asia. The primary compounds identified in *C. citratus* essential oil include α-citral, β-citral, geraniol, nerol, citronellal, myrcene, terpinolene, geranyl acetate, and terpinol methylheptenone. Terpenes, alcohols, ketones, and certain flavonoids and phenolics have also found in the plant [29]. Scientific research has described the antibacterial, anticarcinogenic, anti-inflammatory, antifungal, antioxidant, antiprotozoal, antirheumatic and cardioprotective effects of *C. citratus* [30–32]. It has shown a marked suppression of fungal infections including athlete's foot, itching, ringworm, and yeast infections and has a synergistic effect by suppressing the growth of filamentous fungi by inactivating yeast cells [6, 33]. Citral, myrcene, and citronellal are secondary metabolites that have been isolated from lemongrass and characterized as antimalarials. They showed remarkable activity against *Plasmodium* sp. [34]. In HIV/AIDS patients, oral candidiasis brought on by *Candida albicans* has been demonstrated to be successfully treated with lemongrass essential oil in 1 to 5 days [35].

LGEO's pharmaceutical potential has been reported in rodents in a well-designed trial involving oral administration of EO's key ingredient, citral, in combination with the nonsteroidal anti-inflammatory drug naproxen to experimental rats. The combination of naproxen and citral showed comparable anti-inflammatory effects compared with naproxen alone, but with much less stomach adverse effects [36]. Citral from *C. citratus* is used as an additive in creams and ointments to treat local inflammation as it significantly inhibits inflammatory mediators. It has also been shown to inhibit neutrophil attachment generated by tumor necrosis factor (TNF)-α at a dose of 0.1% and lipopolysaccharide (LPS)-induced nitric oxide synthase (iNOS) and monooxidation-induced signaling pathways co-bind to receptors, thereby blocking the nuclear factor Kappa B (NFƙB) pathway, COX2 and peroxisome proliferators. It suppresses activated receptor alpha (PPARα) by 60–70% and inhibits oral and tissue inflammation [6].

It has been reported to inhibit platelet composition and treat anxiety, gastrointestinal infections, diabetes, malaria, and pneumonia [25]. Tea made from lemongrass essential oil has been proven to have sedative, analgesic, anti-inflammatory, antipyretic, and antispasmodic properties. It has also been used as massage oil for relief of joint and muscle pain [37]. Diarrhea, stomach aches, and digestive issues can all be treated with lemongrass tea [38]. Lipid-lowering and hypoglycemic drugs may also contain lemongrass. In folk remedies and Ayurvedic medicines, it is used to control serum glucose, fat, and lipid levels and prevent obesity and high blood pressure. This plant has been used to keep blood sugar levels stable by secreting insulin (hyperinsulinemia). It lowers blood pressure that may result in hypertension [5]. It has been reported that citral (geranial and neral), the main constituent of *C. citratus* essential oil, is cytotoxic to a number of human leukemia cell lines. This occurs by the activation of procaspase 3. It has also been proven to inhibit the proliferation of pathogenic food-borne bacteria including *Listeria monocytogenes* and *Salmonella* Typhimurium [4].

Essential oils from *C. citratus* have been used to control infections and insects. It is efficient against *Aedes aegypti*, *Phenacoccus solenopsis*, *Dermatophagoides* sp., and *Musca domestica*. *C. citratus* is used in herbal soaps to cure swelling, itching skin, and rashes [6]. It has also been demonstrated that lemongrass essential oil inhibits *Microsporum canis*. Shampoos containing citral were efficient against *Malassezia furfur*, a fungus found in dandruff [39]. Lemongrass essential oil has been noted to exhibit considerable resistance to pathogenic fungi that interfere with the release of mycotoxins during preservation of grains and other food products [40]. *Cymbopogon* is a common herb in tropical regions [30]. It is frequently used as a food ingredient for human consumption. Lemongrass is frequently used in Asian cuisine for its aroma. Industrially, they are important as part of beverages, baked goods, fragrances, pesticides, and preservatives [6, 41]. They can serve as deodorants for perfumes, local samples, candle repellents, and other insect repellents. It has been used as a repellent against snakes and other reptiles in some Asian and African countries [42, 43]. The potential of lemongrass as an effective substitute to antibiotic growth promoters was evaluated [44].

4. Detection of phytochemicals in essential oils in *Cymbopogon* species—lemongrass

An understanding of the chemical components in plants is important for the discovery of beneficial phytochemicals useful in the synthesis of therapeutic agents and other useful chemical compounds. Higher plants are crucial sources of bioactive compounds essential for the maintenance of human health. *C. citratus* contains many pharmacologically active essential oils, flavonoids, phenolic compounds, and other bioactive constituents [6]. The methods of extraction of these essential oils include the following:

1. *Solvent extraction*: The solvent is combined with the plant material and then heated to extract the essential oil (EO). The plant extract is filtered and concentrated by evaporation of the solvent. The oil is extracted from the concentrate by combining it with pure alcohol and distilling it at low temperatures. However, because this method takes a long time to complete, the oils are more expensive than other methods [4, 45, 46].

2. *Distillation methods*: In hydrodistillation, the plant material absorbs water during the boiling process. The oil present in the oil cells diffuses *via* osmosis from the cell walls [46]. During steam distillation, at about 100°C, the combined vapor pressure corresponds to the ambient pressure. This allows volatile components with boiling temperatures between 150 and 300°C to evaporate at temperatures near 100°C [45]. The use of this simple technique for extraction of LGEO has been reported [47, 48]. Hydrodistillation yields less oil because incomplete oil extraction takes place due to a variable rate of distillation caused by heat [49]. Steam distillation has various disadvantages including possible loss of certain volatile chemicals, low extraction efficiency, and unsaturated compound degradation may occur [46].

3. *Supercritical CO_2 extraction*: CO_2 was chosen for use in extraction for various reasons, including inertness, non-toxicity, non-flammability, high solubility, availability at low cost, ideal for thermolabile compound extraction, and ease of removal from the extract. Supercritical CO_2 extraction is strongly suggested at low temperatures to prevent any damage to desirable EO constituents [50]. The use of this technique results in products that are free of toxic waste and of greater quality preserving thermal stability compared with conventional methods. However, the high working pressure necessitates the use of sophisticated equipment, which raises safety risks and costs [51].

4. *Ultrasound extraction (UAE)*: Plant cell walls are disrupted using ultrasound at frequencies greater than 20 kHz, which aid the ability of the solvent to permeate the cells and improve extraction yields. The UAE can process at a low temperature while maintaining good extract quality. The UAE is considered one of the easiest extraction methods due to the use of basic experimental equipment such as ultrasonic baths. The temperature and extraction time are controlled, while the ground sample is mixed with the appropriate solvent and placed in an ultrasonic bath [52]. Unfortunately, there are two major drawbacks to using an ultrasonic probe, both of which are connected to experimental reproducibility [53]. An advantage of this technique is that it is a green technology and its use in extraction of phenolic compounds has risen in popularity in the recent years [54].

5. *Microwave extraction*: Microwave extraction has a similar principle to hydrodistillation except that heating is achieved using microwaves. This method reduces the number of biological components lost during extraction. It reduces time and extraction solvent volume. It has been employed as an alternative to traditional antioxidant extraction techniques by reducing extraction time and being environmentally friendly [4, 54].

6. *Spectroscopy*: Data from variety of spectroscopic techniques are used to determine the structure of substances including ultraviolet (UV)-visible, infrared (IR), nuclear magnetic resonance (NMR), and mass spectroscopy. The fundamental premise of spectroscopy is that electromagnetic energy is passed through organic molecules that absorb part of it. A spectrum can be generated by measuring the amount of electromagnetic energy absorbed. Spectra are unique to each bond in the molecule, and this can be used to elucidate the structure of the organic molecule can be determined using these spectra. Since aromatic compounds are potent ultraviolet chromophores, UV-visible spectroscopy is preferred for quantitative study. When compared with other procedures, this technique takes less time and costs less. Fourier transform infrared spectroscopy (FTIR) is a non-destructive, high-resolution, quick analytical tool for identifying chemical components and determining the structure of compounds [54].

7. *Chromatography*: TLC is a rapid, inexpensive, solid-liquid chromatography technique for determining the presence of components in a mixture. Silica gel (SiO2 x H2O) and alumina (AL2O3xH2O) are the most used solids in chromatography [55]. Bio-autography is a technology that may be used to identify bioactive compounds with antibacterial activity in plant extracts. The bioautographic TLC method combines chromatographic separation with *in situ* activity measurement to facilitate identification and separation of active elements in a compound. High-performance liquid chromatography (HPLC) is an established technology for separating bioactive compounds. For optimal isolation, it is crucial to select the right mobile phase, flow rate, detectors, and columns and other conditions. The identifying peak should have adequate retention time and be well separated from unrelated peaks [56]. Liquid chromatography-mass spectrometry (LC/MS) is also a useful tool for the analysis of phenolic compounds [56]. When pure standards are not available, the combination of HPLC with MS can quickly and reliably identify the chemical composition of herbs [57]. Overpressured layer chromatography (OPLC) is the bridge between thin-layer chromatography (TLC) and high-performance liquid chromatography (HPLC).

OPLC is particularly attractive for separating antimicrobial components from various matrices [58]. Gas chromatography is often combined with a flame ionization detector or an electron capture detector. This technique can quantify and determine the presence of materials present at low concentrations [59]. The gas chromatography section divides sample's compounds into pure chemical pulses based on their vaporization.

8. *Immunoassays*: This method uses monoclonal antibodies to identify pharmaceuticals and natural low-molecular-weight bioactive substances. Monoclonal antibodies are produced by hybridoma technology. They are becoming increasingly relevant in the study of bioactive compounds. They have been shown to have high specificity and sensitivity. MAb-based enzyme-linked immunosorbent assays (ELISA) have been shown to be more sensitive than traditional HPLC procedures in many cases [56].

5. Quantification of phytochemicals of essential oils of *Cymbopogon* species

Refs. [60, 61] have documented the quantitative determination of phytochemicals (total alkaloids, flavonoids, phenol, saponins and tannins) of extracts of leaves using spectrophotometric and Folin-Ciocalteu methods, expressed per gram of the sample dry matter [60, 61]. These were majorly initiated by ethanolic and methanolic of acidified methanolic extractions as necessitated by the phytochemical of interest [62], followed by centrifugation and storage at −20°C until analysis was done.

For instance, the 1,10-phenanthroline method of total alkaloids content (TAC) estimation, as described by Ref. [63], entails the oxidation of alkaloids by iron (III) and subsequent complexation of iron (II) with 1,10-phenanthroline, to form a red-colored complex having the maximum absorbance at 510 nm. The reaction mixture containing 1 ml plant extract, 1 ml of 0.025 M $FeCl_3$ in 0.5 M HCl, and 1 mL of 0.05 M of 1, 10-phenanthroline in ethanol is usually incubated for 30 min in hot water bath with maintained temperature of 70 ± 2°C, before the measurement of the absorbance of red-colored complex at 510 nm against reagent blank. Alkaloid contents are then estimated and expressed as the standard curve of quinine (0.1 mg/ml, 10 mg dissolved in 10 ml ethanol, and diluted to 100 ml with distilled water) and the values expressed as g.100 g^{-1} of dry weight. This simple, sensitive, and economically viable spectrophotometric method has been used in the determination of some *Rauwolfia* alkaloids (ajmaline, ajmalicine, reserpine, and yohimbine HCl) in tablets of pharmaceutical formulations, with reports showing that the common excipients do not interfere with the proposed method. A statistical comparison of these results with the results of the reported approach shows good agreement and no significant difference in accuracy [61–63].

Phenolic compounds inhibit lipid oxidation by scavenging free radicals, chelating metals, activating antioxidant enzymes, reducing tocopherol radicals, and inhibiting enzymes that cause oxidation reactions. These may provide the basis for the rapidly growing interest in the use of natural antioxidants and antimicrobials [64]. Quantification of total flavonoids content (TFC) has been according to the aluminum chloride method reported by [65], involving the dispense of 0.5 ml of extract into test tube, followed by the addition of 1.5 ml of methanol, 0.1 mL of aluminum chloride (10%), 0.1 ml of 1 M potassium acetate, and 2.8 ml of distilled water in a reaction mixture. The absorbance read at 514 nm, after allowing to stand at room temperature for 30 min, is expressed as quercetin equivalent (QE) in mg/g material. In the same vein, the total phenolic content (TPC) quantification of samples extracts can be determined according to the Folin-Ciocalteu method of [62], where 1.5 ml of a 1 in 10 dilution of Folin-Ciocalteu reagent is added to 300 ml of leaf sample extract, followed by 1.2 ml of Na_2CO_3 solution (7.5 w/v). The absorbance read, at 765 nm against a blank after allowing to stand at room temperature for 30 min, is expressed as gallic acid equivalent (GAE) in mg/g material.

Variations in the extraction yields could arise from the different extraction methods and solvents. Other factors could be the evaluated variety, harvest year, processing, and storage [66]. Using walnut leaf as a case study, Ref. [67] investigated several solvents with different polarities such as hexane, chloroform, ethyl acetate, methanol, and ethanol for the evaluation of the cytotoxicity of walnut leaf extract on human cancer cell lines. They reported that the resulting methanol extract had the highest amount of TPC and TFC (120.28 ± 2.32 and 59.44 ± 0.87 mg/g DE, respectively) using colorimetric methods. For the ethanolic extracts, the concentration of the phenolic compounds in young leaves was substantially greater than those in the mature leaves. Employing the response surface methodology of optimization of the ultrasound assisted hydroalcoholic extraction of phenolic compounds of walnut leaves, Ref. [68] tried to establish the optimum conditions and the maximum predicted TPC, using 61% ethanol concentration, 51.28 min extraction time, and the 4.96 v/w liquid-to-solid ratio to obtain 10,125.4 mg GAEs/l, while 2925 mg quercetin equivalents (QEs)/l as maximum TFC was achieved by using ethanol with 67.83% concentration, v/w liquid-to-solid ratio of 4.96, and 49.37 min of extraction time. Under these conditions, the experimental results were reasonably similar to the values predicted by the polynomial response surface model equation. Ref. [69] compared the antioxidant and antimicrobial activities of the prepared ethanol and water extracts from the leaves of three plants, namely *P. aphylla*, Persian walnut, and oleander. They showed that the ethanol extracts had the highest amount of total phenolics and flavonoids in all assays, as well as highest antioxidant and antimicrobial activities when compared with the water extracts.

Initially, the formation of stable and persistent foam on the liquid surface for approximately 15 min represented the presence of saponins [70]. However, the quantification of plant saponins is usually performed by spectrophotometry and chromatography. The major difference in quantitative expression between the two techniques is that spectrophotometry quantifies the total saponin value, while chromatography quantifies specific saponin compound [71].

5.1 Spectrophotometric method of phytochemical quantification

Regarded as a simple, rapid, and inexpensive approach, total saponins assay, also known as vanillin-sulfuric acid assay, involving the reaction of oxidized triterpene saponins with vanillin is one of the spectrophotometric methods used to quantify saponins. It uses sulfuric or perchloric acid as oxidant, resulting in a distinctive purple coloration of the reaction system [71]. Being the most commonly utilized spectrophotometric method for quantifying plant saponins and providing an excellent reference for future experimental design, reports suggest that in order to allow full color development, few criteria, such as choice of standards and wavelength, should be taken into account when selecting this method [70]. However, researchers have reported 544-nm wavelength, with a majority selecting wavelengths in the range of 480–610 nm (excluding 473 nm and 283 nm). This is likely due to the maximum absorption of purple color that falls within this range [72, 73].

Hemolytic method is a spectrophotometric method for determining the saponin concentration of a plant material [74]. This is based on the release of oxy-hemoglobin when saponins react with blood reagent producing measurable color changes detected by the spectrophotometer. Ref. [75] have quantified the saponin content in bitter gourd varieties with the hemolytic technique, with their result showing that white bitter gourd types had much lower levels of saponin (0.25%) than the green varieties (0.67%). In furtherance, the saponin extract was dissolved in distilled water, before incubation of 100 μl of this solution with 1 ml fresh EDTA-blood at 30°C for 30 min. Hemoglobin was measured in the supernatant photometrically at 545 nm, after centrifugation for 10 min, and the result is expressed in hemolytic saponins.

5.2 Chromatographic method of phytochemical quantification

Thin-layer chromatography (TLC), high-performance liquid chromatography (HPLC), and ultra-pressure liquid chromatography (UHPLC) are the most common chromatographic methods employed [73, 76, 77]. Multiple reaction monitoring (MRM)-based UHPLC-ESI-MS/MS technology, according to studies, was developed to provide more precise and sensitive measurement of main saponins and sapogenins in conjunction with chlorogenic acid [78]. Refs. [79, 80] described the use of ultra-high-performance liquid chromatography coupled to single-stage Orbitrap high-resolution mass spectrometry (UHPLC-Orbitrap-MS) to simultaneously detect and quantify phytochemicals in green tea and walnut leaves-derived nutraceuticals. Ref. [81] submit that by combining LC separation and MS detection, a high selectivity could be attained since the MS detector's selectivity allows for more precise identity by confirmation by comparing fragmentation patterns and observing qualifier and quantifier ion transitions.

Additionally, although sensitivity—the connection between analyte signal and concentration—may not be a crucial parameter for evaluation during method validation, it provides information about the instrument signal and might be helpful during method optimization. In light of these, Ref. [78] suggest that this method can be used to quantitatively measure bioactive compounds in crude plant materials and other related products, while also determining the same compounds in other biological sample matrices such as plasma, potentially minimizing matrix effect. Reports of Refs. [82, 83] show chromatographic methods to allow separation and purification of various saponin biotypes from plant materials to identify a specific saponin compound and investigate its pharmaceutical property. Refs. [84, 85] suggest that the main goal of all the studies using HPLC technique is the quantification of specific saponin components. The specific saponin content detected serves as an excellent data reference source to future researchers, in addition to providing a reliable scientific reference to pharmaceutical manufacturers interested in further processing of their respective plant sources [71].

Standardization and purification of complex extracts are still problematic since the mixtures are more toxic than individual components and present more difficulty in detoxification than a single molecule [86, 87]. More so, isolation, synthesis, or formulation processes could be slow and expensive, but relatively inexpensive for plant essential oils. In light of these, Ref. [88] allude that simultaneous quantification and qualitative analyses of phytochemicals could easily be achieved *via* quick and conventional methods such as non-destructive near-infrared spectroscopy and isocratic high-performance liquid chromatography. These improved methods can support rapid and precise content evaluation and confirmation [86].

5.3 Spectroscopic techniques of phytochemical quantification

Spectroscopic techniques such as ultraviolet (UV-visible), infrared (IR), mass spectroscopy, and nuclear magnetic resonance (NMR) provide sufficient information for quantitative as well as qualitative analysis of phytochemicals. Basically, depending on its structure, the organic molecule absorbs electromagnetic radiations in certain regions and produces a spectrum. The obtained spectrum helps in the identification and quantification of the molecule, since they are specific to functional groups [89].

The spectroscopic techniques include the following:

i. Ultraviolet-visible spectroscopy can be used to identify various phytochemicals using maximum absorption (λ max) values that correspond to their structural characteristics, such as total phenolic extract (280 nm), flavones (320 nm), and phenolic acids (360 nm) [54]. This technique is cheaply available and requires less time for observation [90].

ii. Infrared IR spectroscopy also referred to as vibrational spectroscopy exploits the vibrational modifications associated with elongation or bending of certain molecules upon exposure to the infrared region of electromagnetic radiations [89]. The various functional groups or chemical bonds in the molecule have different vibrational frequencies depending on the force constant (bond strength) values and decreased masses [54], which aid structural determination of a bioactive compound by capturing the characteristic frequency absorption bands for the functional groups present in the IR spectrum. Higher resolutions of chemical composition and elucidation of molecular structure can further be achieved with the Fourier transform infrared spectroscopy (FTIR) [91].

iii. Mass spectroscopy involves the conversion of organic molecules to highly energetic charged species through bombardment with electrons or lasers [89]. As such, the molecular formula of the bioactive compound can be predicted by the estimated relative molecular weight of the fragmented ions adjacent to the places of fragmentations [54]. According to the published reports, mass spectroscopy has proven to be a highly effective analytical instrument in the structural elucidation of phenolic compounds when used in conjunction with electrospray ionization (ESI), a preferred method of producing charged species from macromolecules [91].

iv. Nuclear magnetic resonance (NMR) spectroscopy reveals the magnetic properties of some nuclei including 1H, 13C, 19F, and 31P [90]; thus, a signal is produced when the molecule's intrinsic frequency of the active nuclei resonates with the oscillation frequency of the external magnetic field applied. This is determined by measuring the chemical shift arising from variations in the strength of the applied magnetic field, the chemical environment, and changes in the magnetic characteristics of the various nuclei [54, 89]. The quantitative concentrations of the most abundant compounds in the oily fractions of hemp leaves, hurds, and roots were revealed by the NMR method [92–94].

6. Conclusion

This chapter discusses two of the most common species of *Cymbopogon* and their essential oils. The chapter showcases diverse phytochemicals of the oils, economic importance of *Cymbopogon* essential oils, detection, and quantification techniques of the phytochemicals of these essential oils obtained from *Cymbopogon* species. Additionally, different extraction and quantification methods for obtaining the essential oils were explicated. Findings from this chapter portray the different challenges associated with extraction and yields of these essential oils. Although chromatographic methods are commonly employed for extracting and quantifying these essential oils, there is need for improvement *via* the use of combined and more sensitive techniques for greater as well as pure yields of essential oils of lemon grass.

Acknowledgements

We wish to acknowledge Covenant University for sponsoring this publication.

Conflict of interest

The authors declare no conflict of interests.

Author details

Margaret Ikhiwili Oniha*, Eze Frank Ahuekwe and Sharon Oluwatobi Akinpelu
Department of Biological Sciences, College of Science and Technology, Covenant University, Ota, Ogun State, Nigeria

*Address all correspondence to: margaret.oniha@covenantuniversity.edu.ng

References

[1] Wifek M, Saeed A, Rehman R, Nisar S. Lemongrass: A review on its botany, properties, applications, and active components. International Journal of Chemical and Biochemical Sciences. 2016;**9**:79-84

[2] Ojewumi ME, Banjo MG, Oresegun MO, Ogunbiyi TA, Ayoola AA, Awolu OO, et al. Analytical investigation of the extract of lemon grass leaves in repelling mosquito. International Journal of Pharmaceutical Sciences and Research. 2017;**8**(5):1000-1009

[3] Avoseh O, Oyedeji O, Rungqu P, Nkeh-Chungag B, Oyedeji A. Cymbopogon species; Ethnopharmacology, phyto phytochemistry and the pharmacological importance. Molecules. 2015;**20**(5):7438-7453

[4] Majewska E, Kozowska M, Gruczyska-Skowska E, Kowalska D, Tarnowska K. Lemongrass (*Cymbopogon citratus*) essential oil: Extraction, composition, bioactivity and uses for food preservation: A review. Poland Journal of Food Nutrition and Science. 2019;**69**(4):327-341

[5] Shah G, Shri R, Panchal V, Sharma N, Singh B, Mann AS. Scientific basis for the therapeutic use of *Cymbopogon citratus*, stapf (Lemon grass). Journal of Advanced Pharmaceutical Technology and Research. 2011;**2**:3-8

[6] Oladeji OS, Adelowo FE, Ayodele DT, Odelade KA. Phytochemistry and pharmacological activities of *Cymbopogon citratus*: A review. Scientific African. 2019;**6**:1-11

[7] Shendurse AM, Sangwan RB, Kumar A, Ramesh V, Patel AC, Gopikrishna G, et al. Phytochemical screening and antibacterial activity of lemongrass (*Cymbopogon citratus*) leaves essential oil. Journal of Pharmacognosy and Phytochemistry. 2021;**10**(2):445-449

[8] Balakrishnan B, Paramasivam S, Arulkumar A. Evaluation of the lemongrass plant (*Cymbopogon citratus*) extracted in different solvents for antioxidant and antibacterial activity against human pathogens. Asian Pacific Journal of Tropical Disease. 2014;**4**(1):S134-S139

[9] Skaria BP, Joy PP, Mathew G, Mathew S, Joseph A. Lemongrass. In: Peter KV, editor. Handbook of Herbs and Spices. 2nd ed. Vol. 228. Abington Hall Abington, Cambridge CB1 6AH, Cambs, England: Woodhead Pubication Ltd; 2012. pp. 348-370

[10] Debajit S, Sukriti D, Sneha G, Mohan L, Singh BB. RAPD and ISSR based intra-specific molecular genetic diversity analysis of *Cymbopogon flexuosus* L. Stapf with a distinct correlation of morpho-chemical observations. Research Journal of Biotechnology. 2015;**10**(7):105-113

[11] Mukaram M, Masroor M, Khan A, Corpas FJ. Silicon nanoparticles elicit an increase in lemongrass (*Cymbopogon flexuosus* (Steud.) wats) agronomic parameters with a higher essential oil yield. Journal of Harzardous Materials. 2021;**412**(2021):1-13

[12] Abdulazeez MA, Abdullahi AS, James BD. Lemongrass (*Cymbopogon* spp.) oils. In: Preedy VR, editor. Essential Oils in Food Preservation Flavour and Safety. USA: Elsevier, Irish Academic Press; 2016. pp. 509-516. ISBN 978-0-12-416641-7. DOI: 10.1016/B978-0-12-416641-7.00058-4

[13] Srivastava V, Dubey S, Mishra AA. Review on lemon grass: Agricultural and

medicinal aspect. International Resources Journal of Pharmacology. 2013;**4**:42-44. DOI: 10.7897/2230-8407.04807

[14] Dhinesh B, Lalvani JIJ, Parthasarathy M, Annamalai K. An assessment on performance, emission and combustion characteristics of single cylinder diesel engine powered by *Cymbopogon flexuosus* biofuel. Energy Conversation Management. 2016;**117**:466-474. DOI: 10.1016/j.enconman.2016.03.049

[15] Ahmad I, Hanif MA, Nadeem R, Jamil MS, Zafar MS. Nutritive evaluation of medicinal plants being used as condiments in south Asian region. Journal of the Chemical Society of Pakistan. 2008;**30**(3):400-405

[16] Al-Maskri AY, Hanif MA, Al-Maskari MY, Abraham AS, Al-sabahi JN, Al-Mantheri O. Essential oil from *Ocimum basilicum* (Omani Basil): a desert crop. Natural Product Communications. 2011;**6**(10):1934578X1100601020

[17] Arshad Z, Hanif MA, Qadri RWK, Khan MM. Role of essential oils in plant diseases protection: A review. International Journal of Chemical and Biochemical Sciences. 2014;**6**:11-17

[18] Olopade BK, Oranusi SU, Nwinyi OC, Njobeh PB, Lawal IA. Modification of montmorillonite clay with *Cymbopogon citratus* for the decontamination of zearalenone in millet. AIMS Agriculture and Food. 2019;**4**(3):643-657

[19] Omonhinmin AC, Dike IP, Rotimi SO. Phytochemical, cytotoxicity and antioxidant activities of five anti-malaria plants. Research Journal of Medicinal Plant. 2015:1-9. DOI: 10.3923/rjmp.2015

[20] Hasim SF, Ayunda RD, Faridah DN. Potential of lemongrass leaves extract (*Cymbopogon citratus*) as prevention for oil oxidation. Journal of Chemical and Pharmaceutical Research. 2015;**7**(10):55-60

[21] Erminawati NR, Sitoresmi I, Sidik W, Bachtiar A. Antioxidant activity of microencapsulated lemongrass (*Cymbopugon citratus*) extract. In: International Conference on Sustainable Agriculture for Rural Development (ICSARD 2018), Indonesia, October 23-24, 2018. IOP Conference Series: Earth and Environmental Science. Vol. 250. Purwokerto: IOP Publishing Ltd; 2019, 012054-1-012054-6

[22] Ademuyiwa AJ, Grace OK. The effects of *Cymbopogon citratus* (Lemon grass) on the antioxidant profiles wistar albino rats. Merit Research Journal of Environmental Science and Toxicology. 2015;**3**(4):51-58

[23] Huynh KPH, Maridable J, Gaspillo P, Hasika M, Malaluan R, Kawasaki J. Essential oil from lemongrass extracted by supercritical carbon dioxide and steam distillation. The Philippine Agricultural Scientist. 2008;**91**(1):36-41

[24] Mirghani MES, Liyana Y, Parveen J. Bioactivity analysis of lemongrass (*Cymbopogan citratus*) essential oil. International Food Research Journal. 2012;**19**(2):569-575

[25] Manvitha K, Bidya B. Review on pharmacological activity of *Cymbopogon citratus*. International Journal of Herbal Medicine. 2014;**1**(6):5-7

[26] Hartatie ES, Prihartini I, Widodo W, Wahyudi A. Bioactive compounds of lemongrass (*Cymbopogon citratus*) essential oil from different parts of the plant and distillation methods as natural antioxidant in broiler meat. IOP Conference Series: Materials Science and Engineering. 2018;**532**:1-6

[27] Ganjewala D. Cymbopogon essential oils: Chemical compositions and bioactivities. International Journal of Essential Oil Therapeutics. 2009;**3**(2-3):56-65

[28] Joy PP, Skaria BP, Mathew S, Mathew G, Joseph A, Sreevidya PP. Lemongrass. Indian Journal Arecanut Spices Medicine Plants. 2006;**2**:55-64

[29] Trang DT, Hoang TKV, Nguyen TTM, Van Cuong P, Dang NH, Dang HD, et al. Essential oils of lemongrass (*Cymbopogon citratus* Stapf) induces apoptosis and cell cycle arrest in A549 lung cancer cells. Biomedical Research International. 2020;**2020**:1-8. DOI: 10.1155/2020/5924856

[30] Ajayi EO, Sadimenko AP, Afolayan AJ. GC–MS evaluation of *Cymbopogon citratus* (DC) Stapf oil obtained using modified hydrodistillation and microwave extraction methods. Food Chemistry. 2016;**209**:262-266

[31] Chukwuocha UM, Fernández-Rivera O, Legorreta-Herrera M. Exploring the antimalarial potential of whole *Cymbopogon citratus* plant therapy. Journal Ethnopharmacolology. 2016;**193**:517-523

[32] Ntonga PA, Baldovini N, Mouray A, Mambu L, Belong P, Grellier P. Activity of *Ocimum basilicum, Ocimum canum*, and *Cymbopogon citratus* essential oils against *plasmodium falciparum* and mature-stage larvae of *anopheles funestuss*. Parasite. 2014;**21**:33

[33] Boukhatem MN, Ferhat MA, Kameli M, Saidi F, Kebir H. Lemon grass (*Cymbopogon citratus*) essential oil as a potent anti-inflammatory and antifungal drugs. Libyan. Journal of Medicine. 2013;**8**:25431

[34] Tarkang PA, Franzoi KD, Lee S, Lee E, Vivarelli D, Freitas-Junior L, et al. *In vitro* antiplasmodial activities and synergistic combinations of differential solvent extracts of the polyherbal product, Nefang. Biology Medical Research International. 2014;**2014**:1-10

[35] Wright SC, Maree JE, Sibanyoni M. Treatment of oral thrush in hiv/aids patients with lemon juice and lemon grass (*Cymbopogon citratus*) and gentian violet. Phytomedicine. 2009;**16**:118-124

[36] Ortiz MI, González-García MP, Ponce-Monter HA, Castañeda-Hernández G, Aguilar-Robles P. Synergistic effect of the interaction between naproxen and citral on inflammation in rats. Phytomedicine. 2010;**18**(1):74-79. DOI: 10.1016/j.phymed.2010.05.009

[37] Akande IS, Samuel T, Agbazue U, Olowolagba B. Comparative proximate analysis of ethanolic and water extracts of *Cymbopogon citratus* (lemon grass) and four tea brands. JPBMS. 2012;**22**:1-7

[38] Dutta S, Munda S, Lal M, Bhattacharyya PR. A short review on chemical composition therapeutic use and enzyme inhibition activities of *Cymbopogon* species. Indian Journal of Science and Technology. 2016;**9**(46):1-9

[39] Wuthi-UdomlertM,ChotipatoomwanP, Panyadee S, Gritsanapan W. Inhibitory effect of formulated lemon grass shampoo on *Malassezia furfur*: A yeast associated with dandruff. Southeast Asian Journal of Tropical Medicine and Public Health. 2011;**42**:363-369

[40] Nguefack J, Tamgue O, Dongmo J, Dakolea C, Leth V, Vismer H, et al. Synergistic action between fractions of essential oils from *Cymbopogon citratus, Ocimum gratissimum* and *Thymus vulgaris* against *Penicillium expansum*. Food Control. 2012;**23**:377-383

[41] Bossou AD, Mangelinck S, Yedomonhan H, Boko PM, Akogbeto MC, De Kimpe N, et al. Chemical composition and insecticidal activity of plant essential oils from Benin against *Anopheles gambiae* (Giles). Parasites and Vectors. 2013;**6**:1-17

[42] Tajidin NE, Ahmad SH, Rosenani AB, Azimah H, Munirah M. Chemical composition and citral content in lemongrass (*Cymbopogon citratus*) essential oil at three maturity stages. African Journal of Biotechnology. 2012;**11**:2685-2693

[43] Usha R. Fumigant and contact toxic potential of essential oils from plant extracts against stored product pests. Journal of Biopesticides. 2012;**5**:120-128

[44] Alagawany M, El-Saadony MT, Elnesr SS, Farahat M, Attia G, Madkour M, et al. Use of lemongrass essential oil as a feed additive in quail's nutrition: Its effect on growth, carcass, blood biochemistry, antioxidant and immunological indices, digestive enzymes, and intestinal microbiota. Poultry Science. 2021;**100**(6):101-172

[45] Stephane FFY, Jules BKJ. Terpenoids as important bioactive constituents of essential oils. In: de Oliveira M a, Costa WA d, Silva SG, editors. Essential Oils—Bioactive Compounds, New Perspectives and Applications. London: IntechOpen; 2020. DOI: 10.5772/intechopen.91426

[46] Tongnuanchan P, Benjakul S. Essential oils: Extraction, bioactivities, and their uses for food preservation. Journal of Food Science. 2014;**79**(7):1231-1249. DOI: 10.1111/1750-3841.12492

[47] Boukhatem MN, Kameli A, Ferhat MA, Tayebi K, Saidi F. The food preservative potential of essential oils: Is lemongrass the answer? Journal of Consumer Protection and Food Safety. 2014;**9**(1):13-21

[48] Anggraeni NI, Hidayat IW, Rachman SD. Ersanda. Bioactivity of essential oil from lemongrass (*Cymbopogon citratus* Stapf) as antioxidant agent. In: The 1st International Conference and Exhibition on Powder Technology Indonesia (ICePTi) 2017; AIP Conference Proceedings. 1927, art. no. UNSP039997; Melville, USA: AIP Publishing LLC.; 2018: 030007-1-030007-5. DOI: 10.1063/1.5021200

[49] Ameh OE, Achika JI, Bello NM, Owolaja AJ. Extraction and formulation of perfume from *Cymbopogon citratus* (lemongrass). Journal of Applied Science and Environmental Management. 2021;**25**(8):1461-1464

[50] Yousefi M, Rahimi-Nasrabadi M, Pourmortazavi SM, Wysokowski M, Jesionowski T, Ehrlich H, et al. Supercritical fluid extraction of essential oils. TrAC Trends in Analytical Chemistry. 2019;**118**:182-193

[51] Wu H, Li J, Jia Y, Xiao Z, Li P, Xie Y, et al. Essential oil extracted from *Cymbopogon citronella* leaves by supercritical carbon dioxide: Antioxidant and antimicrobial activities. Journal of Analytical Methods in Chemistry. 2019:8192439. DOI: 10.1155/2019/8192439

[52] Corrales M, Toepfl S, Butz P, Knorr D, Tauscher B. Extraction of anthocyanins from grape by-products assisted by ultrasonics, high hydrostatic pressure or pulsed electric fields: A comparison. Innovative Food Science and Emerging Technology. 2008;**9**:85-91

[53] Tabaraki R, Nateghi A. Optimization of ultrasonic-assisted extraction of natural antioxidants from rice bran using response surface methodology. Ultrasonic Sonochemistry. 2011;**18**: 1279-1286

[54] Altemimi A, Lakhssassi N, Baharlouei A, Watson DG, Lightfoot DA. Phytochemicals: Extraction, isolation, and identification of bioactive compounds from plant extracts. Plants (Basel, Switzerland). 2017;**6**(4):42. DOI: 10.3390/plants6040042

[55] Karpagam GN, Gayathri R, Vishnupriya V. Bioactivity analysis of lemongrass oil. Research Journal of Pharmacy and Technology. 2016;**9**(7):903-906. DOI: 10.5958/0974-360X.2016.00172.4

[56] Sasidharan S, Chen Y, Saravanan D, Sundram KM, Yoga LL. Extraction, isolation and characterization of bioactive compounds from plants' extracts. African Journal of Traditional, Complementary, and Alternative Medicines. 2011;**8**(1):1-10

[57] Ye M, Han J, Chen H, Zheng J, Guo D. Analysis of phenolic compounds in rhubarbs using liquid chromatography coupled with electrospray ionization mass spectrometry. Journal of American Society Mass Spectrometry. 2007;**18**(1):82-91

[58] Móricz ÁM, Ott PG, Böszörményi A, Lemberkovics E, Mincsovics E, Tyihák E. Bioassay-guided isolation and identification of antimicrobial compounds from thyme essential oil by means of overpressured layer chromatography, bioautography and GC–MS. Chromatographia. 2012;**75**:991-999

[59] Al-Rubaye AF, Hameed IH, Kadhim MJ. A review: Uses of gas chromatography-mass spectrometry (GC-MS) technique for analysis of bioactive natural compounds of some plants. International Journal of Toxicological and Pharmacological Research. 2017;**9**(1):81-85

[60] Adedeji GA, Ogunsanwo OY, Elufioye TO. Quantifications of phytochemicals and biocide actions of Lawsonia inermis Linn.Extracts against wood termites and fungi. International Biodeterioration & Biodegradation. 2017;**116**:155-162. DOI: 10.1016/j.ibiod.2016.10.026

[61] Adedeji GA, Ogunsanwo OY, Aiyeloja AA. Characterisation of solvent components in Erythrophleum suaveolens (Guill. & Perr.) Brenan stem bark extracts treated Triplochiton scleroxylon K. schum. wood. International Journal of Scientific and Engineering Research. 2013;**4**(10):926-937

[62] Chan EWC, Lim YY, Chew YL. Antioxidant activity of Camellia sinensis leaves and tea from a lowland plantation in Malaysia. Journal of Food Chemistry. 2006;**102**:1214-1222

[63] Singh DK, Srivastva B, Sahu A. Spectrophoto-metric determination of Rauwolfia alkaloids, estimation of reserpine in pharmaceuticals. Analytical Sciences. 2004;**20**:571-573

[64] Jahanban-Esfahlan A, Ostadrahimi A, Tabibiazar M, Amarowicz R. A comparative review on the extraction, antioxidant content and antioxidant potential of different parts of walnut (Juglans regia L.) fruit and tree. Molecules. 2019;**24**(11):2133. DOI: 10.3390/molecules24112133

[65] Kale A, Gaikwad S, Mundhe K, Deshpande N, Salve-kar J. Quantification of phenolics and flavonoids by spectrophotometer from Juglans regia. International Journal of Pharma and Bio Sciences. 2010:1-4

[66] Mundhe KS, Kale AA, Gaikwad AA, Deshpande NR, Kashalkar RV. Evaluation of phenol, flavonoid contents and antioxidant activity of Polyalthia longifolia. Journal of Chemical and Pharmaceutical Research. 2011;**3**(1):764-769

[67] Salimi M, Majd A, Sepahdar Z, Azadmanesh K, Irian S, Ardestaniyan MH, et al. Cytotoxicity effects of various Juglans regia (walnut) leaf extracts in human cancer cell lines. Pharmaceutical Biology. 2012;**50**:1416-1422

[68] Nour V, Trandafir I, Cosmulescu S. Optimization of ultrasound-assisted hydroalcoholic extraction of phenolic compounds from walnut leaves using response surface methodology. Pharmaceutical Biology. 2016;**54**(10):2176-2187

[69] Einali A, Azizian-Shermeh O, Ghasemi A. Phytochemical screening and antimicrobial activities of Periploca aphylla Decne, Persian walnut (Juglans regia L.) and oleander (Nerium indicum mill.) leaf extracts. Journal of Food Measurement and Characterization. 2018;**12**:1350-1359

[70] Ncube B, Ngunge VNP, Finnie JF, Staden JV. A comparative study of the antimicrobial and phytochemical properties between outdoor grown and micropropagated Tulbaghia violacea Harv. Plants. Journal of Ethnopharmacology. 2011;**134**:775-780

[71] Cheok CY, Salman HAK, Sulaiman R. Extraction and quantification of saponins: A review. Food Research International. 2014;**59**:16-40. DOI: 10.1016/j.foodres.2014.01.057

[72] Mostafa A, Sudisha J, El-Sayed M, Ito S-I, Yamauchi N, Shigyo M, et al. Aginoside saponin, a potent antifungal compound, and secondary metabolite analyses from Allium nigrum L. Phytochemistry Letters. 2013;**6**:274-280

[73] Liu Y-W, Zhu X, Lu Q, Wang J-Y, Li W, Wei Y-Q, et al. Total saponins from Rhizoma *anemarrhenae* ameliorate diabetes-associated cognitive decline in rats: Involvement of amyloid-beta decrease in brain. Journal of Ethnopharmacology. 2012;**139**:194-200

[74] Barve KH, Laddha KS, Jayakumar B. Extraction of saponins from Safed Musli. Pharmacognosy Journal. 2010;**2**:561-564

[75] Habicht SD, Kind V, Rudloff S, Borsch C, Mueller AS, Pallauf J, et al. Quantification of antidiabetic extracts and compounds in bitter gourd varieties. Food Chemistry. 2011;**126**:172-176

[76] Serventi L, Vodovotz Y, Chitchumroonchokchai C, Schwartz SJ, Failla ML, Riedl KM, et al. Saponins from soy and chickpea: Stability during breadmaking and in vitro bioaccessibility. Journal of Agricultural and Food Chemistry. 2013;**61**:6703-6710

[77] Ha TJ, Lee BW, Park KH, Jeong SH, Kim H-T, Ko J-M, et al. Rapid characterisation and comparison of saponin profiles in the seeds of Korean leguminous species using ultra performance liquid chromatography with photodiode array detector and electrospray ionisation/mass spectrometry (UPLC–PDA–ESI/MS) analysis. Food Chemistry. 2014;**146**:270-277

[78] Sabaragamuwa R, Perera PO, Fedrizzi B. Ultrasound assisted extraction and quantification of targeted bioactive compounds of Centella asiatica (Gotu Kola) by UHPLC MS/MS MRM tandem mass spectroscopy. Food Chemistry. 2022;**371**:131-187. DOI: 10.1016/j.foodchem.2021.131187

[79] Lopez-Gutierrez N, Romero-Gonzalez R, Plaza-Bolanos P, Vidal JLM, Frenich AG. Identification and quantification of phytochemicals in nutraceutical products from green tea by UHPLC–Orbitrap-MS. Food Chemistry. 2015;**173**:607-618

[80] Su C, Li C, Sun K, Li W, Liu R. Quantitative analysis of bioactive components in walnut leaves by UHPLCQ-Orbitrap HRMS combined with QAMS. Food Chemistry. 2020;**331**:127-180. DOI: 10.1016/j.foodchem.2020.127180

[81] Kruve A, Riin R, Karin K, Maarja-Liisa O, Hanno E, Koit H, et al. Tutorial review on validation of liquid chromatography–mass spectrometry methods: Part I. Analytica Chimica Acta. 2015;**870**:29-44. DOI: 10.1016/j.aca.2015.02.017

[82] Gupta S, Rayehaudhuri B, Banerjee S, Das B, Mukhopadhaya S, Datta SC. Momordicatin purified from fruits of Momordica charantia is effective to act as a potent antileishmania agent. Parasitology International. 2010;**59**:192-197

[83] Adão CR, da Silva BP, Parente JP. A new steroidal saponin with anti-inflammatory and antiulcerogenic properties from the bulbs of Allium ampeloprasum var. porrum. Fitoterapia. 2011;**82**:1175-1180

[84] He H, Xu J, Xu Y, Zhang C, Wang H, He Y, et al. Cardioprotective effects of saponins from Panax japonicas on acute myocardial ischemia against oxidative stress-triggered damage and cardiac cell death in rats. Journal of Ethnopharmacology. 2012;**140**:73-82

[85] Zheng T, Shu G, Yang Z, Mo S, Zhao Y, Mei Z. Antidiabetic effect of total saponins from Entada phaseoloides (L.) Merr. in type 2 diabetic rats. Journal of Ethnopharmacology. 2012;**139**:814-821

[86] Gahukar RT. Evaluation of plant-derived products against pests and diseases of medicinal plants: A review. Crop Protection. 2012;**42**:202-209

[87] Koul O. Phytochemicals and insect control: An antifeedant approach. Critical Reviews in Plant Sciences. 2008;**27**:1-24

[88] Highi G, Hatami A. Simultaneous quantification of flavonoids and phenolic acids in plant materials by a newly developed isocratic high performance liquid chromatography approach. Journal of Agricultural and Food Chemistry. 2010;**58**:10812-10816

[89] Sharma M, Kaushik P. Vegetable phytochemicals: An update on extraction and analysis techniques. Biocatalysis and Agricultural Biotechnology. 2021;**36**:102149. DOI: 10.1016/j.bcab.2021.102149

[90] Rasul MG. Extraction, isolation and characterization of natural products from medicinal plants. International Journal of Basic Sciences and Applied Computing. 2018;**2**:1-6

[91] Ingle KP, Deshmukh AG, Padole DA, Dudhare MS, Moharil MP, Khelurkar VC. Phytochemicals: Extraction methods, identification and detection of bioactive compounds from plant extracts. Journal of Pharmacognosy and Phytochemistry. 2017;**6**:32-36

[92] Liu Y, Liu H, Li S, Ma W, Wu D, Li H, et al. Cannabis sativa bioactive compounds and their extraction, separation, purification, and identification technologies: An updated review. Trends in Analytical Chemistry. 2022;**149**:116554

[93] Salami A, Heikkinen J, Tomppo L, Hyttinen M, Kekalainen T, Janis J, et al. A comparative study of pyrolysis liquids by slow pyrolysis of industrial hemp leaves, hurds and roots. Molecules. 2021;**26**:3167. DOI: 10.3390/molecules26113167

[94] Petrovici AR, Simionescu N, Sandu AI, Paraschiv V, Silion M, Pinteala M. New insights on hemp oil enriched in cannabidiol: Decarboxylation, antioxidant properties and in vitro anticancer effect. Antioxidants. 2021;**10**(2021):738. DOI: 10.3390/antiox10050738

Chapter 13

Effects of the Invasive Alien *Prosopis juliflora* (Sw.) DC and Its Management Options in Ethiopia: A Review

Wakshum Shiferaw and Sebsebe Demissew

Abstract

This paper aimed to review the effects of *P. juliflora* (hereafter *P. juliflora* is represented by *P. juliflora*) on environmental factors in Ethiopia, management options and take up lessons learned elsewhere, and discuss about utilization and management of *P. juliflora*. It addresses relevant scientific information based on the evaluation of data collected from different journals, books, manuals, and various reports using the systematic review method. Invasion of *P. juliflora* had positive effects on soil properties, negative effects on plant diversity, human health, livestock health, and other economic losses. Its negative effects are outweighing its positive effects. The main negative effects of *P. juliflora* are its biotic factors such as allopathic chemicals and active ingredients such as phenolic compounds that are impairing animals and human beings. Mechanical, chemical, management by utilization, fire, disruption of its phenological stages, and biological control methods are among control methods of the species. However, these control methods have their pros and cons for its management. The review was based on limited research findings and sources because there are limitations in research works regarding *P. juliflora* and its management. This review is used to know the invasion of *P. juliflora* and its management options in Ethiopia and other similar tropical countries across the world.

Keywords: allopathic, control methods, management, *Prosopis juliflora*, policy

1. Introduction

Invasive species are either indigenous or exotic which are being taken over a particular habitat [1, 2]. Invasive alien species are species that are introduced intentionally or unintentionally into new areas and cause loss on an environment they invade [2]. In recent decades, biological invasions are increasing [3, 4] and threaten ecosystems, biodiversity, and food security [5]. Invasive alien species are representatives of all taxonomic levels of viruses, bacteria, algae, plants, invertebrates, and large mammals [6].

IntechOpen

In the world database, *P. juliflora* is one of the 100 worst invasive alien species [7]. The report by Ethiopian Biodiversity Institute showed 35 invasive alien plant species (IAPS) in the country [8]. Some of these invasive alien plants include *P. juliflora*, *Parthenium hysterophorus*, *Eichhornia crassipes*, *Lantana camara*, *Acacia drepanolobium*, *Orobanche,* and *Cuscuta* species, which are identified as major plant invaders [2]. It was found that there were emerging plant invaders such as *Cryptostegia grandiflora*, *Parkinsonia aculeate*, *Mimosa diplorotricha*, *Mimosa pigra*, and *Argemone Mexicana,* and *Nicotine glauca* [8]. In Ethiopia, plant species including *P. juliflora*, *P. hysterophorus*, *E. crassipes*, *L. camara,* and *A. drepanolobium* are the worst that threatened biodiversity losses [9]. These species were overtaking other land uses such as woodlands, grazing lands, parklands, urban greening, farmlands, which had reduced their ecosystem services of the land uses.

P. juliflora is a perennial evergreen multipurpose tree or shrub native to the Caribbean, North and South America transported out of its ranges through human activities [10–12]. It was introduced to African countries, for instance, since 1822 in Senegal, South Africa in 1880, Egypt in 1900, Kenya since 1973, and into Eritrea from Sudan probably during the early 1980s [2]. There is contradicting information regarding the introduction of *P. juliflora* into Ethiopia. For example, it was introduced in the late 1970s from India into Goro nursery-Dire-Dawa. On the other hand, *P. juliflora* was first introduced to the Afar region in the late 1970s. The introduction of *P. juliflora* in the Afar region was through coordinated efforts between government and communities to stop desertification, greening up the region, and mitigate the impacts of drought [12]. *P. juliflora* was then planted over large areas up to 1982, and the sector Food for Work Program in 1986–1988 continued to expand its plantation. This species now exists in most regions of Ethiopia for instance, Afar, Oromia, Amhara, Somali, and Southern Nations and Nationalities Region States. It was reported as one of the invasive and problematic invasive alien species in Afar and Somali Regions and expanded to Great Rift Valley toward South Omo in Southern Nations and Nationalities Regional State of Ethiopia [13] and is the dominant invasive tree species in semiarid and arid ecosystems in the tropical regions of Eastern Africa [14]. During 2019 in Afar Region, Shiferaw et al. [15] reveal that there were 1.2 million ha of *P. juliflora* invaded lands that expanded at the rate of 31,127 ha per annual. It established 12.3% of *P. juliflora* land surface invasions in Afar region. Moreover, other information of Pitroff [16] reported by Farm-Africa that *P. juliflora* invaded over 1.8 million ha in Afar region.

Several research works show that *P. juliflora* had environmental effects, which is aggravating and influencing invasions into various ecosystems [10] that significantly had weaken ecosystem services [17, 18]. It reduced palatable grasses of livestock and replaced grass species such as *Chrysopogon plumulosus, Cenchrus ciliaris* and *Setaria verticillata,* and valuable woody species *Acacia tortilis, Acacia Senegal,* and *Acacia nilotica*. *P. juliflora* resulted also in social instability and economic hardship, placing constraints on sustainable development, economic growth, poverty alleviation, and food security in Afar region [19–21]. It caused harm or is likely to cause harm to the environment, people, economy, or human health [22, 23]. *P. juliflora* pods fed by livestock caused tooth decay and death through indigestion in the absence of supplementary feeds in the dry seasons [24, 25]. In arid and semiarid areas, ecosystem services obtained from woodlands, rangelands, livestock production, groundwater, and benefits of conservation areas and tourism were also under threat [26].

In different parts of Ethiopia, so as to minimize and control the invasion of *P. juliflora*, different strategies were applied such as eradication through mechanical methods and burning and cutting of the juvenile plant at 10 cm and adult plant at 40 cm down to the ground were tried by the Ethiopian Institute of Agricultural Research [27]. These methods were costly to manage the species. In addition, controlling *P. juliflora* invasion through utilization such as charcoal production and animal feeds were also tried [28, 29]. However, most of these efforts failed to control the species. Current management practices are also not satisfactory to sustain the rangelands and woodlands. *P. juliflora* management options counter to environmental effects are not sufficient to control its invasion progress toward rangelands and expand into other land use systems in the invaded regions of the country. Unless improved management interventions are adopted, the sustainability of ecosystem services will be at stake in near future. This paper aims to review the effects of *P. juliflora* on environmental constituents in Ethiopia, review the management options and take up lessons learned elsewhere or in Ethiopia, and review about the utilization and management of *P. juliflora*. Thus, this paper addresses relevant scientific information based on the evaluation of information collected from different journals, books, manuals, and various reports.

2. Review method

In this paper, relevant pieces of literatures were selected using the systematic review. Based on the specific objectives of the topic, journals, books, manuals, various reports, and related synthesized ideas were screened. Hence, 81 journals, 9 books, 5 proceedings, and 17 various reports were selected. In addition, except for the concept of some terminologies used, update sources were used for the review.

3. Environmental effects of *Prosopis juliflora*

Dense impenetrable thickets of *P. juliflora* compete with native plant species and harms environment thus disrupt ecosystem functions and services. The species affected soil properties, hydrology, land use and land cover changes, rangelands, quality and availability of animal feeds, threat to fertile agricultural lands and loss of their productivity, reduce biodiversity, invaded wildlife reserves and national parks, affect human and livestock health, the economy of the country and the overall livelihoods of pastoralists, social conflicts and cerate political instability, reduce bird diversity, blocks roads of both animal and inhabitants, reduce urban amenity, but induce carbon sequestration in the invaded areas (see **Table 1**).

3.1 Effects of *Prosopis juliflora* on soil environment

3.1.1 Soil physicochemical properties

Windbreaks, cover crops, and cultivation practices can control loss of soil [64]. Likewise, a shelter belt of *P. juliflora* is planted around fields in many semiarid regions to reduce wind speed and reduce wind induced soil erosion, decrease desiccation by reducing transpiration, and thereby increase plant and animal production [65]. These types of plantation using *P. juliflora* were also practiced in the invaded regions of Ethiopia. The capacity to block the flow of wind depends upon the height, density, and thickness of the stands of plantations. Apart from preventing the loss of fertile

Effects of *P. juliflora*	**Authors**
environmental harms	[21, 30, 31]
Disrupt ecosystem functioning and services	[14, 30, 32, 33]
Soil properties and soil seed bank	[30, 34–39]
Hydrology	[40]
Land use land cover changes	[41–46]
Rangelands, quality, and availability of animal feeds	[11, 21, 30, 32, 43, 47]
Productivity of agricultural lands	[30, 48, 49]
Loss of biodiversity	[30, 43, 48, 50]
Invade parts of wildlife reserves and national parks	[2, 32, 35, 47, 48, 51]
Human and livestock health	[2, 13, 30, 35, 43, 52, 53]
Effect economy and overall communities' livelihoods of a country	[2, 32, 54]
Social tensions or conflict	[29, 30, 36, 43, 55]
Political instability	[13, 17, 56]
Barrier to movements of the animal and human beings	[30, 43, 57]
Reduces urban amenity	[54, 58]
Reduces bird diversity	[43, 59]
Carbon sequestration	[30, 60–63]

Table 1.
Effects of Prosopis juliflora *on environmental properties and its environmental services.*

soil, *P. juliflora* reduced wind damage from crops, reduced loss of soil moisture, and improved microclimate. A report by Patnaik et al. [43] in Sudan shows that wind speed inside 5-year-old *P. juliflora* plantation reduced an average 14%, while potential evaporation reduced by 22% in the same site. *P. juliflora* was growing quickly, and it was a wind-resistant plant, which could be planted successfully to control soil erosion and could serve as shade and shelter that affected water balance by increasing relative humidity but reducing temperature and evapotranspiration [10]. On the other hand, in low land areas of Central Sudan, a study by Al-Amin et al. [66] shows that *Leptadenia pyrotechnica*l provided relatively good protection windward against consequences from erosion than cover of *P. juliflora*. This study shows protection of *P. juliflora* against soil erosion lower than the later species. Another study in Central Sudan by Al-Amin [67] reveals that the growth of *P. juliflora* in clusters could be more effective against wind protection than individual stems. *P. juliflora* had planted where soil fixing or improvement is an important consideration [68], 1). The authors also proved that *P. juliflora* was particularly suitable for stabilizing dunes and easily erodible soils. This is because of its ability to survive and grow on poor sites in which a few other species could tolerate, and its extensive lateral root system could bind soil particles particularly in the upper 60 cm soil depth.

Findings by Giessen et al. [69] show that *P. juliflora* enriched SOC, total P, total N, and available P under its canopies of topsoil in semiarid of Northeast Brazil. In Kenya, results by Mwangi and Swallow [24] reveal that biomass of understory plant species was five times lower under the canopy of *P. juliflora* than open grasslands. SOC and total N concentrations in soils under *P. juliflora* were higher than those under open grassland areas. In Afar region of Ethiopia, a study by Shiferaw et al. [14] shows that invasion of *P. juliflora* changed the physicochemical properties in Teru and Yalow Woredas. Several findings show that positive effects of *P. juliflora* on soil properties overweight the negative ones. For instance, *P. juliflora* significantly affected soil pH, exchangeable Na^+, water-soluble Ca^{2+} + Mg^{2+}, water-soluble Na^+, and exchangeable Na percentages. The invasion of *P. juliflora* significantly increased soil pH but decreased exchangeable Na^+, exchangeable Na percentage, and water-soluble Ca^{2+} + Mg^{2+} than non-invaded open grazing lands. The clay content of *P. juliflora* invaded lands was higher than non-invaded open grazing lands. However, the sand content of the soil was higher under non-invaded grazing lands than *P. juliflora* invaded lands. In this study, though in most of the findings, the invasion of *P. juliflora* had positive effects on physicochemical properties, Shiferaw et al. [2] show negative effects on plant diversity, human and livestock health, economic losses, and it means that negative effects are outweighing its positive effects.

According to Sadeq et al. [35], SOC, total N, available P, total S, and total soluble salts were higher under canopy of *P. juliflora* than outside in soil depth of 0–45 cm, but total Na increased within this soil depth. In Kenya, Muturi *et al.* [70] show that soil characteristics such as %sand, %clay, N, P, K, Mg, Mn, Fe, and Cu under *P. juliflora* species and mixed species of *Acacia* and *P. juliflora* canopies were similar except that pH and calcium were higher under the *P. juliflora* species and mixed species of *Acacia* than under canopy of *P. juliflora*. But, in Turkwel riverine forest, silt and carbon were lower under *Acacia* canopy than under *P. juliflora*. In terms of soil salinity, neutralizing alkaline, sodicity, and soil nutritional status, physical properties such as soil moisture, bulk density, and soil texture, *P. juliflora* have ameliorating effects. These are primarily due to complex interactions between the effects of nitrogen fixation, incorporation of leaf litter, changes in microclimate, and changes in the floral soil fauna and soil microbial populations [10]. In Kenya, research by Kahi et al. [71] reveals that organic matter and total N were higher under *P. juliflora* canopy than under open areas. However, available P, soil pH, soil bulk density were lower under *P. juliflora* canopy than under open areas. Thus, growth of *P. juliflora* implications for creation of suitable soil microclimate probably due to litter turnover and its facilitation of infiltration and draw water from ground to surface soil.

3.1.2 Soil biological properties

Mehadi et al. [72] indicate that invasion of *P. juliflora* increased total mycorrhizal colonization of roots and reduced heavy metals such as Cadmium (Cd) levels in plants. In addition, the intensity of mycorrhization under canopy of *P. juliflora* was significantly higher than under species of native *Acacia*. Results in Saudi Arabia revealed that Cmic under *P. juliflora* was greater in rhizosphere for *P. juliflora* than in rhizosphere of *Acacia ehrenbergiana* and *A. tortilis*. As a result, extracts from parts of *P. juliflora* were used in disinfecting and bio-functions against different bacterial pathogens. But, litter fall of *P. juliflora* inhibited plant growth and their Arbuscular Mycorrhizal Fungi colonization of roots [34]. For instance, the litter and leaf extracts of *P. juliflora* significantly inhibited the germination of *Sorghum bicolor*. On the other hand, *P. juliflora* stimulated soil microbial biomass of carbon, soil metabolic quotient, and activities of soil enzymes.

3.1.3 Hydrology

Moisture and nutrients that were taken from deep soils under *P. juliflora* were beneficial to the herbaceous plants. The removal of trees from the savannah ecosystems during wet and dry seasons supported large numbers of grazers that facilitates the growth of herbaceous plants. But, during the dry periods, the survival of the shallow-rooted herbaceous plants could be endangered by removal of trees [71]. During this season, transportation of moisture from deep soil by trees was encountered in semiarid and arid regions. Extensive lateral root systems of *P. juliflora* capture surface water after rain, but its deep tap roots allowed them to survive prolonged drought through accessing the water table [10]. However, in dry areas, evapotranspiration of plants was escalated than their transpiration. For example, Shiferaw et al. [40] indicate the daily average transpiration of *P. juliflora* lower than the daily average evapotranspiration of a dense *P. juliflora*.

3.1.4 Soil seed banks

Soil seed bank is ground flora of various vegetation ecosystems. It is important for shaping the composition, diversity, structure, and regeneration of plant communities and consequently restoration of vegetation ecosystems. Soil seed bank depends on the spatial distribution of vertical and horizontal seeds of different species and vegetation communities. The spatial distribution of seeds in the soil is primarily a function of the dispersal process [73]. Dense thicket of *P. juliflora* hindered the dispersal of other seeds of other plant species in the invaded area. The seeds of *P. juliflora* were characterized by a seed coat-imposed dormancy and established a huge persistent seed bank in the soil. This character makes it easy and continues germination of seeds of the species. In addition, livestock and wild animals are attracted by the green foliage to eat ripened pods and disperse the seeds. Dispersal and successful germination of the seeds of *P. juliflora* were thus through endozoochory of animals' seed ingestion. Seeds subsequently dispersed away from the parent plant and the pods are easily transported by runoff [36].

Land use dynamics, struggle over resources, and change of climate are key factors that influenced the probability in the expansions of *P. juliflora* [2]. When an invasive species became irresolutely verified, its control can often be challenged and eradication is habitually not possible. Subsequently, its impacts on biodiversity, ecosystem progressions, and ecosystem services can be serious [36]. In vegetation ecosystems, the seed bank of soil has been considered as a promising and cost-effective method for reestablishing of other native plant species, but its influence factors have not been clearly understood [74]. Possibility of vegetation restoration from soil seed bank is usually dependent on its seed density and species composition [75, 76]. The increases in the invasion of *P. juliflora* in certain ecosystem inhibit free dispersal of seeds in other native plants. Therefore, seed dispersal determines species diversity, composition, and density of plant species. Studies show that both livestock and wildlife species played a critical role in the dispersal of *P. juliflora* that enhanced the arrival of its seeds and progress into other land uses [77]. These in turn affected the fate of seeds of other native plant species.

4. Effects on land use and land cover changes

In the introduced areas, *P. juliflora* invasion and expansion increased both in coverage of area and density of the species. At global level, PENHA [78] reported that land covered by *P. juliflora* was 50 million hectares. In Africa alone, for example, the land covered was about 5 million hectares forming dense thickets of *P. juliflora*. In several African countries such as Kenya, Ethiopia, Sudan, Senegal, and South Africa, it had become an invasive species [10, 79]. In the Afar region in Ethiopia, *P. juliflora* is now threatening serious problems on pastoral areas where its invasion existed in four of five zones and 11 of 32 districts of the region. Among the five administrative zones of the Afar region, the Amibara woreda of Zone 3 is thought to be recognized as the starting point for the introduction and spread of *P. juliflora* [55]. This woreda was represented as a degraded semiarid ecosystem in the region [30, 80]. Zone 1 and Zone 3 were the two zones that severely invaded by *P. juliflora*, and it was expanded to the remaining zones [18]. Dubti, Asayita in Zone 1 and Mile, Gewane, Amibara, Gelealu, and Awash Fentale in Zone 3 were the most severely invaded woredas in Afar region. Reports also show that Zone 4 and Zone 5 were partly invaded woredas of Afar region.

EBI [8] reported that *P. juliflora* was threatening vegetation types including *Acacia-Commiphora* woodland, desert, and semi-desert scrublands in Afar Floristic Region. Within these vegetation types, habitats invaded by *P. juliflora* were river banks, irrigated cropland, roadsides, and the settlement areas [2]. *P. juliflora* displaced grazing lands and threatened wildlife conservation areas [47, 50, 81]. According to Helland [42], *P. juliflora* was associated mostly with the loss of pasture and invasion of woodlands. The major factors that aggravated and influenced invasions into various environments are the changes of land use and land cover and climate change [10], and increase in population pressure and overgrazing of pasture lands owing to large herding were also other causes for the increase of invasion of *P. juliflora* in invaded regions of Ethiopia [15].

5. Rangeland quality and availability of animal feeds

P. juliflora replaced local biodiversity in several spots in Afar rangelands and riversides [27]. In such habitats, the grasslands had no more used for grazing and ecosystem functions of rangelands were changed to thickets of *P. juliflora*. These made Afar pastoralists moved further from their home and pasture fields. As a result, these aggravated foods and feed shortage. Mitiku [82] reported that in Amibara woreda of Afar region, *P. juliflora* severely invaded dense *Acacia* woodlands, riverine forests, and agricultural lands. His results indicated that in 16 years (1986–2001) of land use, land cover changes by *P. juliflora* and displaced 9.91 km^2 areas of *Acacia* woodlands. Most households reported that invasion of *P. juliflora* into rangelands was more in Amibara woreda than in Awash Fentale [83]. These could be probability and the adaptability of the species and its first arrival in the earlier woreda than the later. Similarly, in border country Eritrea, study by Harnet [84] shows that the invasion of *P. juliflora* invaded both dry season and wet season rangelands and roadsides in lowlands. A study by Zarga [45] reveals household responses reveal that rangelands had taken over by *P. juliflora*. Moreover, in South Africa, Ndhlovu et al. [85] suggested wider areas of rangelands covered by *P. juliflora* invasions and reduced its grazing capacity.

6. Biodiversity

Invasive species are the second threats to global biodiversity loss next to the land use changes [86]. In the world, biotic invasions by alien plant species are considered as the major factors in biodiversity loss and endangered plant species. The reasons are the natural bio-geographical barriers of oceans, mountains, rivers, and deserts provided that isolation of essential species that ecosystems to evolve have lost their effectiveness due to the increase in global economy [87]. Biodiversity loss aggravated particularly the decline of plants that had associated with deforestation, land degradation, climate change effects, land use dynamics, and spread of invasive alien plants. Some of these invasive species caused considerable disasters in dry vegetation and rangelands of East Africa [88]. In Ethiopia, invasive alien plants and other native invasive plants could affect entire ecosystem services. As a result, natural agro-ecosystems are largely affected by the invasive alien plants [89].

Enormous invasive alien plants indicate in the decline of threatened and endangered native species; because they changed ecosystem processes, change of vegetation structures, and shift native species for the reason that they reached high densities and biomasses [58]. In Afar region, Tessema [13] indicated that *P. juliflora* threatened native plant species such as *Acacia prasinata*, *Boswellia ogadensis*, *Euphorbia doeloensis*, *Euphorbia ogadensis*, and *Indigofera kelleri*. *P. juliflora* blamed for many disaster effects such as replacing grasses, herbs, and shrubs, which were consumed by local livestock, injured livestock with its poisonous thorns, and causing goat teeth to rot and fall out because the small seeds got stuck between the teeth. Thousands of goats had been toothless and died from starvation following teeth loss, which decreased their number and threatening goat breed [88].

7. Effects of *P. juliflora* on the habitat of wild animal

In Afar region, wild animals were endangered due to the disruption of ecosystem integrity. These habitats that harbor threatened plant species also harbored many globally threatened and vulnerable mammal and bird species [88]. It reduced the diversity of birds under the thickets of *P. juliflora* than adjutant non-invaded habitats [43]. But, the invasion of species caused for the loss of agricultural crops adjacent its thicket. This is probability the habitat harbor wild animals to hide in the thicket affected production of crops.

8. Health

8.1 Effect of *P. juliflora* on human health

P. juliflora thorns are dangerous and inflict pain that is like being bitten by a snake [21]. For instance, among human injuries, 8.4% in Awash Fentale and 5.2% household respondents in Amibara woreda reported that pricked by thorns of *P. juliflora* [83]. Furthermore, the studies also highlighted the complexity of the cause of Anopheles' relationship with invasive alien plants such as *P. juliflora* that increased malaria incidences [53, 90]. These have implications to the production of anopheles insect in thickets of *P. juliflora*, which threated human health in the invaded regions of the country.

8.2 Effect of *P. juliflora* on animal health

Due to the loss of quality pasture lands, thousands of goats have been rendered toothless and died from starvation following teeth loss, which has been decreased their number [88]. Nevertheless, households in Awash Fentale and Amibara woredas replied that pods/seeds were the most palatable part of *P. juliflora* by animals [83]. Most of them argued that leaf of *P. juliflora* was the most toxic and killed their animals after consumptions [10]. This might be due to a permanent weakening of the ability to digest cellulose in pods and high sugar content of the pod that depresses the rumen bacterial cellulose activity and finally killed the animal. Feeding of *P. juliflora* pods was leading to neurological disorders, toxicity to neurons of cranial nerve nuclei leading to partial anorexia, depression, salivation, twitching, dehydration, and bloody diarrhea, bile duct hyperplasia, renal tubules, and there is rarefaction of lymphoid tissue, chronic and progressive injury to neurons, and causing denervation atrophy [43]. As a result, various plant parts of *P. juliflora* affected animal health and reduced their production in the invaded regions of the country.

9. Effects of *Prosopis juliflora* on the economy

9.1 Use of *Prosopis juliflora*

Introducing species to new locations had tremendous contributions to societal development [75]. Due to this, human welfare had improved the introduction of its parts out of its native ranges. Trade enabled modern societies to benefit from the unmatched movement and formation of species in the world [91]. Wood sources of *P. juliflora* were also used either as fuel wood or structural material. As a fuel, it could be burned directly or formed as charcoal, and timber, it can be used poles or formed into different types of furniture [24].

P. juliflora pods and seeds are consumed by a wide variety of animals, both in their native range and where it was hosted, and it is an important mammalian diet when trees were existent in large numbers [43]. For instance, a study by Ilukor et al. [92] in the Afar region of Ethiopia shows that households fed their animals with leaves and pods of *P. juliflora*. However, feeding animals that consumed only *P. juliflora* leaves and pods caused acrimonious milk and lost body weight of animals [93]. *P. juliflora* branches were widely used as fencing posts, while its pods high in protein and sugars were important food for human being [24]. *P. juliflora* species had also ameliorating effects on soil under natural and semi-natural systems because of nitrogen fixation and leaf litter incorporated into the soil improved physical and nutritional status of soil. These had reduced the use of inorganic fertilizers commercially purchased by land managers in the invaded regions. High mineral content and rapid decay of small leaves were favorable characteristics for the use of foliage as a soil ameliorant. Compost making could detoxify its allelochemical effects on germination and growth of plants [94]. In addition, when added to agricultural and forest fields, compost made from leaf of *P. juliflora* replaced the cost incurred to commercial fertilizers in various countries. Other uses of *P. juliflora* are flowers for the supply of nectar and pollen as bee forage to produce honey. Flower of *P. juliflora* is small, yellow, and gathered on long inflorescences producing pollen and nectar that is high in protein and sugars [43]. Although *P. juliflora* has diverse economic values, the use of the plant is limited. In Ethiopia, for example, local communities in the invaded areas used the plant only for animal feed, fuel wood, charcoal, and construction purposes [2]. These indicate that *P. juliflora* was underutilized in the invaded areas of the country that made it easy invasion and threatened vegetation ecosystems and affected native plant species.

10. Effects of *Prosopis juliflora* on grazing lands

In animal-rearing areas of lowlands, pressures on environment caused living incomes make dynamics, which improved demands and concentrated supplies of natural resources owing to the upswing of population. Widespread rearing of livestock and crusade in search for feed and water, resource use for charcoal production and construction materials, and shortage of institutional capacity levied influences on present natural resources particularly on woodland vegetation. These caused devastation and deprivation of the vegetation resources in the invaded areas [95]. As a result, the socioeconomic and ecological impacts of *P. juliflora* on grazing lands became severe [13]. *P. juliflora* is the source of income for the poor, but has damaging effects for pastoralists and rich people by slaying their animals and attacking their rangelands [82].

11. *Prosopis juliflora* versus social conflict and political instability

Reports show that invasion of *P. juliflora* might lead to serious food insecurity and might even trigger tribal conflicts for the remaining few pastures and farmlands [13, 19]. This resulted in the taken over of grazing lands by the *P. juliflora* aggravated by climate change and has been caused border conflicts and political instability, for example, between Afar clan and *Issa* of Somali on the border of the two regions. The combination of diminishing grazing areas and population growth both human and animal has contributed to land degradation, competition for pasture and water, and interethnic and intra-ethnic conflict. These factors have favored the invasion of *P. juliflora* in the invasion region [96]. Furthermore, Rogers et al. [97] reported that impacts of *P. juliflora* interact with other drivers of vulnerability. Pastoralists report had broadened conflict, complicated relationships with the state, and increased decentralization within invaded areas of southern Afar of Ethiopia. In Kenya, some individuals, from Chemonke village, claimed to have been displaced from their original settlements by *P. juliflora*. They had to seek alternative settlements elsewhere sometimes imposed them to lease land for cultivation in their new areas. Conflicts might rise as the displaced areas, which seek alternative settlements [24]. A report by FAO [57] further pointed out that invasion of *P. juliflora*, climate, conflict, and economic shocks were among the causes of political instability and the main drivers of food insecurity in Somalia in East Africa.

12. Deterioration of urban amenity by *Prosopis juliflora*

P. juliflora in its native range in Mexico is described as an urban afforestation program that would mitigate air pollution by planting with other native species [98]. But it deteriorates ecosystems in its exotic land ranges in tropics such as Asia and Africa. *P. juliflora* is an aggressive invader of urban fallows and abandoned fields that deteriorate its beauty [99]. Pasiecznik et al. [43] pointed out that *P. juliflora* does not fulfill all the qualities required for urban trees. It disrupts urban amenities in towns (**Figure 1**).

Figure 1.
P. juliflora *disrupts urban beauty in Melka Sedi (left) and Awash Sebat (right), Southern Afar region in Northeast Ethiopia.*

13. Blockade of roadsides by *P. juliflora*

A Report by Patnaik *et al*. shows that invasion of *P. juliflora* creates impenetrable thickets, which could be seen encroaching upon roadsides and blocking the roadsides resulted in human dwellings and blocking pathways (**Figures 2** and **3**) and seriously aggravate road accidents in (**Figures 3** and **4**). As indicated, the growth of *P. juliflora* moderately invades roadside next to wetlands and homesteads [50, 82]. These could be homesteads and wetlands have better organic matter and moisture for its establishment and growth. Due to the thorns of *P. juliflora*, these thickets can seriously aggravate a road accident and human and livestock movement to their homes and water sources (**Figure 5**). The invasions of *P. juliflora* block access to water sources, irrigation canals, homesteads, and blocked paths in Garrisa of Kenya [100]. Findings by Esther and Brent [101] in Kenya and Koyira [50] in Afar of Northeast Ethiopia show that the invasion of *P. juliflora* blocked trails and roads used by humans and livestock.

Figure 2.
Typical scenes of P. juliflora *advancing toward human dwellings (left) and blocking pathways (right).*

Figure 3.
Encroachment of P. juliflora *in a wetland (left), and typically impenetrable thickets formed by the species (right) in Kebena site of Awash Fentale Woreda in Afar region, Northeast Ethiopia. (Source photos: Shiferaw, 2017).*

Figure 4.
Encroachment of P. juliflora *in a wetland (left), and typically impenetrable thickets formed by the species (right). Due to the thorns contained in* P. juliflora, *these thickets can seriously aggravate a road accident.*

Figure 5.
P. juliflora *blocks paths to water sources in Melka Sedi, Southern Afar region in Northeast Ethiopia (Photo: Shiferaw, 2019).*

14. Allelochemical effects of *P. juliflora*

Analysis of chemical extracts showed that allopathic compounds are phenolic or antioxidant capability in nature. Sluggish in breakdown and heavy buildup of leaf litter below *P. juliflora* may result in the increase of toxic substances in the soil layers, hindering the growth of other plant species [102, 103]. For instance, report by Kaur et al. [104] shows that there was a noticeable accrual of litter beneath the thicket of *P. juliflora* compared with *P. juliflora cineraria* in the hyper-arid habitats. Studies by El-Keblawy and Al-Rawai [103] reveal that lower species richness, evenness, and frequency of native species in plots under shades of *P. juliflora*. However, it had no allopathic effect on its seedlings growing below its canopies. But, Muturi et al. [70] show *P. juliflora* canopy hindered regeneration of other native species. Findings by Samuel et al. [105] depict that leaves of *P. juliflora* have greater inhibitory effects than its roots and barks. Muturi et al. [70] also indicated that bark contained the least inhibitory composites. But, shoot and root growth introverted by leaf and root extracts at higher application and concentrations. *P. juliflora* had both positive and negative interactions with plant communities in naturalized areas [2, 13, 23, 24, 50, 105]. Several studies carried out in many parts of the world show that problems of *P. juliflora* were outweighing positive ones ecologically, socioeconomically, and in all health aspects. However, some studies indicate that above-ground biomass, frequency, and cover of understory plant species were significantly higher under non-invaded grazing lands than under the shades of *P. juliflora* [71].

15. Understandings for the management of *P. juliflora*

A study by Shanwad et al. [106] indicates that control of *P. juliflora* is extremely difficult and costly for eradication once it invaded ecosystems. Report by Pasiecznik et al. [43] also confirmed that once *P. juliflora* established into large areas, prevention of further spread is not possible as the species quickly builds up soil seed bank. Thus, it required regular removal of all new seedlings over very many years, as seeds could remain viable for more than 40 years [31]. To control its invasions, different strategies can be applied such as eradication by utilization and mechanical control of *P. juliflora*. The potential irreversibility of the damage of invasion costs and may impose countries economic losses for management. The main management reactions after an alien species invaded are mitigation and adaptation [91].

To reverse the situations, integrated management strategies, participation of all stockholders and multidisciplinary research approaches within and across countries should be designed. To control invasive alien plants, various management options are tried but barriers faced for management in developing counties like Ethiopia [2]. In the invaded areas, experiences indicate that there are several methods of *P. juliflora* management. Chemical, mechanical, biological, management by utilization, and disrupting phenology of *P. juliflora* are among the management methods to control the species. There are three methods for the control of *P. juliflora* namely mechanical, chemical, and biological strategies [30]. Control of *P. juliflora* by utilization is an effective control way and management. In the following subsections, the pros and cons of management and control of *P. juliflora* invasions are discussed.

15.1 Chemical control methods for *P. juliflora*

Several available herbicides were tried to control *P. juliflora*. For example, the study by Shanwad et al. [106] shows combinations of chemicals reduced growth and development of *P. juliflora*. In their study, Mera-71, 2, 4-D ester followed by paraquat was the best in affecting weed recovery. The control of the regrowth of *P. juliflora* was effectively achieved by two times applications of systemic trans-located herbicides like Mera-71 (Glyphosate) and 2, 4- compared to paraquat, diuron. Controlling *P. juliflora* through chemicals at juvenile stages is highly effective than adult stages. Eradication of *P. juliflora* has been attempted in several countries through chemicals but has proved unsuccessful and chemical eradication is not environmentally friendly [107]. In, Ethiopia this management strategy was also tried to control *P. juliflora* but failed to manage it.

15.2 Mechanical control methods of *P. juliflora*

The mechanical control of *P. juliflora* is very labor-intensive, expensive, and economically feasible only for high-value lands. Mechanical control method of *P. juliflora* plant was recommended to cut at 10 cm for young trees and 40 cm for matured stems to control coppicing (**Figure 6**). Control of the spread of the *P. juliflora* is ineffective to eradicate it by mobilizing communities. Management of *P. juliflora* by utilization such as fuel wood, construction, and charcoal production, feed livestock by crushing pods are the best management options [57, 93]. But small-scale households cannot afford to control using mechanical methods. Pastoralists, agro-pastoralists, and urban dwellers should also manage their territory of rangelands, woodlands, homesteads, and compounds through hand weeding of *P. juliflora* seedlings. This method can minimize its invasion before it takes roots in different land uses. However, some of these methods need labor costs for control. Since cutting promotes regeneration, all mechanical methods of control cannot be recommended [107]. This management strangely was largely applied in Ethiopia to control the invasion of *P. juliflora* in Afar and Somale region, but did not succeed.

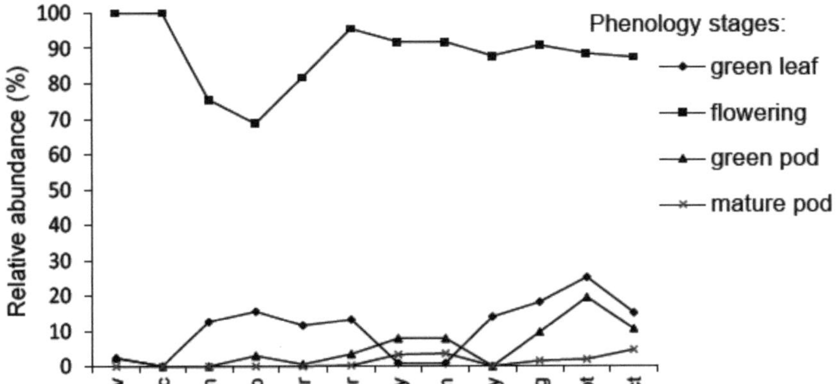

Figure 6.
Phenological patterns of P. juliflora *in the Awash Fentale and Amibara Woredas, Afar region of Ethiopia in 2016/2017 cropping season. (Source: Shiferaw et al. [83]).*

According to Shiferaw et al. [83], disrupting the phenology of *Prosopis juliflora* or aborting the juveniles of the species throughout the year, particularly during the peak time of flowering before seed set could also be used for the management *P. juliflora*. This method controls the species before it disperses seeds into the soil stores as a soil seed bank (**Figure 6**). These needs forced community involvement during its flowering times like yearly voluntary and community participation in soil and water conservation works in Ethiopia. During floral removal, care should be taken due to the allergy to its pollination.

According to Pantaik et al. [10], pollens of *P. juliflora* trigger allergic asthma, rhinitis, and skin allergy. Defoliation of only the leaves of *P. juliflora* also inhibits photosynthesis of the species (**Figure 6**). But, this method of destruction of leaf parts needs care not to cut the branches as *P. juliflora* aggressively propagates by all parts of its stems. Appropriate silvicultural techniques (e.g., thinning) of *P. juliflora* should also be practiced to lessen the invasiveness and effects on other plants [55]. For instance, a research by Singh et al. [108] shows that plant height was recorded to be 20 and14% higher in *P. juliflora* and *A. nilotica* respectively grown in combination with grasses than the sole plantation of these species applying a silvicultural treatment of *P. juliflora*. Moreover, Walter and Armstrong [109] reported that proper silvicultural management increased the quality of wood for maintaining its benefits for livelihood and control the invasion of *P. juliflora*. This management strategy could not be tried in Ethiopia at large to control the progress of *P. juliflora* into various vegetation ecosystems to enhance the values we get from ecosystem services.

15.3 Biological control methods of *P. juliflora*

In recent decades, biotic control has increased recognition in various countries due to lucrative and reliable means of managing large invasiveness of alien plants. It comprises the restrained, strictly administered introduction of one or more species of highly studied alien organisms that blizzard from the original home ranges of invading plant species, and which physiologically are reformed to feeding absolutely on or attacking completely plants of that [30]. In Africa, though the goal of cooperation for biological control of *P. juliflora* in South Africa with other countries, it delayed trendy due to the controversy for the introduction of biological control agents onto *P. juliflora*. Additionally, debates about the relative value and costs of trees continued to hinder progress with the planned increase of biological control [110]. Ravhuhali et al. [111] concluded that managing the spread of invasive species could also be accomplished using livestock as biological control while improving the productivity of the animals (e.g., *P. juliflora* pods utilized for livestock feed ingredients with other feeds). The release of bio-control agents was also considered where these technologies were not feasible because careless release in some species might escape and change into invasiveness inclining to threaten the native organisms [107]. These strategies were not largely tried on research fields of Ethiopian Agricultural Research Institutes, for instance, and had no theories in management of *P. juliflora* in the invaded areas of Ethiopia.

16. Conclusions

P. juliflora has effects on the soil environment in the form of allelochemicals that are ingredients in its plant parts that have effects on living organisms, take over land uses land cover aggressively, economic loss, the health of humans and animals, and reduce the diversity of organisms. On the other hand, the species is used to ameliorate soil properties (soil physical, biological properties), e.g., sequester CO_2 for mitigation of greenhouse effects, and other economic values (construction materials, furniture, medicinal values, sources of pollination for honey production, and animal feeds, etc.) particularly in the semiarid and arid tropics. However, the negative effects of *P. juliflora* exceed its positive values. Mechanical, chemical, management by utilization, fire regime, disrupting its phenological stages, and biological control methods have their pros and cons to control *P. juliflora*. Therefore, prevention, integrated management strategies, and management of the species by utilization are the best measures used to control *P. juliflora* recommended globally and in Ethiopia. Mechanical methods (aborting phenological stage, utilization of the species, others) need awareness of the communities, and forced mobilization is vital for effective and efficient management of the species.

Previously, in several countries, the management methods, namely mechanical, chemical, and biological, were being tried to control the alien *P. juliflora*. But all the methods failed to eradicate the species. As a result, the species is continuing to invade and take valuable farmlands, grazing lands, rural homesteads, wetlands, roadsides, and urban areas, particularly in Afar and Somali regions. In Ethiopia, more research involving multidisciplinary research approaches should be designed to conduct the effects of *P. juliflora* on the environment, economy, and health of animals and communities. Moreover, other teams should also design other effective and efficient management designs that are different from previous researches that is vital to alleviate the invasion of the species and thus improve the livelihood of inhabitants in the region.

Acknowledgements

The authors express their sincere thanks to Addis Ababa and Arba Minch Universities that had taken attention to review on this topic to manage invasive alien plant species *Prosopis juiflora* in the invaded areas of the country.

Conflict of interest

The authors declare no conflict of interest.

Author details

Wakshum Shiferaw[1]* and Sebsebe Demissew[2]

1 Department of Natural Resources Management, College of Agricultural Sciences, Arba Minch University, Arba Minch, Ethiopia

2 College of Natural and Computational Sciences, The National Herbarium, Addis Ababa University, Addis Ababa, Ethiopia

*Address all correspondence to: waaqs@yahoo.com

References

[1] Burgiel SW, Muir AA. Invasive Species, Climate Change, and Ecosystem Based Adaptation: Addressing Multiple Drivers of Global Change. Washington DC: Global Invasive Species Programme (GISP); 2010

[2] Shiferaw W, Demissew S, Bekele T. Invasive alien plant species in Ethiopia: Ecological impacts on biodiversity a review paper. International Journal of Molecular Biology Open Access. 2018;**3**:169-176

[3] Pysek P, Richardson DM. Invasive Species, environmental change and management, and health. Annual Review of Environment and Resources. 2010;**35**:25-55. DOI: 10.1146/annurev-environ-033009-095548

[4] Thapa S, Chitale V, Rijal SJ, Bisht N, Shrestha BB. Understanding the dynamics in the distribution of invasive alien plant species under predicted climate change in Western Himalaya. PLoS One. 2018;**13**(4):e0195752

[5] Turbelin AJ, Malamud BD, Francis RA. Mapping the global state of invasive alien species: Patterns of invasion and policy responses. Global Ecology and Biogeography. 2017;**26**:78-92

[6] Neill PE, Arim M. Human health link to invasive species, Encyclopedia of Environmental Health, 2nd edition, 2011; 3:116-123.

[7] Lowe S, Browne M, Boudjelas S, De Poorter M. 100 of the world's worst invasive alien species a selection from the global, invasive species database. ISSG. 2000;**6**:1-11

[8] EBI (Ethiopian Biodiversity Institute). Ethiopian's Fifth National Report to the Conservation Biodiversity. Addis Ababa, Ethiopia; 2014

[9] Shiferaw W, Bekele T, Demissew S, Aynekulu E. Socio-ecological impacts of invasive Plant Species in Ethiopia: A review paper. Journal of Agricultural Forest Meteorological Research. 2020a;**3**(2):292-296

[10] Pasiecznik NM, Felker P, Harris PJC, Harsh LN, Cruz G, Tewari JC, et al. The *Prosopis juliflora - Prosopis pallida* Complex: A Monograph. UK: HDRA; 2001. p. 172

[11] Shiferaw W, Bekele T, Demissew S, Aynekulu E. Relationship between *Prosopis juliflora* invasion and livelihood diversification in the South Afar region, Northeast Ethiopia. Regional Sustainability. 2020c;**1**(1):82-92. DOI: 10.1016/j.regsus.2020.09.002

[12] Wakie T, Laituri M, Evangelista H. Assessing the distribution and impacts of *P. juliflora* through participatory approaches. Applied Geography. 2016;**66**:132-143

[13] Tessema AY. Ecological and economic dimensions of the paradoxical invasive species– *Prosopis juliflora* and policy challenges in Ethiopia. Journal of Economics and Sustainable Development. 2012;**3**(8):2222-2855

[14] Shiferaw W, Demissew S, Bekele T, Aynekulu E. Effects of *Prosopis juliflora* invasion on native Species diversity and Woody Species regenerations in rangelands of Afar National Regional State, Northeast Ethiopia. Journal of Resources and Ecology. 2023;**14**(1):1-16. DOI: 10.5814/j.issn.1674-764x.2023.01.004

[15] Shiferaw H, Schaffner U, Bewket W, Alamirew T, Zeleke G, Teketay D, et al. Modeling the current fractional cover

of an invasive alien plant and drivers of its invasion in dry land ecosystems. Scientific Reports. 2019a;**9**:1576. DOI: 10.1038/s41598-018-36587-7

[16] Pittroff W. Invasive alien species: The threat to sustainable livelihoods and ecosystems health. In: Policies and Realities -Needs for Environmental Rehabilitation in Ethiopia. 29thAnnual Conference of the Biological Society of Ethiopia. Addis Ababa, Ethiopia; 2019. pp. 1-58

[17] Rettberg S. The spread of *Prosopis juliflora* in the wetlands of the Middle Awash Basin. In: Managing *Prosopis juliflora* for better (agro-) pastoral Livelihoods in the Horn of Africa. In: Proceedings of the Regional Conference. Addis Ababa, Ethiopia; 2014

[18] Shiferaw W, Bekele T, Demissew S, Aynekulu E. Effects of *Prosopis juliflora* invasions on land use/cover change in South Afar region, Northeast Ethiopia. Advance Research Journal of Multidisciplinary Discoveries. 2019d;**34**(5):26-43

[19] Admasu D. Invasive Plants and Food Security: The Case of *Prosopis juliflora* in the Afar Region of Ethiopia, Addis Ababa. 2008

[20] Haile MZ. Invasion of *Prosopis juliflora* (SW.) DC and Rural Livelihoods: The Case of Afar Pastoralists at Middle Awash Area of Ethiopia. Ethiopia: Norwegian University of Life Sciences (UMB). 2008

[21] Seid O, Haji J, Legesse B. Rural households' perception on the effects of *Prosopis juliflora* invasion: The case of Amibara District of Afar National Regional State, Ethiopia. Pastoralism: Research, Policy, and Practice. 2020;**10**:21. DOI: 10.1186/s13570-020-00174-1

[22] Richardson DM, Cambray RA, Chapman WRJ, Dean CL, Griffiths DC, Le Maitre DJ, et al. Vectors and pass ways of biological invasions in South Africa: Past, present, and future. In: Ruiz G, Carlton JT, editors. Invasive Species. Vectors and Management Strategies. Washington, DC: Island Press; 2003. pp. 292-349

[23] Shiferaw H, Bewket W, Alamirew T, Zeleke G, Teketay D, Bekele K, et al. Implications of land use/land cover dynamics and *Prosopis* invasion on ecosystem service values in Afar region, Ethiopia. Science of the Total Environment. 2019b;**675**:354-366

[24] Mwangi E, Swallow B. *Prosopis juliflora* invasion and rural livelihoods in the Lake Baringo area of Kenya. Conservation and Society. 2008;**6**(2):130-140. DOI: 10.4103/0972-4923.49207

[25] Preeti K, Avatar SR, Mala A. Pharmacology and therapeutic application *of Prosopis juliflora*: A review. Journal of Plant Sciences. 2015;**3**(4):234-240. DOI: 10.11648/j.jps.20150304.20

[26] Wilgen BWV, Forsyth GG, Maitre DCL, Wannenburgh A, Kotze JDF, Berg EVD, et al. An assessment of the effectiveness of a large, national-scale invasive alien plant control strategy in South Africa. Biological Conservation. 2012;**148**:28-38

[27] Berhanu A, Tesfaye G. The *Prosopis juliflora* dilemma, impact on dryland biodiversity and some controlling methods. Journal of the Dry Lands. 2006;**1**(2):158-164

[28] Sertse D, Pasiecznik N. Controlling the spread of *Prosopis juliflora* in Ethiopia by its utilization. Programs for International Development. 2014:1-2

[29] Wakie T, Evangelista P, Laituri M. Utilization Assessment of *Prosopis*

juliflora in Afar Region, Ethiopia. US Forest Service, an Office of International Programs USAID Pastoral Livelihoods Initiative II Project (PLI II). 2012

[30] Abdullah MM, Ute AJ, Regasa T. *P. juliflora* (Sw.) DC distribution, impacts and available control methods in Ethiopia. Tropical and Subtropical Agroecosystems. 2017;**20**(1):75-89

[31] Wise RM, van Wilgen BW, Le Maitre DC. Costs, benefits, and management options for an invasive alien tree species: The case of mesquite in the Northern Cape, South Africa. Journal of Arid Environments. 2012;**84**:80-90

[32] Belay M, Goshu A. Impact of *Prosopis juliflora* on pastoral communities of bale zone Oromia region, Ethiopia. World Rural Observation. 2020;**12**(3):85-94. DOI: 10.7537/marswro120320.07

[33] Hussien A, Seboka S, Shiferaw W. Assessment of irrigation water quality of lowlands in the Bale Zone, South Eastern Oromia, Ethiopia. International Journal of Water Resources and Environmental Engineering. 2017;**9**(12):264-269. DOI: 10.5897/IJWREE2017.0747

[34] Mehadi M, Tounekti T, Khemira H. Effects of *Prosopis juliflora* on germination, plant growth of Sorghum bicolor, mycorrhiza, and soil microbial properties. Allelopathy Journal. 2019;**46**(2):121-132. DOI: 10.26651/allelo.j/2019-46-2-1214

[35] Sadeq AMM, Abido SM, Salih AA, Alkhuzai AJ. The effects of Mesquite (*Prosopis juliflora*) on soils and Plant communities in the deserted rangelands of Bahrain. International Journal of Forestry Research. 2020;**2020**:1-8. DOI: 10.1155/2020/8810765

[36] Shiferaw H, Teketay D, Nemomissa S, Assefa F. Some biological characteristics that foster the invasion of *Prosopis juliflora* (Sw.) DC. at middle awash Rift Valley Area, North-Eastern Ethiopia. Journal of Arid Environments. 2004;**58**(2):135-154

[37] Shiferaw W, Bekele T, Demissew S, Aynekulu E, Pitroff W. Invasion of *Prosopis juliflora* and its effects on soil physicochemical properties in Afar region, Northeast Ethiopia. International Soil and Water Conservation Research. 2021b;**9**(4):631-638. DOI: 10.1016/j.iswcr.2021.04.003

[38] Shiferaw W, Bekele T, Demissew S, Aynekulu E. *Prosopis juliflora* invasion and effects of environmental factors on the density of soil seed bank in Afar Region, Northeast Ethiopia. Journal of Ecology and Environment. 2019c;**43**(1)

[39] Tiedemann AR, Klemmedson JO. Effect of mesquite on physical and chemical properties of the soil. Journal of Range Management. 1973;**26**(1): 27-29

[40] Shiferaw H, Alamirew T, Sebinasi D, Bewket W, Zeleke G, Urs S. Water use of *Prosopis juliflora* and its impacts on catchment water budget and rural livelihoods in Afar Region, Ethiopia. Scientific Reports. 2021;**11**:2688. DOI: 10.1038/s41598-021-81776-6

[41] FAO. FAO Invasion of *Prosopis juliflora* in India. In: Problems Posed by the Introduction of *Prosopis juliflora* Species in Selected Countries. Rome: FAO; 2007. p. 13

[42] Helland J. Afar Resilience Study for Feinstein International Center (Tufts University) and Afar Region Disease Prevention, Preparedness and Food Security, Coordination Office. Ethiopia; 2015

[43] Patnaik P, Abbasi T, Abbasi SA. *Prosopis (Prosopis juliflora)*: Blessing and bane. Tropical Ecology. 2017;**58**(3):455-483

[44] Syomiti M. How *Prosopis juliflora* can be economically rewarding to pastoral communities in Kenya's rangeland. East Africa TIRI Research, TRB-13. Kenya Agricultural Research Institute, Kenya. 2015.

[45] Zerga B. Degradation of rangelands and rehabilitation efforts in Ethiopia: The case of Afar rangelands. Journal of Advances in Agricultural Science and Technology. 2015;**3**(6):81-94

[46] Zergaw YA, Anke J, Detle M. Ecosystem engineer unleashed: *Prosopis juliflora* threatening ecosystem services? Regional Environmental Change. 2015;**15**:155-167. DOI: 10.1007/s10113-014-0616-x

[47] Kebede TA. Sustaining the Allideghi Grassland of Ethiopia: Influence of Pastoralism and vegetation change. 2009

[48] Shiferaw W, Bekele T, Demissew S, Aynekulu E. Phenology of the alien invasive plant species *Prosopis juliflora* in arid and semi-arid areas in response to climate variability and some perspectives for its control in Ethiopia. Polish Journal of Ecology. 2020b;**68**:37-46. DOI: 10.3161/15052249PJE2020.68.1.004

[49] Tegene S, Ayele N, Belay L. Assessment of *Prosopis juliflora* infestation and management options at Tendaho Sugar Estate. International Journal of Advanced Research and Biological Science. 2019;**6**(3):191-200

[50] Koyira TS. Impact of *Prosopis juliflora* L. (Fabaceae) on Plant Biodiversity at Alledeghi Wildlife Reserve and Surrounding Local Community, Ethiopia. Ethiopia: Addis Ababa University; 2015

[51] Mukherjee A, Dilip AV, Nagaraj HK. Invasive *Prosopis juliflora* replacing the native floral community over three decades: A case study of a world heritage site, Keoladeo National Park, India. Biodiversity and Conservation. 2017:1-18. DOI: 10.1007/s10531-017-1392-y

[52] Hussain IM, Shackleton TR, El-Keblawy A, Perez TMM, González L. Invasive Mesquite (*Prosopis juliflora*), an allergy and health challenge. Plants. 2020;**9**:141. DOI: 10.3390/plants9020141

[53] Stone MC, Witt BRA, Walsh CG, Foster AW, Murphy TS. Would the control of invasive alien plants reduce malaria transmission? A review. Parasites & Vectors. 2018;**11**:76. DOI: 10.1186/s13071-018-2644-8

[54] Haji J, Mohammed A. The economic impact of *Prosopis juliflora* on agropastoral households of Dire Dawa administration, Ethiopia. African Journal of Agricultural Research. 2013;**8**(9):768-779. DOI: DOI, 10.5897/AJAR12.014

[55] Shiferaw W. Invasion of *Prosopis juliflora* (Sw.) DC and its Ecological Impacts in Afar National Regional State, Northeast Ethiopia. Ethiopia: Addis Ababa University; 2020

[56] Eriksen S, Lind J. Adaptation as a political process: Adjusting to drought and conflict in Kenya's drylands. Environmental Management. 2009;**43**:817-835

[57] FAO. Innovative ways of managing *Prosopis juliflora* trees in Somalia, safeguarding agricultural and pastoralist livelihoods by transforming a longstanding threat into a sustainable resource for women and youth. 2020

[58] Surendra B et al. Invasive alien Plant Species assessment in urban ecosystem: A case study from Andhra University, Visakhapatnam, India. International Research Journal of Environmental Science. 2013;**2**(5):79-86

[59] Khera N, Mehta V, Sabata BC. Interrelationship of birds and habitat features in urban green spaces in Delhi, India. Urban Forestry & Urban Greening. 2009;**8**:187-196

[60] Archer SR, Andersen EM, Predick KI, Schwinning S, Steidl RJ, Woods SR. Woody plant encroachment: Causes and consequences. Rangeland Systems. 2017;**2**:25-84

[61] Bogale AG, Tolossa TT. Climate change intensification impacts and challenges of invasive species and adaptation measures in eastern Ethiopia. Sustainable Environment. 2021;7:1875555

[62] Dangles O, Malmqvist B. Species richness–decomposition relationships depend on species dominance. Ecology Letters. 2004;7(5):395-402. DOI: 10.1111/j.1461-0248.2004.00591.x

[63] Hejda M, Pysek P, Jarosik V. Impact of invasive plants on the species richness, diversity, and composition of invaded communities. Journal of Ecology. 2009;**97**:393-403. DOI: 10.1111/j.1365-2745.2009.01480.x

[64] Ogunsola AO, Adeniyi DO, Adedokun AV. Soil management and conservation: An approach to mitigate and ameliorate soil contamination. Soil Contamination. 2020

[65] Tewari JC, Mathur BK, Tewari P, Singh Y, Singh M, Ram M, et al. Prosopis: A miracle Species of hot arid and semi-arid regions of India. Popular Kheti. 2013;**1**(2)

[66] Al-Amin NKN. Semi-arid vegetation pattern, stability and suitability to suppress sand movement in Central Sudan. Asian Journal of Agricultural Sciences. 2011;**3**(1):44-50

[67] Al-Amin NKN, Stigter CJ, Mohammed AE. Wind reduction patterns around isolated biomass for wind-Erosion control in a Desertified area of Central Sudan. Research Journal of Environmental and Earth Sciences. 2010;**2**(4):226-234

[68] Zeila AD. Mapping and Managing the Spread of Prosopis juliflora in Garissa County, Kenya. Kenya: Kenyatta University; 2005. pp. 1-81

[69] Giessen H, Menezes RSC, Salcedo IH, Wick B. Organic matter transformations and soil fertility in a tree pasture in semiarid NE Brazil. Plant and Soil. 2003;**252**:195-205

[70] Muturi GM, Poorter L, Mohren GMJ, Kigomo BN. Ecological impact of *Prosopis juliflora* species invasion in Turkwel riverine forest, Kenya. Journal of Arid Environments. 2013;**92**:89-97

[71] Kahi CH, Ngugi RK, Mureithi SM, Ngethe JC. The canopy effect of *Prosopis juliflora* (Sw.) DC. and *Acacia tortilis* (Hayene) trees on herbaceous plants Species and soil Physico-chemical properties in Nejemps Flats, Kenya. Tropical and Subtropical Agroecosystems. 2009;**10**:441-449

[72] Mehadi M, Taieb T, Habib K. Invasive character of *Prosopis* facilitated by its allelopathy and a wide mutualistic interaction with soil microorganisms. Journal of Biological Sciences. 2017;**18**(3):115-123

[73] Shiferaw W, Demissew S, Bekele T. "Ecology of soil seed banks: Implications for conservation and restoration of natural vegetation": A review. International Journal of Biodiversity and Conservation. 2018b;**10**(10):380-393

[74] Li CD, Xiao B, Wang Q, Zheng R, Wu J. Responses of soil seed Bank and vegetation to the increasing intensity of human disturbance in a semi-arid

region of northern China. Sustainability. 2017;**9**:18-37. DOI: 10.3390/su9101837

[75] Duncan RP, Diez JM, Sullivan JJ, Wangen S, Miller AL. Safe sites, seed supply, and the recruitment function in plant populations. Ecology. 2009;**90**(2129-2138):e0195752

[76] Gonzalez SL, Ghermandi L. Comparison of methods to estimate soil seed banks: The role of seed size and mass. Community Ecology. 2012;**13**(2):238-242

[77] Mworia JK, Kinyamario JI, Omari JK, Wambua JK. Patterns of seed dispersal and establishment of the invader *Prosopis juliflora* in the upper floodplain of Tana River, Kenya. African Journal of Range and Forage Science. 2011;**28**(1):35-41

[78] Penha Tsegay B, Livingstone J, Zeremariam F, editors. Proceedings of a Regional Conference, Exploring *Prosopis juliflora* Management and Policy Options in the Greater Horn of Africa. Addis Ababa, Ethiopia; 2014

[79] Shackleton RT, Le Maitre DC, Pasiecznik NM, Richardson DM. *Prosopis*: A global assessment of the biogeography, benefits, impacts and management of one of the world's worst woody invasive plant taxa. AoB PLANTS. 2014;**6**:plu027. DOI: 10.1093/aobpla/plu027

[80] Ilukor J, Rettberg S, Treydte A, Birner R. To eradicate or not to eradicate? Recommendations on *Prosopis juliflora* management in Afar, Ethiopia, from an interdisciplinary perspective. Policy and Practice. 2016;**6**:4. DOI: 10.1186/s13570-016-0061-1

[81] Zeray N, Legesse B, Mohamed HJ, Aredo KM. Impacts of *Prosopis juliflora* invasion on livelihoods of pastoral and agro-pastoral households of Dire Dawa Administration, Ethiopia. Pastoralism: Research, Policy, and Practice. 2017;**7**(7):1-14. DOI: 10.1186/s13570-017-0079-z

[82] Mitiku GS. The Ecological and Socio-Economic Impacts of Invasive Plant Species: The Ecological Distribution and Socio-Economic Impacts of *Prosopis juliflora* in the Afar Regional State of Ethiopia. Müller: VDM Verlag Dr; 2010

[83] Shiferaw W, Demissew S, Bekele T, Aynekulu El. Community perceptions towards invasion of Prosopis juliflora, utilization, and its control options in Afar region, Northeast Ethiopia. Plos One; 2022;**17**(1): e0261838

[84] Harnet M. The ecological and socioeconomic role of *Prosopis juliflora* in Eritrea. In: An Analytical Assessment within the Context of Rural Development in the Horn of Africa. Gutenberg University; Gutenberg. 2008

[85] Ndhlovu T, Milton-Dean SJ, Esler KJ. Impact of *Prosopis* (mesquite) invasion and clearing on the grazing capacity of semiarid Nama Karoo rangeland, South Africa. African Journal of Range and Forage Science. 2011;**28**:129-137

[86] Miranda JD, Armas C, Padilla FM, Pugnaire FI. Climatic change and rainfall patterns: Effects on semi-arid plant communities of the Iberian southeast. Journal of Arid Environments. 2011;**75**:1302-1309

[87] Borokini TI, Babalola FD. Management of invasive plant species in Nigeria through economic exploitation: Lessons from other countries. Management of Biological Invasions. 2012;**3**(1):45-55

[88] Obiri JF. Invasive plant species and their disaster–effects in dry tropical forests and rangelands of Kenya and Tanzania. JAMBA. 2011;**3**(2):417-428

[89] Fessehaie R, Tessema T. Alien plant species invasions in Ethiopia: Challenges and responses. In: International Workshop on *Parthenium* Weed in Ethiopia. Addis Ababa; 2014

[90] CBD. WHO Biodiversity and health. State of Knowledge Report. 2015;**2015**:1-4

[91] McNeely JA, Mooney HA, Neville LE, Schei P, Waage JK. A Global Strategy on Invasive Alien Species. Switzerland: IUCN; 2001

[92] Ilukor J, Birner R, Gelaye MT, Negusse GY. Economic Impact of *Prosopis juliflora* on Animal Health and Production: A Case of Afar Region, Ethiopia. Tropentag. Prague, Czech Republic; 2014

[93] Tegegn GG. Experiences *on Prosopis juliflora* Management Case of Afar Region. South Africa; 2008

[94] Asrat G, Seid A. Allopathic effect of meskit (*Prosopis juliflora* (Sw.) DC) aqueous extracts on tropical crops tested under laboratory conditions. Momona Ethiopian Journal of Science. 2017;**9**(1):32-42. DOI: 10.4314/mejs.v9i1.3

[95] Dong S, Wen L, Liu S, Zhang X, Lassoie JP, Yi S, et al. Vulnerability of worldwide pastoralism to global changes and interdisciplinary strategies for sustainable pastoralism. Ecology and Society. 2011;**16**(12):10

[96] Wakwaya K, Yirga M, Yesehak E. In: Stark J, Terasawa K, Ejigu M, editors. Climate Change and Conflict in Pastoralist Regions of Ethiopia: Mounting Challenges, Emerging Responses. Ethiopia: USAID; 2011. pp. 1-64

[97] Rogers P, Nunan F, Fentie AA. Reimagining invasions: The social and cultural impacts of *Prosopis juliflora* on pastoralists in southern Afar, Ethiopia. Pastoralism: Research, Policy, and Practice. 2017;**7**(22):1-13. DOI: 10.1186/s13570-017-0094-0

[98] Munoz-Pizza MD, Villada-Canela M, Rivera-Castaneda P, Reyna-Carranza AM, Osornio-Vargas A, Martínez-Cruz LA. Stated benefits from air quality improvement through urban afforestation in an arid city – A contingent valuation in Mexicali, Baja California, Mexico. Urban Forestry & Urban Greening. 2020;**55**:26854

[99] Mude A, Ouma R, Steeg VJ, Kariuki J, Opiyo D, Tipilda A. Kenya Adaptation to Climate Change in the Arid Lands: Anticipating, Adapting to and Coping with Climate Risks in Kenya - Operational Recommendations for KACCAL. Nairobi, Kenya: International Livestock Research Institute; 2007. p. 135

[100] Maundu P, Kibet S, Morimoto Y, Imbumi M, Adeka R. Impact of *Prosopis juliflora* on Kenya's semi-arid and arid ecosystems and local livelihoods. Biodiversity. 2009;**10**:2-3. DOI: 10.1080/14888386.2009.9712842

[101] Esther M, Brent S. Invasion of *Prosopis juliflora* and Local Livelihoods: Case Study from the Lake Baringo Area of Kenya. Nairobi: World Agroforestry Center: ICRAF; 2005

[102] El-Keblawy A. Impacts of native and exotic *Prosopis juliflora* species on native plants in arid lands of the UAE. In: International Conference on Ecology, Agriculture and Chemical Engineering, (ICEACS'2012). Phuket, Thailand; 2012

[103] El-Keblawy A, Al-Rawai A. Impacts of the invasive exotic *Prosopis juliflora* (Sw.) D.C. on the native flora and soils of the UAE. Plant Ecology. 2007;**190**:23-35

[104] Kaur R, Gonzales WL, Llambi LD, Soriano PJ, Callaway RM, Rout ME,

et al. Community impacts of *Prosopis juliflora* invasion: Biogeographic and congeneric comparisons. PLoS One. 2012;**7**(9):e44966. DOI: 10.1371/journal.pone.0044966

[105] Samuel G, Sebsebe D, Tadesse W. Allelopathic effects of the invasive *Prosopis juliflora* (Sw.) DC on selected native Plant Species in middle awash, southern Afar rift of Ethiopia. Management of Biological Invasions. 2012;**3**(2):105-114

[106] Shanwad UK, Chittapur BM, Honnalli SN, Shankergoud I, Gebremedhin T. Management of *Prosopis juliflora* through chemicals: A case study in India. Journal of Biology, Agriculture and Healthcare. 2015;**5**(23):30-38

[107] FAO. Invasives. In: Sankaran KV, editor. Newsletter of the Asia-Pacific Forest Invasive Species Network (APFISN). Kerala, India: FAO; 2007

[108] Singh YP, Singh G, Sharma DK. Performance of pastoral, Silvipastoral, and silvicultural Systems in Alkali Soils of Indo-Gangetic Plains. Journal of Soil & Water Conservation. 2015;**14**(2):1-6

[109] Walter JK, Armstrong VK. Benefits, Threats, and Potential of *Prosopis* in South India, Abingdon. 2014.

[110] Zachariades C, Hoffmann JH, Roberts AP. Biological control of mesquite (*Prosopis* species) (Fabaceae) in South Africa. African Entomology. 2011;**19**(2):402-415

[111] Ravhuhali KE, Mudau HS, Moyo B, Hawu O, Msiza NH. *Prosopis* Species—An invasive Species and a potential source of browse for livestock in semi-arid areas of South Africa. Sustainability. 2021;**13**:7369